The GIS Book

Fifth Edition

Updated and Expanded

George B Korte, P.E.

ONWORD PRESS
THOMSON LEARNING™

Australia • Canada • Mexico • Singapore • Spain • United Kingdom • United States

NOTICE TO THE READER

Publisher does not warrant or guarantee any of the products described herein or perform any independent analysis in connection with any of the product information contained herein. Publisher does not assume, and expressly disclaims, any obligation to obtain and include information other than that provided to it by the manufacturer.

The reader is expressly warned to consider and adopt all safety precautions that might be indicated by the activities herein and to avoid all potential hazards. By following the instructions contained herein, the reader willingly assumes all risks in connection with such instructions.

The publisher makes no representation or warranties of any kind, including but not limited to, the warranties of fitness for particular purpose or merchantability, nor are any such representations implied with respect to the material set forth herein, and the publisher takes no responsibility with respect to such material. The publisher shall not be liable for any special, consequential, or exemplary damages resulting, in whole or part, from the reader's use of, or reliance upon, this material.

Trademarks

The products mentioned herein are trademarks or registered trademarks of their respective manufacturers.

OnWord Press Staff

Publisher: Alar Elken
Executive Editor: Sandy Clark
Acquisitions Editor: James Gish
Managing Editor: Carol Leyba
Development Editor: Daril Bentley
Editorial Assistant: Fionnuala McAvey
Executive Marketing Manager: Maura Theriault
Executive Production Manager: Mary Ellen Black
Production and Art & Design Coordinator: Cynthia Welch
Manufacturing Director: Andrew Crouth
Cover Design by Cammi Noah

SAN 694-0269
Fifth Edition, 2001
10 9 8 7 6 5 4 3
Printed in Canada

Library of Congress Cataloging-in-Publication Data

Korte, George.
The GIS book / George B. Korte.—5th ed.
p. cm.
ISBN 0-7668-2820-4 (alk. paper)
1. Geographic information systems. I. Title.
G70.212 .K67 2000
910'.285—dc21
00-061154
CIP

For more information, contact OnWord Press.
An imprint of Thomson Learning
Box 15-015
Albany, New York USA 12212-15015

About the Author

George Korte has nearly thirty years of experience as an engineer and manager. These include twenty years in GIS applications. He is the Vice President of Information Technology Services for InfoTech, working in their Alexandria, Virginia, headquarters. Prior to this, he was an executive manager for Intergraph Corporation; the Director of GIS for a large computer systems integrator; an independent GIS consultant; a Senior Associate and the director of GIS, cartography, photogrammetry, and CADD operations for a Top 50 engineering firm; and was Assistant City Engineer for the City of Fairfax, Virginia. He has done extensive consulting in the planning, selection, implementation, and management of GIS systems.

Mr. Korte received a B.S. in Civil Engineering from George Washington University and an M.B.A. from the University of Virginia, and is a registered professional engineer. Mr. Korte has been invited to speak at numerous industry conferences, and his articles on GIS have been featured regularly in several trade journals.

Contact for further information:

George Korte, P.E., Vice President
InfoTech
4900 Seminary Road, Suite 850
Alexandria, VA 22311-1811
Phone: 703-671-1900
Email: gkorte@infotech.org

Acknowledgments

It was the founding publisher of OnWord Press, Dan Raker, who first suggested the idea of and title for this book. I also want to acknowledge the encouragement and guidance I received from several others regarding the book in the ten years since then. These include Terry Stringer, publisher of *Civil Engineering News*; Dr. Joel Orr, internationally recognized expert in computer graphics; my sister Treena Rinaldi; Jackie Headapohl, Victoria Dickinson, and Walt Walkowski, editors of *P.O.B.* magazine, where much of the material for this book first appeared; and WBH Associates' Brad Holtz, a national CAD consultant and fellow Washingtonian. Beyond his encouragement, Brad's book *The CAD Rating Guide* inspired me to think, "I bet I could do that, too!" I would also like to acknowledge Neil Hohmann of GIS USA in Annapolis, MD, for his technical review of both the fourth and fifth editions.

Grateful acknowledgment is made to Intergraph Corporation for permission to reproduce many of the illustrations that appear in this book. Thank you to the following Intergraph employees for their assistance: David Kingsbury, Terry McClure, Brenda Pearce, and Dan Weston. Special thanks also go to Intergraph's Robert "RMAC" McIntyre, who created the introductory GIS exercises presented in Appendix A.

I also acknowledge the work of Harold L. Thurman of Madsen, Okes, and Associates of Littleton, Colorado, whose paper titled "How to Ruin Your Project" initially identified many of the issues discussed in Chapter 16, "The Pitfalls of GIS."

I want to give special acknowledgment to Daril Bentley, who has edited the last two editions of this book. Daril's become a good friend in the process.

Finally, I want to acknowledge my wife Deborah, who has been so very patient nearly every weekend for the last several months as I worked on this latest edition of the book. She's the Vice President of Administration at Citizens for a Sound Economy in Washington, D.C. She's bright and articulate; a talented seamstress, decorator, and cook; and the most beautiful lady in town. More importantly, she's a woman of strength, faith, and integrity who brought healing into my life. In the words of the country songwriter, "I feel real sorry for all the fellas who ain't me."

I dedicated the third edition of this book to my three children: Mary Ellen, Christopher, and Kathleen. I love them dearly and I am very proud of them. (Mary Ellen has since married Brian Coppage, and they have been blessed with the birth of their son, Brian Spencer.) I dedicated the fourth edition to my parents, Mr. and Mrs. George B. Korte, Sr.

I dedicate this edition to my wife, Deborah. I love you, Darlin'.

Contents

Introduction

I hate reading introductions. I always figure that the good stuff, the stuff I bought the book for, is not going to be in the introduction. The introduction is the place where the author gets too sentimental or preachy or abstract to hold my attention. In fact, I don't recall ever reading an introduction all the way through. Of course, this habit is one I now regret. Because I have always played mental hooky while reading introductions, I don't have the foggiest idea what they are supposed to say. But I can think of several important things you should know about this book.

Who Should Read This Book

The GIS Book was written for people who want to know about the selection, implementation, uses, benefits, and management of geographic information systems (GIS), but do not need to know all of the technical details of how a GIS actually "works."

The people who want to know about GIS are generally involved with operations, assets, and/or systems disbursed over some geographic area. This would include, for example, students, employees, and managers in the following areas: civil, electrical, communications, and environmental engineering; land planning, land surveying, and public health and safety; geology, geography, forestry, agriculture, and other natural resource areas; environmental sciences such as oceanography and meteorology; the military; politics; and business operations such as marketing, advertising, sales, delivery services, insurance, and the like. You need no technical background to gain an understanding of GIS from this book.

Focus and Organization

The GIS Book is a nontechnical guide to help you understand what a GIS is, fundamentally how it works, and how it benefits the user. This book also deals with how to evaluate the economics of a GIS investment and how to select, implement, and manage a GIS.

This book is divided into three parts. The first part explains what a GIS is, how it differs from other types of computer mapping systems, the benefits it can produce, and how it can be used. It includes an overview of the GIS industry and its key players, and discusses technology trends; GIS data sources, formats, and standards; the types of analyses you can perform with a GIS; and how GIS is used on the Internet.

The second part of the book provides specific guidance in planning, selecting, and implementing a GIS. It also presents descriptions of the core products offered by the top four GIS software developers. It discusses the use of a GIS consultant to help you select and implement a GIS, pitfalls to watch out for, and GIS management and staffing issues.

The third part presents other considerations to help make decisions on GIS selection, implementation, and management. It discusses the economics of base map accuracy, GIS data quality, and translating CADD data to GIS, and includes a detailed model for the financial analysis of a GIS program. It also describes government and trade association involvement in GIS, as well as legal aspects of GIS. Finally, it presents three case studies in GIS implementation and management.

Content

Unlike previous editions, this edition of *The GIS Book* incorporates hands-on, practical exercises (see Appendix A, which includes instructions for use). In addition to new and updated illustrations, all existing content has been updated and supplemented, including, among other areas, advances in technology, changing market forces and business paradigms, and the changing legislative and legal landscapes. New and significantly revised material includes the following.

- GIS data sources, collection, and entry
- GIS data formats and standards
- GIS on the Internet
- Managing and maintaining a GIS database
- GIS data quality
- Legal aspects of GIS, including land surveying
- Government and trade association involvement in GIS

There are also three extensive appendices. The first presents a series of self-paced introductory GIS exercises using real GIS data and software that can be downloaded over the Internet for free. The second lists resources for more information on GIS, organizations to join, and conferences to attend. The third appendix lists sources of GIS data. The appendices are followed by a glossary of GIS terms and an index.

Where This Book Came From

This book actually started out as a series of articles on GIS I wrote in 1989 for Dan Raker, then publisher of the monthly newsletter *Design Systems Strategies*. After the first year of articles had been published, Dan suggested they be collected to create a handbook titled *A Practitioner's Guide to GIS*. Soon after that, Dan turned from *Design Systems Strategies* to focus on book publishing, going on to establish High Mountain Press, OnWord Press, and GeoBooks.

Meanwhile, I continued to write articles on GIS for other periodicals. These furnished the material for an expanded second edition, released in 1991 and titled *The GIS Book*. Around that time I also started writing feature articles on GIS for *P.O.B.* magazine, a leading trade publication for the land surveyor. (P.O.B. is a land surveyor's term that stands for "point of beginning.") Many of these articles provided the raw materials for an expanded third edition, released in 1993, the further expanded fourth edition in 1997, and this fifth edition, likewise updated, expanded, and including new and additional illustrations.

I have received many compliments on *The GIS Book*, mostly for its readable style. I am also aware of one criticism, that it is "not technical enough." Purposely, I have done nothing in response to this. Although this edition fills some technical holes in previous editions (such as the discussion of GIS data quality), I have stayed away from the treatment of advanced GIS concepts (such as quadtree data models and map conflation) for fear that I would lose the intended audience's interest.

The technical and technological aspects of GIS are the subject of numerous books. For readers who are interested in the "nitty gritty" of GIS, the appendix of GIS resources includes a bibliography of fine works that treat the more complex technical aspects of GIS, as well as other titles that provide in-depth discussions of the features and functions of specific GIS products.

Understanding GIS

Part One explains what a GIS is and how it can be used. It also discusses the GIS industry and technology trends. The chapters in Part One are:

- *Chapter 1, Introduction to GIS,* describes many of the uses of GIS, the benefits it produces, and the fundamentals of how it works.
- *Chapter 2, Defining GIS,* shows the difference between a GIS and other computer-based mapping systems.
- *Chapter 3, The Uses of GIS,* describes the use of GIS in a small city government to illustrate how it can benefit an entire organization.
- *Chapter 4, An Overview of the GIS Industry and GIS Software,* discusses the industry in North America, including its size, growth rates, and key players, including software developers. It also discusses the primary types of GIS users and applications.
- *Chapter 5, Why Implement a GIS?,* is a discussion of the principal reasons for implementing a GIS, showing the advantages of a GIS over traditional mapping methods.

- *Chapter 6, GIS Technology Trends,* reviews current trends in data communication networks, computer hardware operating systems, and GIS software.
- *Chapter 7, GIS Data Sources, Collection, and Entry,* examines the collection and entry of the data needed to make a GIS useful.
- *Chapter 8, GIS Data Formats and Standards,* talks about two principal types of GIS data commonly used, as well as standards for GIS data transfer and storage.
- *Chapter 9, Types of GIS Analysis,* discusses the basic types of analysis a GIS can perform, as well as several specialized functions.
- *Chapter 10, GIS on the Web,* takes a look at the uses of GIS on the World Wide Web (the Internet) and the types of resources available on the Internet.

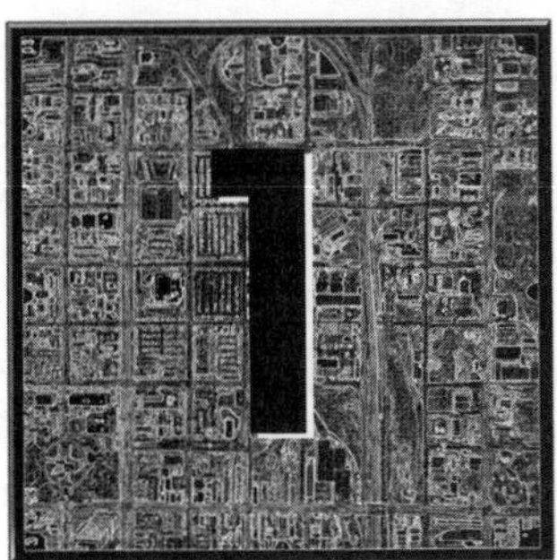

Introduction to GIS

How GIS Is Affecting Our Lives

In This Chapter...

Any organization interested in assets or activities distributed over some geographic area can use a GIS. That would include a lot of organizations. This chapter presents examples of the many uses of GIS, and the benefits it produces. The chapter describes the origins of GIS and how it works, and includes the perspectives of an industry leader on the future of GIS.

A Dramatic Story

Responding to the robbery of an armored truck, Metropolitan Toronto Police spot the getaway car and give chase, eventually following it into the suburbs. Back at MTP headquarters, a dispatcher coordinates the pursuit using a new computerized dispatch system. She sits in front of two computer screens. One shows the status of this and other incidents, as well as the status of available police units. The other shows a computerized map of the city, red and yellow "snowflakes" pinpointing the location of both the incidents and the police cars. She can view the entire city, or zoom in to see details. The system can even show her which unit is closest to an incident, and trace its fastest route to the scene.

When the robbery suspects abandon their vehicle and flee on foot to a nearby golf course, MTP officers find themselves in very unfamiliar territory. "Our people really didn't know the rolling topography of the golf course very well," says Richard Coulis, MTP staff sergeant and acting inspector. "But the dispatcher could see all the roads, cart trails, creeks, and even the water hazards right on her computer screen." Windowing the golf course on her map display, the dispatcher instructs some police cruisers to take up strategic positions at possible exit points, then directs the officers pursuing on foot.

According to Coulis, "An officer would be running along and say, 'I'm going to turn and head west up this hill,' and she'd answer, 'No, don't do that. There is a creek up there and you won't be able to get across.'" All of the suspects are eventually rounded up, most of them right on the golf course. "The information we had was so thorough that our officers returning to the station asked, 'Where were the video cameras?'" Coulis adds. "With an old fashioned paper map this kind of operation would have been impossible." (The following illustration shows a computer-aided dispatch station.) A GIS is the foundation of Toronto's computerized dispatch operation.

Computer-aided dispatch station.

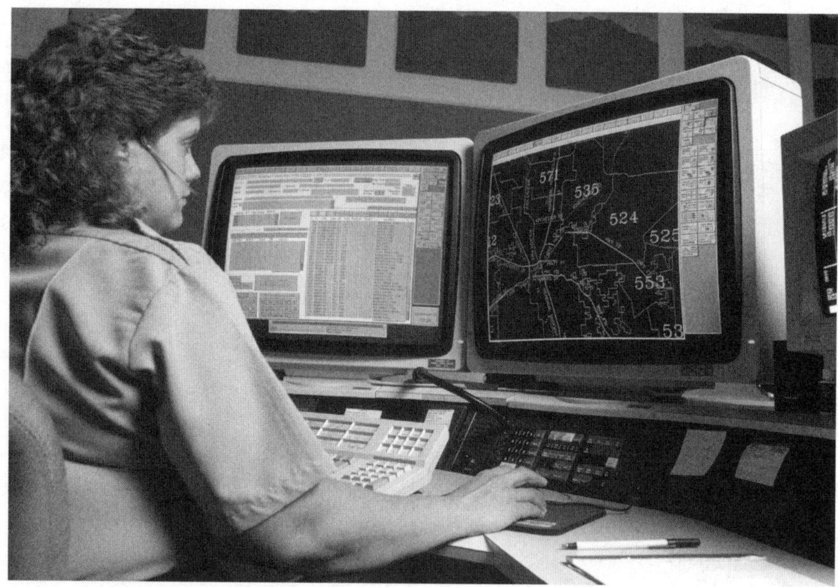

A GIS is a computer system that can store virtually any information found on a paper map. But it can be much more helpful than a traditional map. Whereas the GIS can display maps on a computer screen, it can also provide detailed information about their features, including roads, buildings, streams, and so forth. Moreover, the computer can quickly search and analyze these map features and their attributes in ways that are not possible with paper maps.

For instance, suppose you are looking at the GIS display of a state highway map. Point at a particular road and click the computer's mouse button to find out that the roadway is state route 29, four lanes wide, paved with asphalt, and has a 45-mph speed limit. Click again to see the location of all traffic accidents that have occurred on that stretch of the highway. Click once more to view a picture of the pavement and road signs at that location. Or, ask the GIS to highlight all parts of the state highway system that serve more than 5,000 vehicles per day. In just seconds you see the results.

GIS are being used to help manage forests, utilities, petroleum exploration, and even the taking of the U.S. Census. Corporations, public safety officials, and the military use them. They have been linked to many other technologies, using satellite images to help farmers manage their crops, using global positioning system (GPS) data to help dispatchers locate an available taxi, using seismic data to locate the epicenter of an earthquake, and using bathymetric data gathered by a deep sea submersible to map the ocean floor. The following illustration shows satellite imagery analyzed by a GIS.

Satellite imagery analyzed by a GIS.

The Origins of GIS

GIS was pioneered in the 1960s by the Canadian forestry mapping initiative and continued to develop as Canadian, U.S., and other government and university researchers sought to represent the earth's geography using a computer database, display it on a computer terminal, and plot it on paper. They also developed computer programs to quickly search and analyze this data. Several corporations were founded in the 1970s to develop and sell systems for computer mapping and analysis. Today's two leading GIS software developers trace their roots to these early days, although at first each one emphasized two different aspects of the technology.

Intergraph Corporation of Huntsville, Alabama, focused on efficient input and storage of GIS data, as well as the preparation of computer-generated maps that rivaled traditional maps for their cartographic quality. The Environmental Systems Research Institute (ESRI) of Redlands, California, focused on providing a "tool kit" of computer commands for the analysis of GIS data. Over the years, both companies have rounded out their systems' respective capabilities.

At first, only the largest government agencies, utilities, and corporations could afford GIS because of their expense. Based on mainframe and minicomputers, a typical GIS workstation cost the owner more than $100 thousand, considering all of the hardware, software, and training involved. Nonetheless, the market for GIS expanded steadily in the early 1980s, as trade magazines, conferences, and professional associations spread the word of its benefits. It mushroomed, however, with the advent of the personal computer. GIS software was quickly adapted to this new, less expensive "platform," and the cost fell within the reach of many more users.

Even then, GIS was used primarily by a select few who had both the GIS software on their computer and the training to use it. Although tens of millions of dollars had been invested in GIS databases built from paper maps, aerial photographs, and satellite imagery, this data was still largely inaccessible until GIS specialists plotted it out on paper to distribute it. Then in the 1990s the Internet threw the doors to this valuable GIS data wide open to the public.

Today there are hundreds of web sites that publish "live" GIS data on the World Wide Web. Literally anyone who can use a web browser can now access and view GIS data. As a result, the worldwide market for all GIS products and services was about $7 billion in 1999, split almost evenly between sales in North America and the rest of the world, and is growing nearly 13 percent per year, according to Daratech, a computer market research firm.

The typical GIS is founded on several basic concepts. First, real-world features on the earth's surface are related to a map grid coordinate system and recorded in the computer. The computer stores the grid coordinates of these features to show where they are, and the attributes of these map features to show what they are. Second, map features can be displayed or plotted in any combination and at virtually any map scale, making computerized mapping data far more flexible to use than traditional paper maps. Third, the GIS can analyze the "spatial" (locational) relationships among map features. Thus, the GIS can determine how many acres of land zoned for commercial use are located within a town's flood plain.

The Many Uses of GIS

The Willapa Bay area, located in the southwest corner of Washington state, boasts abundant natural resources that support thriving timber, oyster, and fishing companies, as well as fertile soils for farming and an ideal environment for cranberry bogs. In 1992, local residents formed the Willapa Alliance to find an effective means of managing and preserving these natural resources on which they depend for their livelihood. The alliance used a GIS to map and analyze an area roughly the size of Rhode Island. The GIS provided data about the health and status of the region's ecosystem in a format that enabled the members of the alliance to understand the environmental issues affecting the bay area.

Data from dozens of sources were consolidated on 80 GIS overlays, and then used to examine the various factors that affect local industries. These include geological, environmental, biological, and developmental factors, as well as fishing, forestry, and farming practices, such as the use of pesticides. Moreover, changes in these factors can be monitored over time, providing a tool for measuring the impact of policy decisions.

"All this data would have been impossible to analyze without GIS," says Allen Lebovitz, natural program resource director of the Willapa Alliance. "We're also using it to present very complex and technical data to the average person," he adds. "So far, we have learned that there are definitely some (areas) that will never be highly productive regardless of management. That's important in deciding where you want to spend your conservation and restoration resources." The following illustration shows GIS data for a forested area.

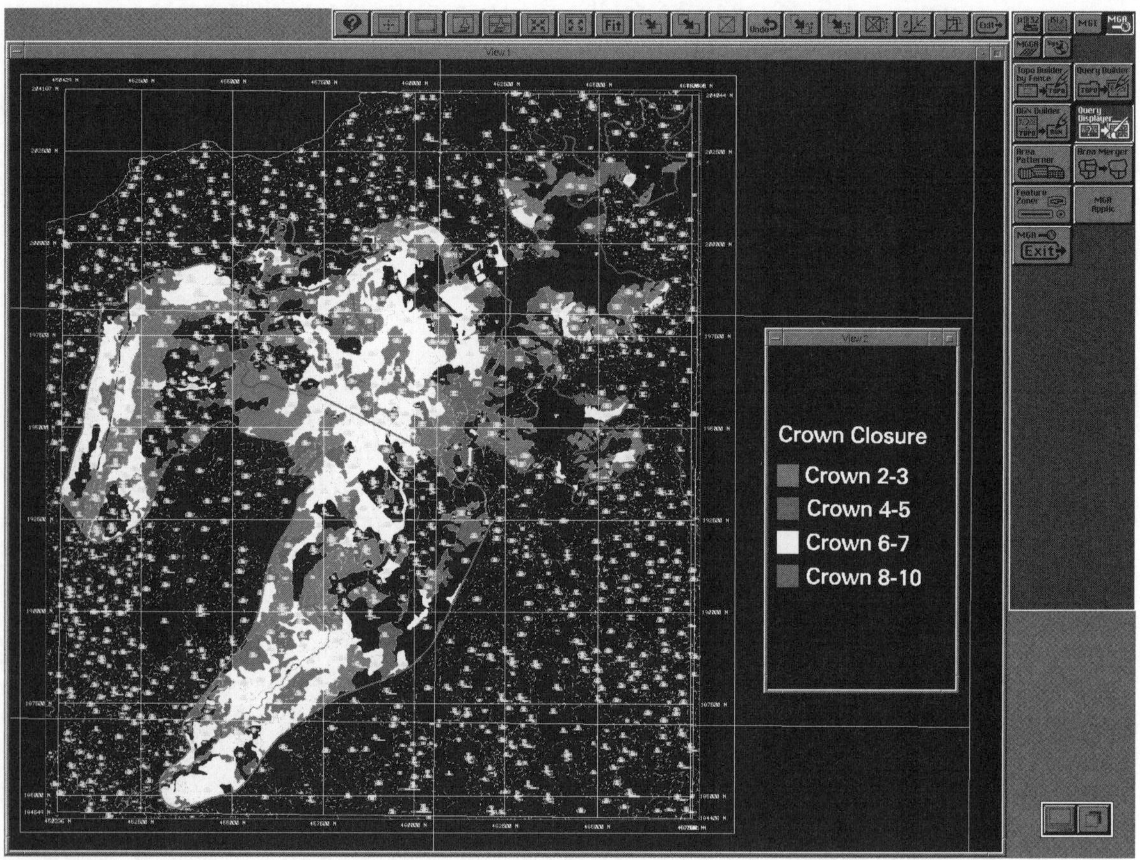

Forest data in a GIS.

In 1984, a propane truck accident caused a four-hour shutdown of the Cross-Bronx Expressway in New York City. The New York Port Authority police were first on the scene, followed by the city's fire and police departments, then the New York State Department of Transportation; however, the ensuing disagreements between the agencies about jurisdiction and disposition of the accident slowed the overall cleanup and road-reopening effort. This and similar incidents prompted the formation of TRANSCOM, a consortium of 14 New York, New Jersey, and Connecticut transportation and public safety agencies located within a 50-mile radius of Manhattan.

The coalition transcends territorial boundaries to solve regional transportation problems, acting as an information clearinghouse for its members. TRANSCOM uses GIS as a tool to coordinate and speed the process. Member agencies notify the TRANSCOM operations center in Jersey City about traffic incidents, construction, and special events. This information is disseminated to other agencies, which view the TRANSCOM GIS data over a communications network to aid their planning and emergency response efforts.

At the local level, the New York City Department of Transportation uses GIS to help manage and maintain 6,500 miles of roadway, traffic signals at 10,000 intersections, 63,000 parking meters, 300,000 streetlights, and 1.3 million street signs. Traffic engineers use GIS to map accidents and pinpoint problems, such as intersections with poorly timed traffic signals. The Division of Parking has used GIS to analyze the use and location of parking meters, as well as the most efficient routes for meter readers.

The Bureau of Bridges maps bridge locations, and engineers can point to a bridge displayed on the GIS to retrieve close-up photographs of its structural details. The Highways Operations Division tracks potholes and planned utility construction projects to help schedule roadway paving projects. "The department has few doubts about the benefits if the GIS effort," reports Fred Cohn, the deputy director of the city's Department of Transportation responsible for the GIS. "The system has paid for itself." The following illustration shows a detail from a Virginia Department of Transportation highway map generated from a GIS.

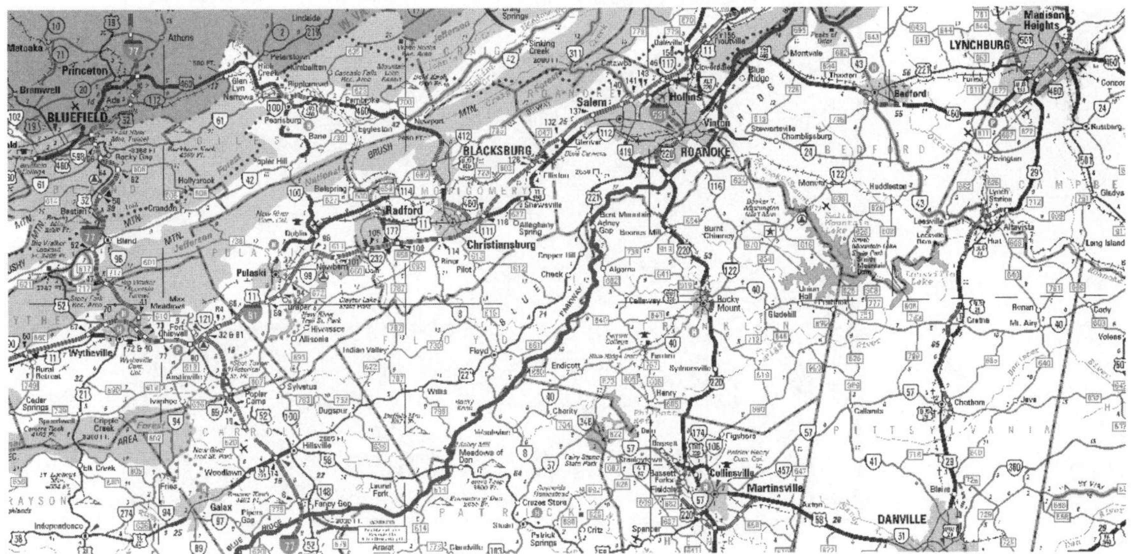

Highway map generated by a GIS.

Business has also turned to GIS to help develop market strategies, determine the best locations for new retail stores, fine tune product delivery routes, dispatch taxis and service trucks, and analyze sales territories. For example, direct mail firms use GIS to help find those households most receptive to the products they market. According to the Direct Marketing Association, advertisers spend more than $30 billion on direct marketing yearly, with direct mail accounting for a large percentage of this figure. Atlanta-based Cox Enterprises is one of the largest media companies in the nation.

One subsidiary, Cox Direct, conducts cooperative mailings (multiple advertisers sharing the cost of printing, postage, and production) in brown envelopes to 30 million homes 10 times a year. The envelopes contain promotional offers for household products, including medications, shampoo, cleaning products, and food. Cox Direct breaks down its large address database into subsets based on demographic data available from the U.S. Census Bureau.

This data is summarized for each census block, and includes household data on marital status, family size, age, income, education, race, and so forth. Cox Direct analysts help clients draw profiles of their ideal consumers, then determine the addresses that best fit those profiles. "One of the ways we use GIS is to simplify the process of making those matches," says David Zeph, Cox information technology manager. The following illustration shows census data analyzed by a GIS.

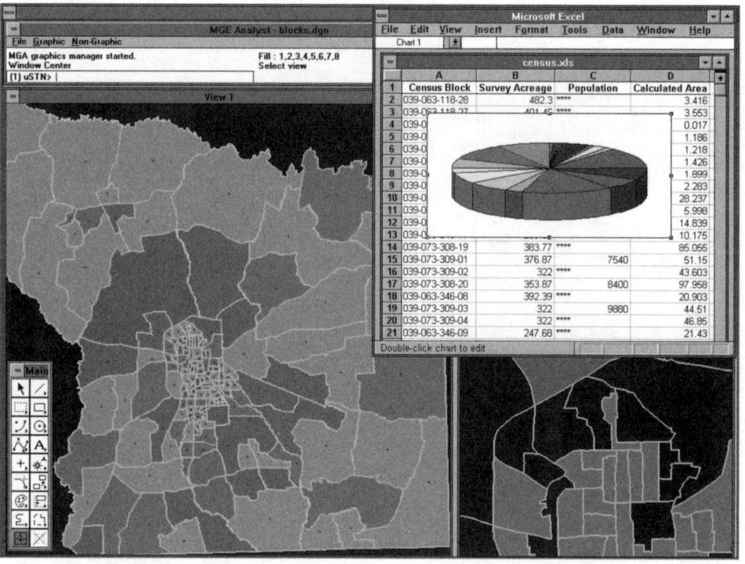

Census data analyzed by a GIS.

When Hurricane Andrew came ashore in southern Florida on August 24, 1992, it left behind not only human suffering but property losses and damages estimated at $15.5 billion. Property and business owners who were insured filed claims in record numbers. For insurance companies with policyholders concentrated in the hurricane's path, the settlements were staggering and seriously affected their financial stability.

Therefore, insurance companies have turned to GIS to help predict and model catastrophes, to avoid concentrations of policies in areas of high risk exposure, to determine the amount of risk to transfer through reinsurance (selling policies to a secondary underwriter), and to help in the pricing of both the original policy and the reinsurance contract. They can analyze natural risk factors such as flood-, fire-, tornado-, and earthquake-prone areas, as well as crime rates, accident frequency, and the availability of fire protection. They can use this data to help evaluate the risk of insuring a property.

For example, an application for property insurance from a chemical factory located near a major geologic fault and across the street from an elementary school would probably be turned down. According to David Langdon, a systems analyst in the Weatogue, Connecticut, office of Tillinghast, a worldwide actuarial and risk management consulting firm, "With the help of GIS, insurance companies are finding it easier to analyze, visualize, and prepare for the many issues in today's insurance market."

Today, municipalities and utilities throughout the nation have generally embraced the concept of GIS, hampered only by the tremendous cost of converting mountains of paper map data to a computer format. Wake County, North Carolina, encompasses 12 municipalities within 864 square miles. The county tracks more than 184,000 tax parcels representing a population of 460,000. The county's planning and assessment departments first implemented GIS to manage property records, including tax parcel, zoning, and land-use maps. An example of tax assessment data retrieved by a GIS is shown in the following illustration.

Tax assessment data retrieved by a GIS.

Today, the GIS database is used by numerous other departments and agencies, including the Fire Services, Inspections, and Emergency Management departments, as well as the county school system, the Board of Elections, and the Soil and Water Conservation Service. Several cities and towns within the county, including the city of Raleigh, use its GIS data as an accurate foundation upon which they can add specifics about their own infrastructure, including water and sewer lines, roadways, utility networks, administrative districts, and so forth.

Charles Friddle is the Wake County GIS director: "Overall, the system is improving the county's responsiveness to our municipalities and the public we serve. With GIS, we not only save money by consolidating information, but we are now poised to provide those who need geographic information with a means to access it."

Likewise, utilities such as West Ohio Gas in Lima, Ohio; Pacific Bell of San Francisco, California; the Washington Suburban Sanitary Commission of Laurel, Maryland; and Consumers Power Company of Jackson, Michigan, are using GIS to automate the "outside plant" records of their telephone, cable TV, gas, electric power, water, and sanitary sewer systems.

These GIS are not only mapping the location and attributes of the utility networks, they are used in the planning, design, and construction of new lines, to dispatch repair crews, to route meter readers, to schedule maintenance and repair work, and to answer customer inquires. "West Ohio Gas must reduce costs to become more efficient, while simultaneously offering new services and improving customer service," says Ty Lotz, the company's superintendent of engineering. "GIS is helping us to do this." The following illustration shows details from an electric utility's GIS.

Electric utility data in a GIS.

An Industry Leader's Comments

Larry Ayers is Executive Vice President of Intergraph Corporation's Mapping Sciences Division. Asked about the most interesting GIS application he's been involved with, he replied, "A military base is like a small city, but with only one landowner, one landlord, and one owner of all the utilities. The management of one of these bases is complicated enough, but is made even worse by the very unusual, sometimes dangerous, things that go on there. The military, Patrick Air Force Base in Florida is one example, has used GIS to bring together all types of information about its property and assets and put it right at the base commander's fingertips."

Asked about the future of GIS, he replies, "So far, GIS has been a specialty in the world of computer science, used by a small group of experts. In the future it will be just another computer function to aid in decision-making. I suppose the average homeowner might use it to plan the itinerary for Saturday morning errands."

This prediction might be so far-fetched, because the forerunners of "GIS for the common man" are already here. For example, a program from Road Scholar of Houston, Texas, (called City Streets) provides street addresses and locations in more than 170 metropolitan areas. It gives details down to one-tenth of a mile, including streets, intersections, and addresses. Using the Washington, D.C., database, a tourist marks the sights she wants to visit and the route she wishes to take. The program calculates the distances from point to point, shows which routes can be traveled by car or taxi, and which ones require walking.

Rental car leaders such as Hertz and Avis provide vehicles equipped with "in-vehicle navigation" systems. The systems combine GPS technology with GIS to pinpoint the vehicle's location on a map displayed by a small computer screen mounted on or under its dashboard. They provide an address finder, place finder, and city finder to help the driver locate his destination. Hotel, restaurant, and entertainment locations are also available. Ordinary citizens can even take advantage of large, sophisticated GIS databases.

Kevin Byrne, a teacher at the Minneapolis College of Art and Design, visits the Ramsey County, Minnesota, Public Works Division. After a brief visit with a county GIS technician, he leaves with a computer file on a small diskette that was extracted from the county's massive GIS database, one of the largest in the state. He manipulates the file on his Macintosh computer to produce maps of his Midland Hills neighborhood in Rosedale, Minnesota, for its Crime Watch committee.

GIS applications are also appearing on the Internet. Web sites such as MapQuest (*http://www.mapquest.com*) and MapBlast (*http:/www.mapblast.com*) provide quick and simple maps. Users can enter a postal address to recall a map of the area around that location, and view "points of interest," including landmarks and public services.

Summary

One problem facing GIS users today is the many different formats in which GIS data is stored. This can make it difficult to merge data from different sources, or for someone using one GIS software program to read the data created using another program. David Schell, President of the Open GIS Consortium (OGC) of Wayland, Massachusetts, is leading an effort to standardize the means of reading and using GIS data: "The exciting thing [about GIS] is that geographic information really is valuable in many human endeavors, and with GPS, high-resolution satellite images, the Internet, and the simple fact that layers of data for the earth's sphere accumulate, we're rapidly entering an era in which geographic information will play a more important role."

Another dramatic example of that role is the use of GIS as a tool for disease control. Recent outbreaks of Ebola virus in Zaire; cholera in India, Bangladesh, and Latin America; dengue fever in Australia and Central America; and Hantavirus in the American Southwest have focused public attention on emerging and reemerging diseases. Researchers have found that epidemics of new diseases, as well as the resurgence of age-old afflictions, are partly linked to the environment.

GIS helped researchers in Mexico study the relationship of the landscape to malaria in isolated areas plagued by the disease, and then create maps showing malaria risks over large regions. This enables public health agencies to target their resources they are most needed. Likewise, a UNICEF-sponsored study in West Africa used GIS to map villages with high rates of Guinea-worm disease and evaluate the effectiveness of policies designed to combat its spread. The analysis demonstrated the power of GIS to determine areas with a high disease prevalence and a large number of cases.

It was also useful for identifying and locating areas that are most in need of assistance. "Areas of high infection can be identified, and programs such as health education, Guinea-worm intervention, and water supply treatment can be targeted for these areas," says Barbara Tempalski, a GIS technician in the Department of Geology and Geography at Hunter College in New York City that worked on the project. "The opportunities for using this technology in epidemiology and disease-related fields are immeasurable."

Larry Ayers believes that in the future GIS will be almost as prevalent as computers themselves. "Employees who learned to use word processing and spreadsheet programs over the past ten years will be learning to use GIS in the next ten."

Defining GIS

The Essential Differences Among CADD, AM/FM, and GIS

In This Chapter...

The commonly accepted definition of a GIS is "a computer-based system used to capture, store, edit, analyze, display, and plot geographically referenced data." However, this broad definition applies to three principal types of computer systems, each of which has distinctly different characteristics and applications: CADD, AM/FM, and GIS. This chapter discusses the key concepts of data structure in each of these three technologies, points out the important differences, and describes their typical applications.

What Is CADD?

Computer Aided Design and Drafting (CADD) technology is widely used by many professionals (including engineers, architects, and planners) to help them design and produce design drawings. The following illustration shows a CADD drawing.

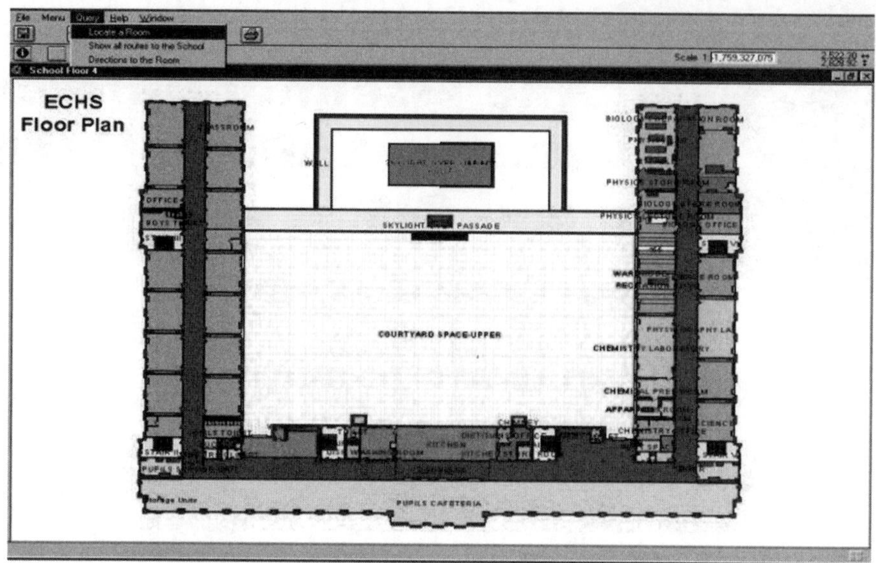

A CADD drawing.

However, CADD can also be used to produce maps. It is an effective replacement for the traditional manual cartographic process. As shown in the following four illustrations, CADD data include all of the graphical elements needed to draw a map: lines, line strings, text, and symbols. These are referenced to a coordinate system, which can be used to represent a mapping grid, such as state plane coordinates.

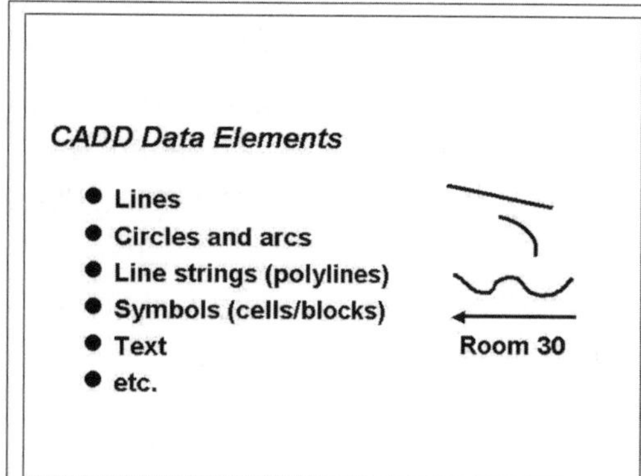

CADD data elements.

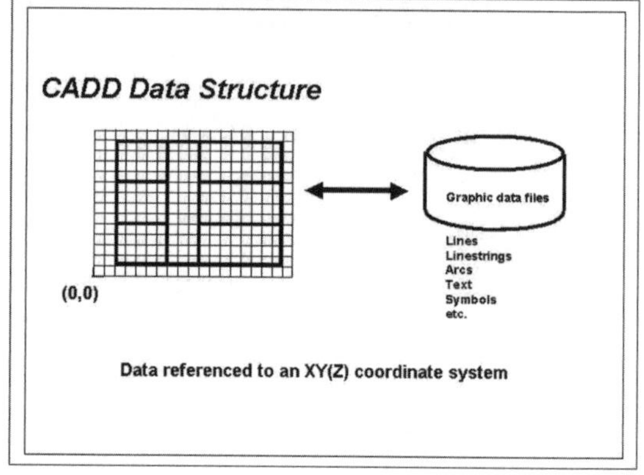

CADD data structure.

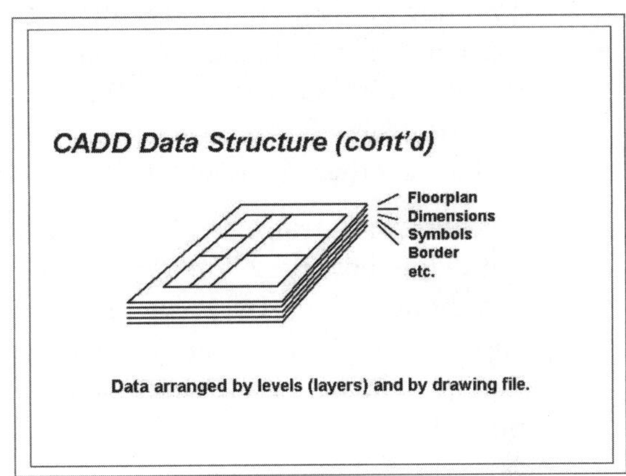
CADD Data Structure (cont'd)
Floorplan
Dimensions
Symbols
Border
etc.
Data arranged by levels (layers) and by drawing file.

CADD data structure (continued).

CADD data records.

CADD Data Records

No.	Type	Level	Color	Style	Weight	Font	Name	Geometry
1	Line	Floorplan	Yellow	Solid	1	-	-	XY,XY
2	Line	Floorplan	Blue	Dotted	1	-	-	XY,XY
3	Linestring	Site plan	Red	Dashed	2	-	-	XY...XY
4	Text	Text	White	-	1	Times	Room 30	XY
5	Symbol	Symbol	Green	-	2	-	Arrow	XY
etc.								

Data in a CADD system is organized on layers that are conceptually like registered overlays. The layers can be used to organize map features by theme, such as streams versus roads versus structures, or by type of data, such as line work versus text. The table represents the elements stored in a CADD data file. Plotted CADD data rivals scribed cartographic products in graphic quality and precision.

CADD can greatly reduce map production time and save money over the traditional cartographic process. For instance, corrections are much easier to make. To make changes to the map, the manual process requires erasing and re-inking, or applying opaque and rescribing. A CADD system allows the user to quickly modify a single element without affecting other features.

CADD offers many other benefits over traditional film-based, manually prepared mapping techniques; the data is better organized, easier to store and retrieve, and so on. For these reasons, many civil engineers use CADD to store and manipulate mapping data used in the design process. Similarly, map atlas producers and other cartographers can use CADD-based systems to create maps.

However, CADD is not suited for analyzing map data. In a CADD system, map features are associated by theme, using layers, and the features are all referenced to a common geographic coordinate system, but further relationships among the data elements are not defined. Although the CADD database can describe the geometry of two roads that cross each other, the fact that they intersect is not identified. Lines may be assigned to a level called "water system," but the CADD database does not define how they are linked to form a network.

Likewise, line strings can be assigned to a level called "soils," but the database does not define how these line strings are connected to form areas of uniform soil type, commonly called soil polygons. Once again, these line strings are related to one another only by layer and by reference to a common coordinate system. The fact that they define a closed area cannot be determined without processing the data to inspect for this condition.

Resource planners and managers often ask questions that require the analysis of *spatial* relationships ("spatial" refers to *location*): What is nearby? How many of these do we have in this area? What areas are both this type and that type? The following are some examples of problems in spatial analysis.

- How many acres of the forest, by age and type of tree, will be within 100 feet of this proposed power line?
- Who are the owners of properties that lie within a mile of this hazardous materials storage site?
- Which areas of the county are more than five minutes away from a fire station?

A CADD system is not well suited for answering these questions because these spatial relationships are not defined in the data structure.

What Is AM/FM?

Automated Mapping/Facility Management (AM/FM) is used by utilities to manage mapping and attribute data regarding their physical plant. For example, an electric utility would use an AM/FM system to store the location and attributes of its power lines, poles, transformers, and so on. Like CADD, AM/FM uses graphic data elements to represent map features. These are likewise referenced to a map coordinate system and organized on layers by map theme. However, AM/FM goes a step further by defining relationships among utility system components as *networks* (see the following illustration).

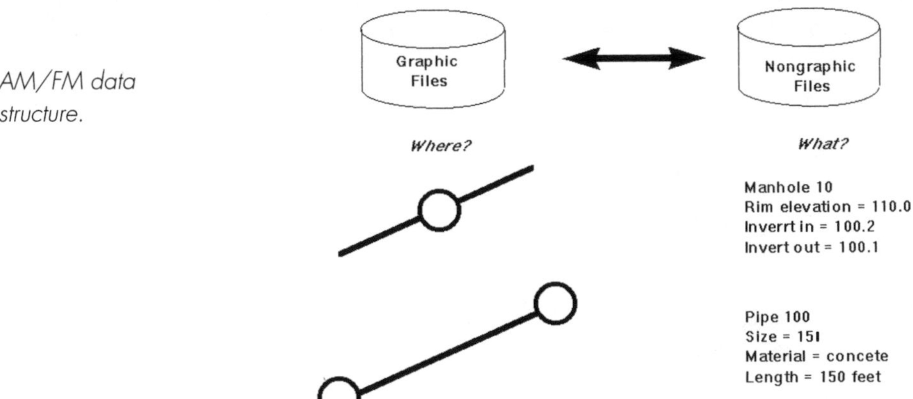

AM/FM data structure.

A network identifies which components are connected to each other. These connective relationships are often defined in a separate data file describing the geometry of the system. Unlike a CADD system, intersections among all network components must be preserved in order to define the connections. Therefore, in an AM/FM system, no two lines of the same type of utility may cross except at an intersection.

Another important feature of AM/FM is that the attributes of the utility system are also stored in a separate data table. As shown in the previous illustration, a unique identification number, such as a utility pole number, links these records to the graphic data elements. These attribute data describe the characteristics of utility system components such as sizes, capacities, materials, and so on. With these two features, a networked data structure and associated attribute data, it is possible to model and analyze a utility system's operation.

What Is GIS?

A GIS is best suited for the analysis of geographic data. GIS is similar to CADD and AM/FM in that it references graphic data elements to an XY coordinate system, and it separates map features by layer (also referred to as a map "theme" or "coverage"). Although the GIS may divide the entire area being mapped into separate files, much like map sheets, it handles all of the data in these files as though they were in one large "seamless" map file.

In addition to graphic (often referred to as "spatial") data, a GIS also stores attribute data. These are associated with the spatial data and provide further descriptive information about them, similar to an AM/FM system. For instance, in a GIS used for municipal tax mapping, a tax parcel would be defined as an area, and its descriptive data might include the lot number, owner's name, acreage, zoning, and so on. This attribute data is placed in a database separate from the graphics data (see the following illustrations).

GIS coordinate system and data structure

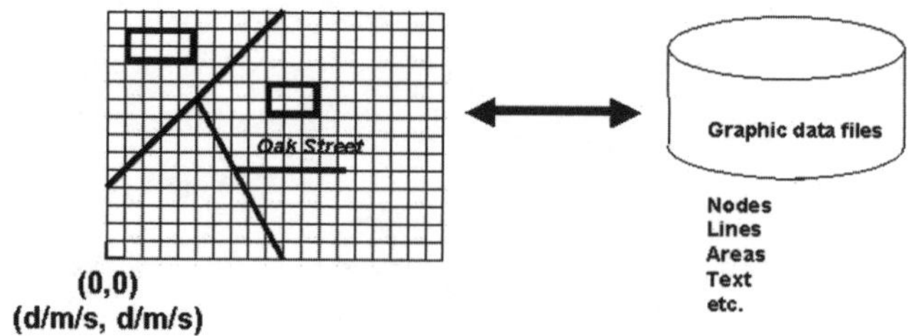

Data referenced to a mapping coordinate system

GIS theme data structure.

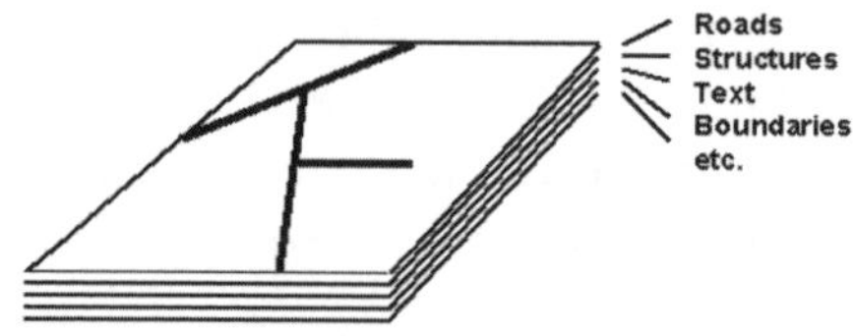

Data arranged by themes and by map sheet file.

GIS spatial and attribute data.

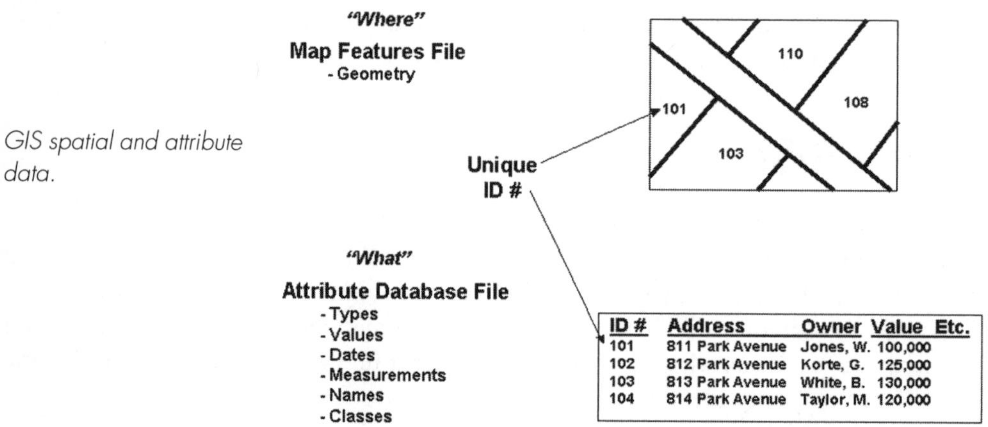

ID #	Address	Owner	Value	Etc.
101	811 Park Avenue	Jones, W.	100,000	
102	812 Park Avenue	Korte, G.	125,000	
103	813 Park Avenue	White, B.	130,000	
104	814 Park Avenue	Taylor, M.	120,000	

Most importantly, GIS differs from CADD and AM/FM in that the spatial relationships among all data elements are defined. This convention, known as data topology, goes beyond merely describing the location and geometry of map features. Topology also describes how linear map features are connected, how areas are bounded, and which areas are contiguous. To define map topology, a GIS uses a special database structure.

As in a CADD system, all map features are related to a geographic coordinate system. However, unlike a CADD system (which defines map features simply as lines, line strings, and symbols), a GIS defines map features as *nodes*, *lines*, and *areas*. (Other terms—such as *points*, *arcs*, and *polygons*—are also frequently used.) The following illustration shows the basic data elements of a GIS.

The basic elements of a GIS.

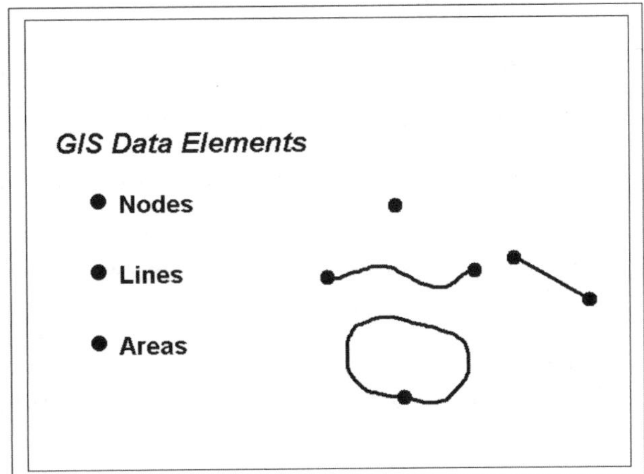

- *Nodes* represent the intersection points and the end points of lines. Each node is uniquely numbered and is located by a pair of XY geographic coordinate values.
- *Lines* are also uniquely numbered. Their geometry is described by a series of coordinate pairs. A straight line is defined by only two coordinate pairs (representing the beginning and the end), whereas additional coordinate pairs are needed to represent curvilinear features. The more coordinate pairs used, the more precise the geometric definition of the line.
- *Areas* are bounded by one or more lines and identified by a centroid, a point that must be located anywhere within the area.

Lines are encoded with their beginning node number and their ending node number, as well as the area to their left and the area to their right. Map features depicted as an isolated symbol, such as a tower or stream gauging station, are referred to as point features. In a GIS, points are represented as a special type of line element, known as a *degenerate* line. This is a line described by two identical coordinate values. Thus, it is a line having its beginning and end points at the same location.

The following illustration and table describe a topologically structured map and its corresponding GIS data records. There are several items of interest to note about them.

- All map features are uniquely identified.
- The nodes are labeled N1 to N10, the areas are labeled A1 to A5, and the lines are labeled L1 to L13.
- No node or line is duplicated.
- No line crosses another line without an intersection.
- Nodes and lines can be referenced to more than one area.
- The area outside the map is conventionally labeled A1.
- There are several lines with only two XY coordinate points because they are straight lines. Other lines have numerous coordinate pairs because they are curvilinear.
- The designation of the starting node and ending node of a line is arbitrary, as long as the area left and the area right correspond to the chosen "direction of travel."

Topologically structured map.

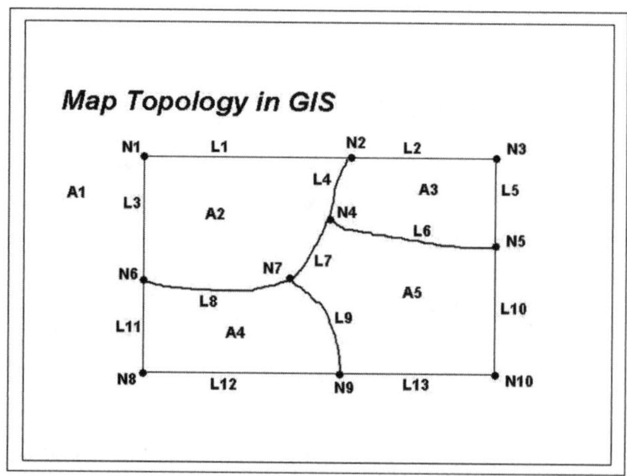

GIS data records.

GIS Data Files and Records

Node File

Node	
N1	XY
N2	XY
N3	XY
N4	XY
N5	XY
N6	XY
N7	XY
N8	XY
N9	XY
N10	XY

Area File

A1 XY (Outside)
A2 XY L1, L4, L7, L8, L3
A3 XY L2, L5, L6, L4
A4 XY L8, L9, L12, L11
A5 XY L6, L10, L13, L9, L7

Line File

Line	Start Node	End Node	Area Left	Area Right	Geometry
L1	N1	N2	A1	A2	XY,XY
L2	N2	N3	A1	A3	XY,XY
L3	N1	N6	A2	A1	XY,XY
L4	N2	N4	A3	A2	XY...XY
L5	N3	N5	A1	A3	XY,XY
L6	N4	N5	A3	A5	XY...XY
L7	N4	N7	A5	A2	XY...XY
L8	N6	N7	A2	A4	XY...XY
L9	N7	N9	A5	A4	XY...XY
L10	N5	N10	A1	A5	XY,XY
L11	N6	N8	A4	A1	XY,XY
L12	N8	N9	A4	A1	XY,XY
L13	N9	N10	A5	A1	XY,XY

Given these three data tables, the GIS software can start anywhere and quickly determine which lines are connected, which lines make up the boundary of an area, and which areas are adjacent. This capability is the foundation of spatial analysis in GIS.

In addition to the geometric and spatial data of the foregoing table, a GIS also contains attribute data. These are associated with the topologic elements (nodes, lines, and areas) and provide further descriptive information about them. For instance, in a GIS used for municipal tax mapping, a tax parcel would be defined as an area, and its descriptive data might include the lot number, owner's name, acreage, zoning, and so on. This attribute data is placed in a database separate from the graphics data.

GIS software is designed to permit the routine examination of both spatial and attribute data at the same time. The user is able to search the attribute data and relate it to the spatial data, and vice versa. For instance, in our municipal tax mapping example, the city planner may ask where are all the lots in the west end, larger than one acre, and zoned for industrial use. The GIS can respond by either listing their lot numbers or plotting their location on the city street map. This is a capability that neither a CADD nor an AM/FM system will normally provide.

Typical Applications

All three types of computer mapping systems have a specific role to play. For instance, a commercial map atlas company may use a CADD system. Its mapping applications are primarily for cartographic products. A telephone company will use an AM/FM system to support its telephone system operations and maintenance. It must be able to quickly trace a cable network and retrieve its attributes.

A wood and paper products company will use a GIS to manage its timber reserves. It needs to map tree stand polygons and perform many types of spatial analysis. Some organizations find they need all three capabilities. A state department of transportation, for instance, needs to create base maps to support highway design, make highway maps, analyze highway networks, and analyze spatial relationships among transportation, natural, and cultural resources.

Summary

In this chapter you looked at three principal types of computer systems that can be used to store maps and related data. CADD and AM/FM systems can be used to store, manipulate, and retrieve these data and are well suited for particular uses. However, a GIS is needed to fully analyze mapping data because it is specifically designed for spatial analysis.

A CADD system can be used for mapping, but it does not model the spatial relationships among map features. An AM/FM system models the network relationships of utility lines and equipment, as well as their attributes. A GIS models spatial relationships for all types of map features, representing them as nodes, lines, and areas. Moreover, a GIS can also store the attributes of map features.

The Uses of GIS

A Typical Municipal Government Application

In This Chapter...

The best way to begin to understand how GIS can be used is to visit a typical organization using GIS. This chapter looks at how a GIS might be used in a typical municipal government. It also describes the ways things were done before the GIS was installed, so that you can also understand some of the benefits it produced.

Consider a typical town of approximately 20 square miles in area with a population of roughly 30,000. You will take a look at the operations of the municipal government, assuming it installed a GIS several years ago. You will also visit the town hall, and take a tour of several departments using GIS in the following areas.

- Tax assessment
- Planning
- Registrar
- School board
- Public works
- Public safety

Tax Assessment

The first stop on our tour is the tax assessor's office. This department is responsible for assessment and collection of taxes on the town's 11,000 tax parcels. In the past, the tax assessor maintained two principal types of tax records: a set of tax maps and a set of real estate files. The tax maps were drawn in ink on Mylar at a scale of 1inch = 200 feet. Thirty map sheets were required to cover the town at this scale. These maps showed the city boundary, the rights-of-way, street names of all roads, the boundary lines of all property parcels, street addresses, and lot numbers, block numbers, and section numbers for all parcels. Parcel dimensions had been penciled in on a few of the larger and more prominent parcels.

These tax maps were first compiled about twenty years earlier by the state Department of Taxation from property records. The drafting section of the town's engineering department kept the tax maps up to date as new subdivisions were recorded, roads were built, or the town limits changed. After the revisions were made, the originals were copied and a new set of paper prints was distributed around town hall. Revisions were made on an irregular basis, but usually about every three months.

When the town bought a mainframe computer in the early 1970s, its real estate records were one of the first applications tackled by the new data processing department. The tax records had been maintained on assessment cards filed in the tax assessor's office. When the town installed a minicomputer in the 1980s, and then a client-server system in the 1990s, the tax assessment system and records migrated to these new systems. The tax assessment office has a full-time data entry clerk, who updates these real estate files daily. These updates are made necessary by new subdivision recordings, property transactions, and the annual reassessment. The updates are made at a computer terminal in the tax assessment department linked to the minicomputer.

When the town installed a GIS, the tax maps were among the first to be digitized. The town hired a service bureau to take its thirty tax maps and digitize them in a format compatible with the GIS that was being purchased. As each of the tax parcels was digitized, it was also tagged with the parcel identification number.

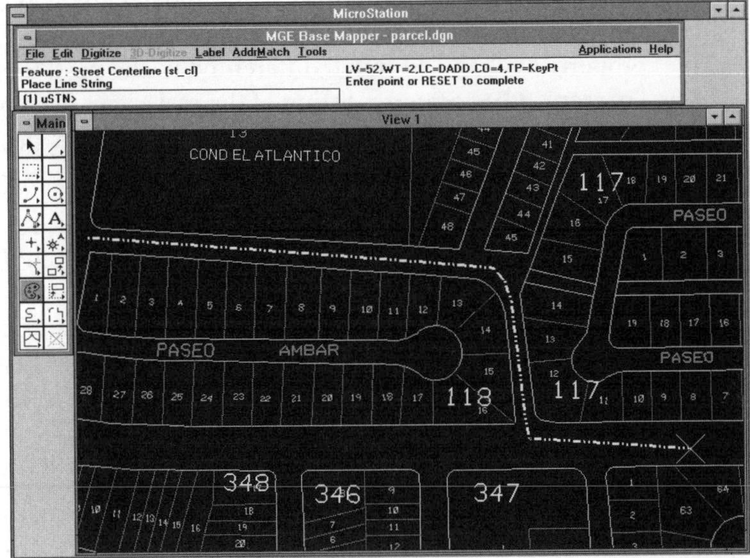

Tax parcel map on a GIS.

After the tax maps were loaded into the GIS, the software was able to link the graphic files describing parcel boundaries with the real estate files containing tax parcel data. The parcel identification number made this linkage possible, even though these two sets of files are maintained on separate computers.

Today, the engineering department still updates the tax parcel boundary data in the GIS similar to the way it updated tax maps in the past. However, now the updated information is immediately available to the tax assessor and other town departments. It is no longer necessary to make sets of paper copies and distribute them around town hall.

Very few people now require paper copies of tax maps. Instead, when someone calls or drops in for tax map information, they can call it up on a GIS terminal at the public information desk. Town employees can access the GIS data from the computers on their desks, and two more computers are located on the public information counter in the tax assessment department.

Tax map and assessment data is also available over the Internet. Citizens can use a PC equipped with a web browser in their home or office, or at one of the branches of the town's library to gain access to the town's home page. The tax assessment office maintains a web page and provides tax maps and assessment data through a simple query screen. Town employees and citizens alike are particularly pleased that map information is no longer divided into map sheets. The GIS gives them a continuous map of the entire town.

The tax assessment department still updates the real estate files that reside on the town's first computer, just as it did before. However, the annual reassessment of all properties is handled much differently. In the past, the tax assessor would sit with a tax map and a computer listing of the real estate files to conduct his reassessment. It was a very tedious process for her to correlate tax parcel data shown on the maps with tax records in the computer printouts.

Today, she sits at a GIS terminal and obtains the same information in a fraction of the time it used to take. She can ask the GIS to search the real estate files for properties with certain characteristics, and the GIS will not only find them in the real estate files but identify them on the tax maps. For instance, she may want to know all properties in a neighborhood that have sold in the past year, as well as their selling prices. The GIS can quickly find these properties in the real estate files and display them on the screen, along with their sales prices. The assessor uses this data to help with her reassessment of the properties in the neighborhood.

Even routine daily questions are answered much faster with the GIS. When a citizen, developer, or real estate agent comes into town hall seeking information about a particular tax parcel, the GIS finds and displays the tax map and the real estate data on that parcel much faster than it could have been retrieved before.

Planning

Now let's go down the hall to the planning department. In the past, the town planner used the engineering department's draftsmen to update the town land-use plan and zoning maps. The originals were drafted in ink on Mylar at a scale of 1inch = 1,000 feet. These maps showed the town boundary, streets, street names, and major landmarks, such as the local shopping mall. The land-use map also showed planned land uses, and the zoning map showed the current zoning of all parcels in the town.

One of the problems with these two maps was that, at this scale, it was not possible to show parcel numbers or street addresses. Therefore, the planning department also kept a set of paper tax maps and added the zoning boundaries to it using a felt-tip pen. The zoning boundaries were updated from time to time as the town council approved zoning changes. Unfortunately, there was not only a duplication of effort to keep these two zoning maps up to date but a problem of conflicting information between the annotated tax maps and the "official" town zoning map.

Since the town's GIS was installed, the planning staff has added the zoning boundaries and land-use plan boundaries to the GIS database. This has given them several important benefits. When searching for routine information about a parcel, a planning department employee simply types in the parcel number or street address instead of trying to locate the parcel on the official plan and zoning maps, or searching through the annotated tax maps. Moreover, special planning studies can now be done much faster than before.

For instance, the town manager frequently asks the town planner to research questions such as where are all parcels larger than five acres and zoned for industrial use? Or a member of the town council may ask him how much undeveloped land is planned for residential development. He now uses the GIS terminal located in the planning department to answer these questions. He completes his research in a fraction of the time it took him to do so in the past.

The town planner can also produce a variety of custom maps and reports with relatively little effort. Moreover, plotting the maps in color using the town's color electrostatic plotter makes very attractive presentation materials for public hearings. The planning department's web site also offers a variety of maps showing the town's land usage and trends from a variety of viewpoints.

Registrar

The last stop on the first floor of town hall is the registrar's office. The registrar uses the GIS in two ways. When new citizens come in to register to vote, she can quickly determine in what voting district they reside. She simply types in their address at the GIS terminal and the system searches for the address in its database and then compares this location with the overlay of voting precincts.

Before the GIS was implemented, she frequently was unable to locate the citizen's house. This was usually because he was moving into a new subdivision and the new streets had not yet been added to her paper copy of the town street map. Today, these street-map changes are immediately available to her as soon as the engineering department updates the GIS database.

The registrar also uses the GIS to analyze voting districts. State law requires that voting districts be reexamined after every census. The GIS can compare precinct boundaries with the Census Bureau population data and count the number of voters in each district. The registrar can try new precinct boundaries and immediately receive a recount of the population in the newly defined precincts. She can play this "what if" game to arrive at an even distribution of voters with a fraction of the effort required in the past.

School Board

Now let's walk upstairs to the office of the town's school board. The school board's facility engineer uses his GIS terminal to retrieve engineering and architectural drawings of the schools. All of the paper drawings were scanned into a digital format about fifteen years ago. Since then, all new drawings have been submitted in CAD format, as required by town code. The engineer can display a map of the town, click on the location of a school, and see a list of all drawing sets available for that building. He then selects the drawing set and the drawing of interest, and it is immediately displayed on the computer screen.

This system has replaced the large sets of flat files and hanging files he used to have in his office. Moreover, he does not have to worry that someone in Public Works has borrowed the drawing set he needs. They have access to the same drawings over the town hall computer network.

The school facility engineer also works with the town's public transportation manager to plan school bus routes. Because the GIS includes an overlay of U.S. Census Bureau demographic data, they have access to the most recent census data on school-age children. This data helps them plan school bus routes that evenly distribute the student passengers on each bus. These types of studies were not even attempted in the past. Instead, bus routing was done largely by intuition and trial and error.

The school board looks to the town planning office to help them plan new schools or consider the consolidation of schools. The planner uses the GIS for studies to support these actions. He can quickly determine the distribution of school-age children throughout the town, or report the number of students that live within a given distance, say two miles, of a potential school site. These studies are done in a fraction of the time they took when everything was done using paper maps and records.

Engineering and Public Works

The largest department of the second floor of town hall is public works. The director of public works is responsible for the town survey crew, the drafting operation, building inspection and permitting, maintenance and operation of the town's water and sewer systems, public transportation, solid waste removal, public buildings, and general engineering design. He is also responsible for the maintenance of all graphic data in the GIS database.

Survey Crew

The town survey crew uses an automated survey data collector for its field survey operations. This device allows the crew to record measured angles and distances automatically. In the past, these data were recorded by hand in field survey notebooks. Instead, the data collector records the point being occupied, the backsight station, the foresight station, the angle turned, the distance to the station, and a code describing each of these points. The survey crew uses a keypad on the side of the instrument to record station identification numbers and other descriptive codes.

When the survey is complete, the crew brings the data collector into the office, connects it to a data port in their PC, and uploads the survey notes from the data collector to a CAD system. The CAD program reads the data and calculates X,Y coordinate values for all points surveyed, as well as their elevations. The program also plots the data in the CAD file, using symbology appropriate to the type of point surveyed. Thus, if a highway was surveyed, the curb lines, manholes, sidewalks, poles, and so forth are automatically plotted by the system using the proper drawing symbols.

This automated process eliminates the time-consuming and tedious process of "breaking down" field notes by hand, calculating locations, and plotting a drawing by hand. When the file has been checked and the drawing finished, this CAD data is used to update the mapping data in the GIS.

The survey crew also uses a GPS survey system. This system was first used in the 1990s to establish a network of ground control points around the town, all referenced to the state's mapping coordinate system. Ever since, it has been relatively easy to tie all new survey work into the state mapping system. Because the GIS uses the same coordinate system, these surveys can be registered to the data in the GIS. Whenever digging permits are issued to utility contractors, the county "Miss Utility" service is called in.

This service uses electromagnetic sensors to locate any underground utilities in the excavation area. These locations are then spray painted on the ground. The town survey crew uses this opportunity to survey the exact location of the utility line. Using its GPS system and employing a procedure similar to that used with the automated survey data collector, this survey data is automatically recorded and plotted in the CAD system.

This data is then used to update the location of utility lines shown in the GIS. These lines had been traced into the GIS from paper utility maps, but their locations were only approximately correct. About half the town's utility line locations have been verified and refined in this manner. As a result, both municipal and local engineers have come to rely on the utility locations shown in the town's GIS when designing new utility systems or improvements in the public right of way.

Public Utilities Department

The public utilities department uses the GIS to manage the town's water and sewer systems. In the days before the GIS, the department used two sets of maps: one showed the location of water mains and fire hydrants; the other showed the locations of sanitary sewer mains and manholes.

These two utility-system maps were drawn by hand at a scale of 1inch = 1,000 feet. They showed pipe sizes and slopes where appropriate, but they did not have more detailed information such as pipe inverts, date of installation, pipe materials, manhole rim elevations, water pressures, and so on. This detailed information had to be found by searching through construction drawings.

Unfortunately, there was often confusion, even conflicts, in these drawings. This was because not all as-built information had been collected or kept up to date. Therefore, it was often necessary to verify utility data in the field. This required the survey crew to spend half a day or more gathering or verifying utility data.

Today, utility department personnel can use their GIS terminal to research utility data. For information about a particular manhole, they might call up the sanitary-sewer system map for the town and window the area in which the manhole is located. They then identify the manhole in question by pointing to it using the workstation's mouse. The system responds with a report on the manhole's attribute data, including invert elevations, rim elevation, date of installation, and so on.

Similarly, they can find information about a section of pipe, including its size, material, slope, capacity, and date of installation. Moreover, they can ask the GIS to search for information using questions such as where are all storm sewer lines larger than 24 inches in diameter? Or, where are all water valves installed before 1950? The system will search its attribute files to locate these facility items, and then display them on the workstation screen.

The public utilities department found that one of the great unexpected benefits of implementing the GIS was that it required a total cleanup of all utility data. To load the GIS database, all unknown information had to be researched and all conflicts in the existing data had to be resolved. This was a major effort, but today the operation runs much more smoothly because the data is readily at hand and reliable.

In the future, the public utilities department plans to add programs that link the engineering and topographic data in the GIS with standard engineering analysis programs. These include hydrology programs for performing rainfall runoff computations and hydraulics programs for analyzing the flow in pipe systems and for performing flood-plain computations.

Engineering Department

The engineering department is responsible for utility and roadway design. This department was successful in convincing the town to get relatively detailed topographic mapping at 1 inch = 100 feet, with a 2-foot contour interval. Although this cost more than the less-detailed mapping required by other departments, it has been very beneficial. The detailed topographic data provides a much better base for registering the utility and tax map data added to the GIS. In addition, less field survey work is required for engineering designs.

Moreover, the engineering department sells topographic information from the GIS to outside engineering and surveying firms for a nominal fee. The department's engineers use the GIS to aid in roadway and utility design. The system provides base sheet information for engineering design, including topographic and planimetric map data, as well as digital orthophotography. The topographic information in the GIS is compatible with their CAD system's engineering design programs. These include programs for roadway design, site planning, and utility system design.

The engineering department was also successful in convincing the town to require that all plan submissions to the town, such as for new subdivisions or for new utility systems, be made in a digital format compatible with the town's GIS. This has made it much easier to incorporate new data into the database and keep it up to date.

GIS Department

The public works department first established a GIS department when the system was installed. The GIS manager is responsible for overall GIS promotion, maintenance, and operations. The GIS system manager is responsible for supporting the GIS database. This includes daily file backup and archiving, contact with hardware and software vendors, and establishing user access privileges.

A GIS cartographer is responsible for maintaining the graphic files in the GIS database. This job involves updates to the tax parcel boundaries, topographic and planimetric data, utility data, and planning and land-use data. The town recognizes that as it grows and the GIS database grows, some or all of these database maintenance functions may have to be decentralized to the user departments.

The GIS system manager also provides user training for new employees, advanced training for existing employees, and technical support to users. This technical support includes troubleshooting, answering routine questions, referring problems to the vendor, and minor programming tasks. Programming is done chiefly to customize the basic capabilities of the GIS for GIS users. This programming is done principally using the GIS macro programming language.

Public Safety

Our final stop on this tour takes us to the police department's public safety center, which is across the street from town hall. The town's chief dispatcher is responsible for the operation of the public safety center. In the past, his dispatchers used a variety of reference materials to direct responses to police, fire, and rescue incidents. These included a collection of United States Geological Survey (USGS) maps, a town street map, and a collection of hand-drawn subdivision map sketches.

The reference materials also included a directory of street address ranges and an index card file. This card file contained important phones numbers, property owners' names and addresses, and an inventory of hazardous materials and buildings with invalid occupants. New dispatchers required several months on the job to master this eclectic set of references.

Shortly before the GIS was installed, the town installed an E911 telephone capability. When the GIS was installed, it was linked to the E911 system. Today, when a call is received through the E911 system, it transmits the caller's address to the GIS. The GIS then automatically locates the address and displays the tax map for the surrounding area on a console in front of the dispatcher. A target in the middle of the display pinpoints the caller's location. Thus, while the dispatcher is responding to the incident and directing response units, he is also presented with a map of the surrounding area. Information regarding the address, such as the presence of hazardous materials, is provided from the E911 database.

The GIS can also assist the dispatcher in routing the responding unit. The dispatcher points to the location of the available unit or units, and the GIS calculates and displays the route each unit should take to report to the scene. Either the shortest distance traveled, the shortest time required to make the trip, or the fewest number of left-hand turns, can determine these routes. The GIS will even display and print out directions for the responding unit to follow to the incident.

The police and fire departments also use the GIS for incident analysis. The GIS can read a file of police or fire incident locations, and then locate and display them on the town street map. This has proven to be a great aid to the analysis of trends in criminal activities.

Summary

The GIS database is stored on a server connected to the town hall's computer network, which serves all departments. GIS users have access to its database over the network. Before the GIS, many departments kept their own mapping records. Much of the data on these various map sets was duplicated, inconsistent, conflicting, or out of date.

Having a single GIS database eliminates these problems. All GIS users now have access to the latest and best mapping data available. Moreover, the GIS map data is linked to several databases that contain attribute data, including the tax assessment, utility system, and building and zoning permits databases. GIS users can ask questions of the map data and get answers involving attribute data, and vice versa.

As you can see, this typical town has used its GIS for a wide variety of applications. Because land data is so widely used in a municipality, utility, or other government agency, many different departments can use a GIS. The typical benefits of using GIS are discussed in a later chapter.

4 An Overview of the GIS Industry and GIS Software

How Big Is the GIS Industry, Who Are the Key Players, Who Uses GIS, and What Do They Use It For?

In This Chapter...

Before looking at GIS as a technology, let's first set the stage with an overview of the GIS industry. We will look at the software vendors, the users, the applications, and the service providers.

GIS Software Vendors

Most GIS software vendors are relatively small compared to major corporations, ranging in size from a few dozen to a few hundred employees. These companies usually sell indirectly to the customer, relying on dealers and value-added resellers (VARs) to demonstrate and deliver their products. The "value added" by the reseller may be training, software customization, implementation support, or other support services. A dedicated direct-sales force may cover some large market segments, such as utilities and the federal government. The GIS software vendors rely heavily on trade magazine advertising, direct mailings, and trade shows to advertise and promote their products.

Most GIS software programs offer the full range of commands for data input, manipulation, analysis, and presentation. The basic product may be a single program that offers nearly all needed GIS commands, or it may be the first of several modules needed to provide a complete suite of GIS commands. These are distinct from "desktop mapping" programs, which offer many of the same features as a GIS but are limited in their ability to support spatial analysis. These programs were developed to support individual user needs for mapping presentations, as opposed to supporting the entire organization's needs for managing, analyzing, and presenting geographic data.

Some GIS programs have been developed by governments and universities and are available for a nominal cost. These account for a small portion of all GIS software in use.

The GIS Software and Overall Products and Services Markets

Daratech, a computer market research company, annually compiles statistics on the GIS industry. The statistics cited in this chapter are taken from the Daratech report "GIS Markets and Opportunities 2000." The full report can be purchased by contacting Daratech at 617-354-2339, and more information is available at their web site, *www.daratech.com.*

The GIS market in North America enjoyed double-digit annual growth rates in the 1980s, but these dropped below 10 percent during the 1990-1992 recession. Since the middle of the last decade, the GIS market has been surging along with the national economy. Daratech reports that the GIS market grew a robust 12.8 percent in 1999. Worldwide, GIS software sales were more than $845 million in that year. The following illustration shows the top GIS software vendors in terms of 1999 company revenue throughout the world. This does not include sales figures for dealers and resellers.

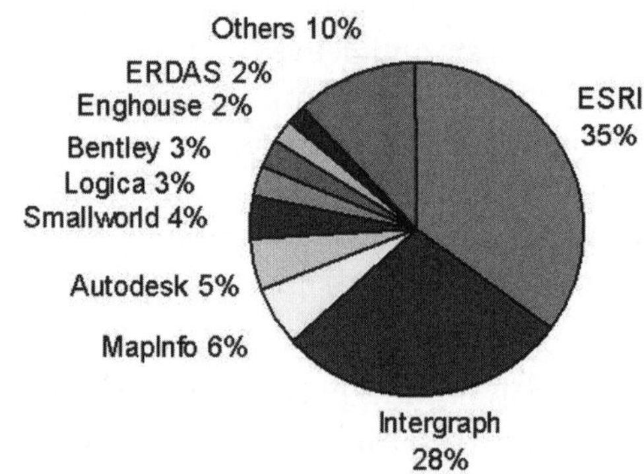

Worldwide GIS market shares.

As previously mentioned, Daratech estimates the total worldwide market for all GIS products (hardware and software) services at about $7 billion in 1999, split about evenly between North America and the rest of the world.

GIS Users

The following conceptual way of looking at GIS users is based on the author's own observations, and conversations with others who have consulted with and served large organizations that use GIS. These include municipalities, utilities, and military installations. The following illustration shows four general categories of the users of an organization's GIS data, and the relative numbers of each. GIS software vendors typically provide tools tailored to each type of user's needs.

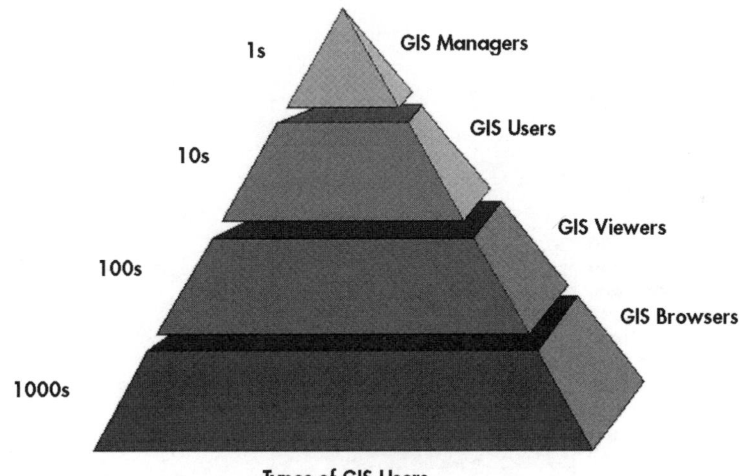

Types of GIS users.

- *GIS managers* are those who set data access privileges, manage the system and the GIS database, and maintain primary contact with vendors. They will frequently perform the functions of a GIS user, as well use "one-of-a-kind" GIS software tools on behalf of all GIS users (e.g., GIS software to change a map projection from a state plane grid map projection to a Universal Transverse Mercator projection).
- *GIS users* are those who maintain the GIS data and use it to perform sophisticated analyses. They may also need to create quality cartographic maps.

- *GIS viewers* are those who view and query GIS data, but do not edit it. They may need to create and plot simple thematic maps using GIS data.
- *GIS browsers* are those who occasionally view GIS data on the World Wide Web (the Internet) or the organization's internal LAN (an intranet). They are generally looking only for information content or for links to other web sites.

People in a wide variety of professions use GIS. The following illustration shows the primary types of GIS users, and indicates the percentage of all users that each type represents. This data was compiled in a survey of 3,800 readers of *GEOWorld* and *Business Geographics* magazines, of which 442 responded, or 11.6 percent.

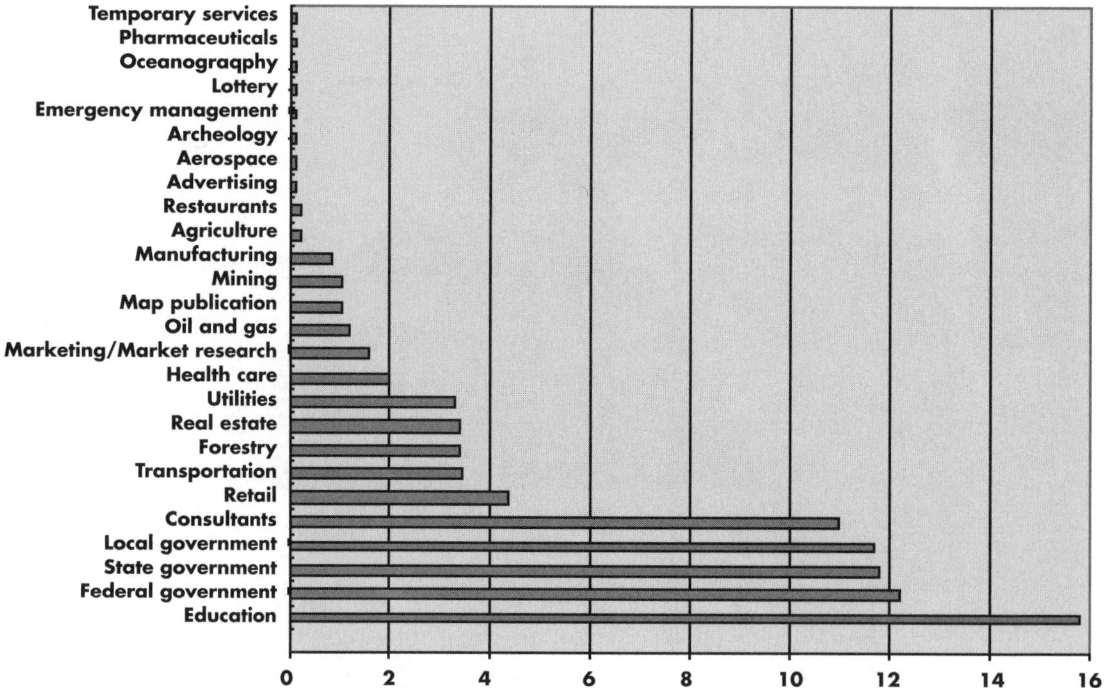

GIS users (percentages of all GIS users).

GIS Applications

Dataquest's market survey subdivided GIS applications into the following nine categories.

- Base data includes the creation of vector data and raster images representing physical features on the earth's surface. This information is used to register and digitize other types of GIS data, both tangible and intangible features. Such data is derived from aerial photography, digital orthophotography, photogrammetry, surveying, satellite imagery, and hard-copy maps. It is often generated under direct contract to the GIS user or for open market resale.
- Land information includes the creation and maintenance of the data for land records, land planning, and land use.
- Biological uses include environmental, public health and safety, forestry, and agricultural.
- Geoscience applications include oil, gas, and mineral exploration.
- Infrastructure management includes transportation, logistics, emergency services, and dispatch management.
- Utilities include water, sewer, storm water, electric, gas, telephone, CATV, data communications, and steam systems.
- Business marketing and sales involves demographic, sales and locational analysis, as well as providing travel directions.
- Geopolitics involves the military or other defense use, as well as political districting.
- Cartography, or mapmaking, is the final category.

The following illustration shows the distribution of the worldwide software sales according to these nine categories of GIS applications.

Worldwide GIS software sales by application.

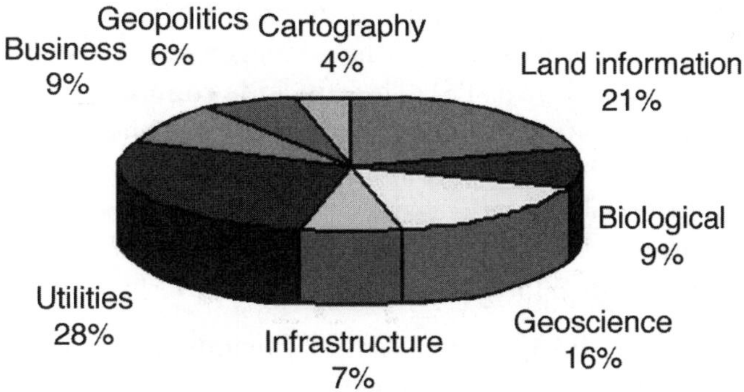

GIS Services

Now let's look at the types of services needed by organizations that implement a GIS. These roughly correspond to industry segments. As you will see, however, there is a great deal of overlap among the services provided by these industry segments.

GIS Consulting

The GIS consultant assists the organization in examining its needs for GIS, and then in planning the implementation of the new system. He may also be called upon to prepare specifications for the GIS hardware, software, mapping, and data conversion services, and assist in the evaluation of vendor proposals or the review of contractor submissions.

According to *GEOWorld*'s *1999 GIS Sourcebook*, there are about 500 companies offering GIS consulting services worldwide, roughly two-thirds of these based in the United States. Consultants have assisted a significant portion of the nation's larger cities, counties, and utilities in planning their GIS programs. Some firms specialize in GIS consulting, but most mapping and data conversion firms, as well as many of the hardware manufacturers and software vendors, also offer this service.

Mapping and Database Conversion

Purchasing a GIS is like buying a refrigerator in that it is "empty" when delivered. Someone must "stock it" with data before the system is useful. This requires creating a digital base map to which all other GIS data themes can be registered. This often means creating a new topographic base map from aerial photography using photogrammetric mapping techniques.

Photogrammetric mapping is a specialized field requiring expensive, specialized equipment and highly trained mapping technicians. There are over 200 private photogrammetric mapping firms in the United States, according to John Pallatiello, Executive Director of the Management Association of Private Photogrammetric Surveyors (MAPPS). These include companies that specialize in this service, as well as consulting engineers, mapping firms, and surveyors that offer this service among a broad range of capabilities. According to a recent MAPPS survey, most of these firms had experienced a growth in business during the past year. "Much of this growth has resulted from increased demand for digital orthophotography to serve as a base map for GIS," says Pallatiello.

Other GIS database themes (including tax parcels, environmental data, district boundaries, and utilities) are usually digitized from existing hard-copy maps. Attribute data for these other GIS database features are usually keyed in and attached to the corresponding graphic data features. Providing these services also requires special equipment, software, and training. Photogrammetric mapping firms, consulting engineers and surveyors, and specialized service bureaus provide these services. According to the *GEOWorld 1999 GIS Sourcebook*, about 350 companies offer GIS data conversion services around the world. As with GIS consultants, roughly two-thirds of these are based in the United States.

The decision whether to perform data conversion in-house or use contractor personnel is not a straightforward choice. Most GIS users, especially public agencies and utilities, only account for departmental salaries and (sometimes) fringe benefits. Contractors and consultants must charge for general and administrative costs, as well as a profit, in addition to labor and labor overhead. Therefore, on an hourly basis, in-house personnel are far less expensive than outside contractors and consultants. On the other hand, the outside contractor is a specialist in the conversion process and usually has sufficient equipment and resources to accomplish the work in a fraction of the time required by in-house personnel.

Training

The new GIS user requires training for the employees who will use the system. There are different levels of training required, depending on the type of user the employee happens to be. The GIS software vendor usually provides the initial user training using both their own training departments and authorized dealers and training centers. Subsequent training is often provided through in-house training programs, especially for casual users. Training in a particular GIS software package can also be purchased from software consultants and mapping firms that use the program.

Dr. Jay Morgan, Associate Professor of Geography at Towson State University near Baltimore, Maryland, says that colleges and universities are concentrating on educating students in GIS concepts and theory, not particular GIS software programs. Instead, this is being done by an increasing number of community colleges. Many of these local schools have already developed courses for training in specific word processing, spreadsheet, and computer-aided drafting (CAD) programs, and are adding classes for specific GIS packages.

Surveys conducted by Dr. Morgan have identified more than 900 schools that teach GIS courses around the world. Appendix B presents a list of such schools in North America, as well as further information on GIS education and training. According to Morgan, employers are willing to pay a premium for new employees that have basic training in the same GIS package the firm uses.

Software Customization and Integration

GIS vendors have developed programs that are useful to the largest possible number of customers. They cannot possibly deliver products that will meet all of the needs for every customer. Therefore, they have provided software tools that permit the user to customize the GIS software for his particular applications. This development work greatly enhances the efficiency of the GIS operations. Moreover, the GIS need not be a standalone, or "stovepipe," system.

It is often integrated with other data processing systems and databases, including systems used for computer-aided drafting (CAD), electronic document management (EDM), computer-aided facility management (CAFM), and administrative databases. The GIS software vendor will almost always offer programmer training in its software development languages. Many of the software vendors, as well as numerous small software consulting firms, provide software development services. Sometimes the GIS or mapping consultant will offer this service as well.

Database Maintenance

The GIS is a model of the real world, but the real world is constantly changing. These changes must be reflected in the GIS database. The GIS loses credibility if the data is not kept current. This work involves changes to both the graphic and non-graphic (attribute) data in the GIS database. Like the initial database creation effort, it is usually very labor intensive. However, in-house users are most familiar with the changes to the GIS database. Moreover, in-house personnel are less costly to the individual department in a public agency or utility than outside contractors, as previously discussed. Therefore, in-house personnel usually perform database maintenance.

Summary

In 1999, the worldwide market for all GIS products and related services was nearly $7 billion, according to Daratech. There appears to be plenty of opportunity for the technology to penetrate more deeply in two ways. Many organizations that fully embraced GIS some years ago, such as utilities, have still not completed the implementation of their systems. Across other market segments, such as municipalities and businesses, there are still many organizations that have not made a full commitment to GIS.

Moreover, the technology is supporting entirely new applications, including vehicle navigation systems, precision farming, and battlefield simulation. Recent new computer system technologies, such as the World Wide Web, will undoubtedly make GIS easier to use and more widely available. This may spawn even more GIS applications. If so, the GIS market forecast will likely be revised upward in the future.

GIS is one of the fastest growing segments of the computer industry. GIS has not only proliferated in North America, it is also gaining widespread acceptance throughout the world. The next chapter describes the benefits of GIS.

Why Implement a GIS?

Advantages of GIS Over Traditional Methods of Land Data Management

In This Chapter...

A GIS represents a major commitment of time, money, and organizational energy. Before getting into it, you need good reasons for *investigating* the use of a GIS in your organization. This chapter reviews six of the major benefits of a GIS.

Map Data Is More Secure and Better Organized

A common problem with traditional map files is lost, misplaced, or misfiled sheets. In many cases, there is also a problem with not being able to use a map that someone else has borrowed. And there are many map filing systems that are so poorly organized as to be nothing more than a place to get maps out of the way. The GIS is a central computer database of all map data.

Ideally, one group or person is responsible for maintaining the database. Although others may be permitted to update the content of the database, the GIS manager is responsible for security, organization, and access to the data. If he or she is doing the job properly, the data is consistently available to all authorized users.

Redundancy and Other Problems of Multiple Map Sets Are Eliminated

The amount of data a paper map can portray is limited by the map's size and scale. This usually means that a number of map sets are required to meet all of an organization's map requirements. These map sets have varying themes, scales, and levels of detail.

Much information gets duplicated in these map sets. *Base map* information such as roads, drainage, and boundaries is usually repeated in each set. Thematic information is often repeated at varying scales to show different levels of detail. As a result, the same revisions must often be made to several different map sets. Also, not all map sets are updated at the same time. One set may be current while another has out-of-date, and therefore conflicting, information. Moreover, the quality of the content and appearance of map sets produced by different departments often varies considerably.

A GIS contains only one set of each type of map data theme. Whether the entire GIS database is stored on a single file server, or it is distributed on several file severs around the network, each data theme is stored only once. This eliminates redundant data. Nonetheless, all GIS users can access all of the themes across the network, as though all of them were on his computer. This "virtual database" can be used to present numerous maps at varying scales, showing different levels of detail. In fact, the GIS can handle much more map detail than was normally feasible in the past.

The GIS can also produce maps with different combinations of information to cover a variety of mapping themes. This means that the base information on all types of maps can be updated by a single revision. Because all maps are derived from the same database, the quality of their content and graphic presentation is more consistent. And, when a revision is made in the GIS database, all users immediately have access to the most current data.

Mapping Revisions Are Easier and Faster

Modern manual cartographic techniques are much improved over earlier processes. Copper plate engraving, ruling pens, and linen sheets have given way to scribing, drafting pens, and Mylar. Yet revisions to traditional maps still require erasing and re-inking, or opaqueing and rescribing, the map original. A trip to the photo lab is often required as well.

Similar to a CADD system, a GIS dramatically affects mapping production. Changes are made by simply identifying an element for modification. All nearby or intersecting elements remain unchanged. A 3:1 improvement in the time required to make map revisions is typical. Less time needed for revisions means fewer people and less cost to keep maps up to date. This is very conducive to keeping the map database current. A GIS can produce multiple color plots or color film separations directly from its plotter, thus eliminating the need for a trip to the photo lab.

Map Data Is Easier to Search, Analyze, and Present

Map users often ask questions like these:

- What is nearby?
- What is in this area?
- What else can you tell me about this?
- What areas have both of these characteristics?

These questions usually require extensive research and analysis of maps and related data. With traditional paper maps and manual filing systems, this analysis involves comparing maps sheets of different scales and different themes, as well as researching stacks of card files to correlate the graphic and nongraphic data. This is a tedious and time-consuming process. (One of Murphy's Laws must be: Important map information always falls on the match line between two sheets. Critical information always falls at the intersection of four sheets.)

Even where nongraphic data such as real estate records has been computerized, the researcher may still have to comb through sheets of computer printouts while visually comparing the data with the maps. A GIS provides the researcher with powerful automated tools to answer these questions. These tools make it much easier to analyze the data for special studies and reports. In fact, new types of analyses that were not feasible before are now possible.

The GIS can quickly search through map data, looking for features with certain characteristics, or inspecting spatial relationships among features. Moreover, graphic data and attribute data are explicitly linked. Thus, the GIS user can search for map data using attribute data as a criterion, and vice versa.

For example, the GIS can automatically and quickly answer a question such as where are the locations of all vacant parcels larger than one acre and zoned for commercial use? With traditional maps, this question would require a search through the real estate files, followed by a visual review of the tax maps, and possibly a review of the zoning map if the real estate files did not include zoning data.

The GIS can also greatly reduce the cost of producing customized maps and reports. The user can go directly from the GIS display of the results of an analysis to a plotted map, graphic, or report. Maps can be produced at any scale, covering any area, and in any combination of data.

The problem of map data falling on the match line between sheets is eliminated because the GIS contains a continuous map. It is not only easier to conduct special studies of map data, but retrieving routine data is much easier and faster. This can be a big help to the municipal employee providing general information to the public. It represents an even greater benefit to the police or fire dispatcher directing emergency response operations.

Employees Are More Productive

The GIS greatly increases the productivity of employees who collect, manage, analyze, and distribute land data. They are able to produce more because they can accomplish these tasks in less time. Although the GIS may not eliminate the need for existing staff positions, it usually reduces the need for additional staff. These may be positions that have not been filled because of budget constraints.

Using the GIS, the existing staff becomes more productive and is able to accomplish work that was previously left undone. Moreover, employees are able to produce better work. Because individual tasks can be done faster, employees are able to complete a larger scope of work and do a better job.

Map Data Is Integrated Throughout the Organization

Most organizations have departments that maintain their own independent filing systems. These departmental systems often include maps of various types, typically each one referenced to a different geographic coordinate system. For instance, a municipal tax assessor may reference real estate records to a parcel number, whereas the public utilities department references water and sewer customers to an account number, the public works department references sewer data to a manhole or sewer line number, the building inspector references construction permits to an address, and the police dispatcher references building information in his E911 system to a telephone number.

These reference systems may work well within each department, but it is very difficult to correlate the data from different departments. Each system exists as an independent island of information. A GIS provides the opportunity to key all map and map-related data to a common reference system. Usually this is an X,Y coordinate system, such as the state plane grid.

All data used for reference purposes before is still available to the users, so they can continue to use the retrieval process with which they are most familiar. But the data for all departments is now linked to a common geographic reference system. This permits all departments to have access to a common mapping database. Therefore, the operations of all departments that use GIS can be more fully integrated. Moreover, GIS data can provide a link to other types of corporate data.

For example, a GIS user in the public works department might access tax assessment information on a property parcel through the GIS. Conversely, someone in the tax assessor's office might find out about the availability of water service to a parcel through the GIS. This integration of GIS among departments and with other information systems is more than simply building bridges between islands of files. The GIS represents a mainland to which all users can go for information (see the following illustration).

The GIS concept of data sharing.

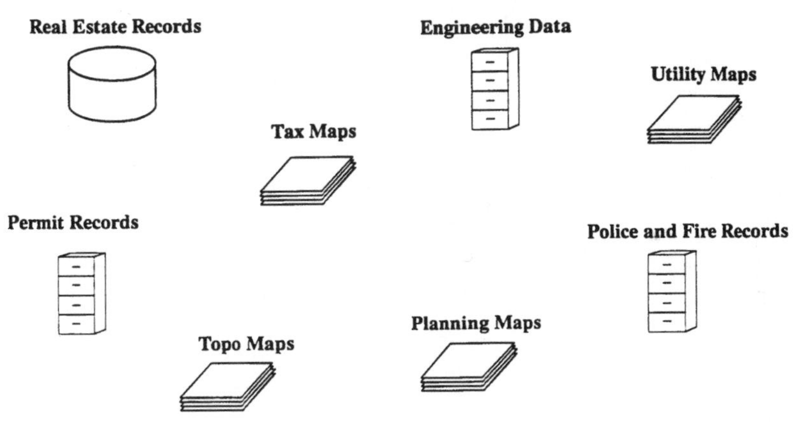

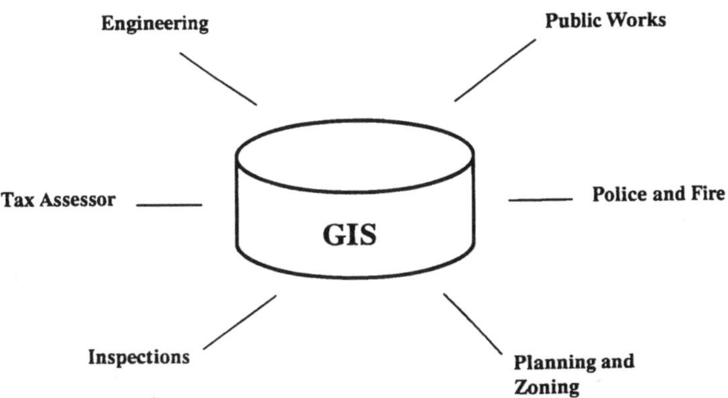

Summary

GIS is revolutionizing the way in which maps and map-related data are stored. The impact on cartographers, planners, engineers, surveyors, and others is comparable to the impact computers first had on accountants and administrators in the 1970s, the impact word processing had on secretaries and typists in the 1980s, and the impact the Internet had on marketing and sales managers in the 1990s.

6 GIS Technology Trends

How Current Developments in Information Technology Are Affecting GIS Applications

In This Chapter...

Information technology (IT) has been developing rapidly for the last 30 years. Because GIS relies on computer systems, it is obvious that changes in the broader field of information systems technology will have an effect on GIS. GIS also relies on a database of graphic information. This spatial database can represent the largest single component of cost for a GIS. Because of the importance of this database, Chapter 7 is devoted to discussion of trends in the area of GIS data collection and entry. This chapter looks at developments in four important areas of IT.

- Data networks and data communications
- Computers
- Computer operating systems
- GIS software

Because technology is changing so quickly, there is a risk that any discussion of GIS technology trends will seem ridiculously out of date by the time it is in the hands of the reader. Therefore, discussion here is limited to well-defined, general trends observed over the past several years. There are several peripheral systems that affect GIS, including video displays, data storage, multimedia, scanning, and plotting. These technologies have also been developing quite rapidly.

Data Networks and Data Communications

From the advent of commercial computer systems in the late 1960s until the early 1980s, mainframe computers and minicomputers were the core of virtually all data processing systems. The dominant trend in computer technology today is toward *distributed computing*. Other terms, such as *client-server architecture* and *network computing*, are also used to refer to this type of computer architecture.

In the past, a central computer was the host to a variety of terminals and peripheral devices. All data was stored on disk drives connected directly to this central computer. Thus, the mainframe or minicomputer was required to execute instructions, perform computations, and control data inputs, outputs, and storage, as well as to direct the flow of data for all users. A *homogeneous* computer environment was required—one in which all computer resources were supplied by one principal vendor and its closely allied manufacturers of peripheral devices.

For GIS users, one of the principal drawbacks of centralized computing was that the computer became a virtual bottleneck for the massive amounts of graphic data to be processed. A common joke among mainframe and minicomputer GIS users was that after returning from lunch, you log into your file, then go for a cup of coffee while the computer retrieves and displays it, along with all of the files needed by five or ten other users returning from lunch (a wait of up to twenty minutes).

However, since the early 1990s, the trend has clearly been toward distributed, *heterogeneous* computing. Computer users have been demanding the ability to network all of their existing systems. They have also rejected the "one vendor" approach. Vendors have responded by downplaying their proprietary solutions and embracing industry standards, a concept often referred to as *open systems*. This gives users the flexibility to select the best combination of solutions for their applications. Individual procedures within an application can be processed on computers and devices best suited for the task.

The concept of distributed computing is shown in the following illustration. It depicts three local area networks (LANs) supporting three different types of applications: CADD, GIS, and management information systems (MIS). The MIS might include financial, personal, logistical, and operational computer systems.

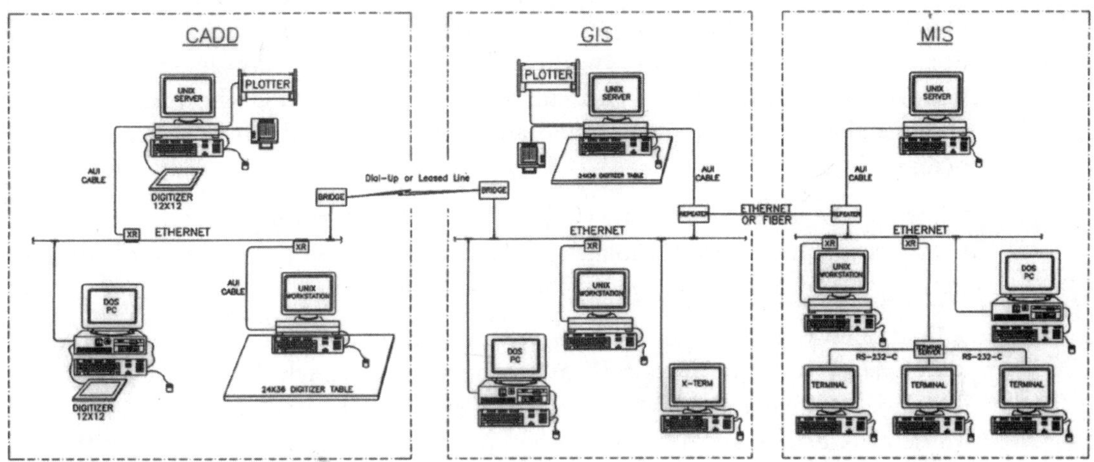

A distributed computing system.

Each of these LANs is built upon a *backbone*, a single cable that interconnects all devices on the network and transmits the data among them at very high transmission rates. Several types of network cabling can be used, including coaxial, twisted pair, and fiber optic. A protocol for data transmission is also required. Several networking protocols are commonly used today, including TCP/IP and IPX/SPX. The network protocol allows all devices on the LAN to communicate and exchange data using a common data format.

The previous illustration also points out the fact that several *types* of computers can be added to the LANs. Both PCs and engineering workstations can be linked over the network. A mainframe or minicomputer could also be connected to the LAN. The primary requirement is that each of these computers is able to support the network's protocols for data transmission.

Some of these computers can be used as file servers for other users on the network. Other network protocols, such as Network File System (NFS), allow a user to have access to files that reside on computers other than her own. This makes it possible to set up a common database of information, as well as to share application software.

Sharing a common database is very important in many multi-user GIS environments, such as municipalities and utilities. Many GIS vendors offer network licenses that permit a designated number of user workstations (*clients*) to have access to the application software residing on a file server. The chief advantage to this concept is that only one copy of the software need be installed and maintained, making system administration much easier.

An obvious advantage of distributed computing is that each user has access to his own computer, whether it is a PC or a workstation. Because the user does not have to share this computer with other users, he can count on consistent response and performance. Again, this is particularly important in GIS, where the size of the graphics files to be manipulated and analyzed can be in the range of several megabytes each.

Numerous other devices can be installed on a LAN, including printers, plotters, and scanners. Because the LAN makes it possible for all users to have access to any of these devices, it is far less expensive to provide them.

A final point to note about the previous illustration is the linkage between the three LANs. There are several ways to interconnect LANs, including coaxial cable, fiber lines, dial-up telephone lines, microwave, and leased (permanent) telephone lines. These LAN *bridges* allow users on one LAN to access devices and data on the other LANs.

This is particularly important in GIS applications, where GIS software must have access to databases that are typically stored on other computer systems. In a municipality, for example, a GIS user in the planning department may need to access data contained in the tax roll database maintained on the computer serving the tax assessor's department. Similarly, this planning department user may need to access the utility data maintained by the public works department on its CADD system. Likewise, there is usually a need to share data between GIS and MIS users.

The previous illustration also indicates a connection to the Internet. The Internet is a worldwide network, or "web," of data servers and telecommunications links. It is operated largely by public and educational organizations on a purely cooperative basis.

Part of the phenomenal success of the Internet, particularly in the case of the World Wide Web, is a user interface that is very easy to learn, use, and navigate, and that is common to several computer operating systems. It has placed tremendous amounts of information instantly at the fingertips of computer users around the world, to say nothing of the fact that millions of Internet users have been linked for sharing information. As will be discussed further under the topic of "Software," which follows, GIS data can now be made available over the Internet.

Distributed computing architecture and open systems standards offer several significant advantages to the organization using GIS. One is the fact that it is possible to increase the size of the network and serve additional users or applications in an incremental fashion. *New computers, new peripherals, and new software products can be incorporated while still protecting the investment in the original network architecture, hardware, software, and databases.* All that is required is that the new hardware and software comply with the standards and protocols of the existing distributed network.

Another significant advantage lies in the fact that individual departments responsible for processing and updating a particular data layer or theme can restrict control of that data to their LAN server. Other departments can still have access to the data, but can be prevented from editing or adding to that data layer.

Computers

GIS users have unique computer requirements the typical office computer user does not. They work with large amounts of data and need to see intricate drawings, plans, and maps on the computer screen. They perform several operations that require a lot of computer power all at once, such as running an analysis while plotting a map. GIS projects can involve dozens of files totaling hundreds of megabytes of data storage.

During the 1980s and early 1990s, many organizations found that the typical PC did not offer the processing power (computational and input/output speed) to satisfactorily handle their GIS applications, especially if their spatial database was large. As the GIS database grew in size, the response time of the PC slowed down significantly, to a point where the user not only became frustrated but lost precious production time.

For this reason, GIS users frequently turned to engineering workstations for their technical work. These systems offered greater processing power than the PC. This was partly due to the fact that these workstations employed a 32- or 64-bit data path, meaning that they process data in chunks two or four times as large as those of the typical PC, which ran on a 16-bit operating system. It was also due to the fact that the microprocessors used in engineering workstations were simply much faster than those used in mass market PCs. Finally, workstations offered some important options, such as high-resolution graphics and large display monitors to view those intricate drawings and maps.

Although engineering workstations offer many advantages to the technical user, there are some disadvantages as well, the principal one being a higher price than the typical PC. Another is the fact that engineering workstations are generally based on the UNIX operating system. UNIX is a more powerful operating system than DOS or Windows, but also more difficult for the average computer user to learn. Moreover, choices for office software running on engineering workstations are somewhat limited because software developers found a much larger market for Windows applications, so they naturally concentrated on PC products.

Some UNIX workstations offer the ability to run PC applications by emulating the Windows operating system, but the programs generally run more slowly in the emulation mode. As a result, one finds many GIS users with two machines on their desktop: a PC for office automation tasks such as word processing and e-mail, and an engineering workstation for technical tasks.

A third type of desktop computer appeared in the 1990s: the personal workstation. The personal workstation is a single computer that supports both processing-intensive technical applications as well as office applications. Based on the Windows NT operating system, personal workstation users can more easily share the same personal productivity tools with PC users: they can exchange files, and can use the same e-mail system, the same word processors, and the same spreadsheet programs "right out of the box." The personal workstation includes a number of standard features for the technical user that the typical PC may not have, including faster data buses and large-screen, high-resolution monitors.

Personal workstations also offer processing speed comparable to engineering workstations. Like the vast majority of PCs, they are built around one of the Intel family of microprocessors. Until the introduction of the Pentium microprocessor, Intel microprocessors had not offered the processing speed and capabilities of the microprocessors used in engineering workstations. The following illustration shows the dramatic growth in the processing speed of Intel microprocessors.

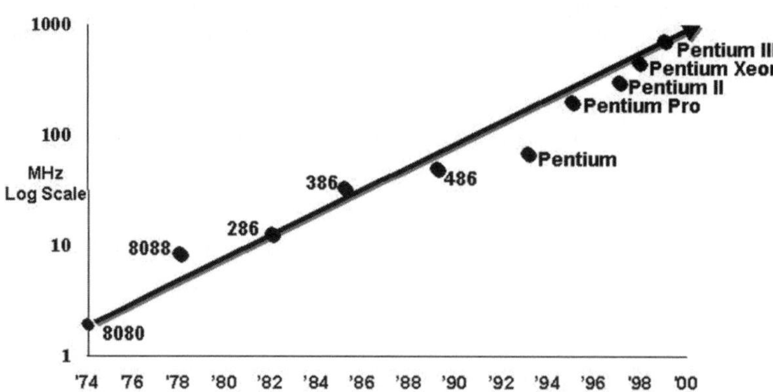

Intel microprocessor speeds.

Thus, the personal workstation offers engineering workstation speed, features, and high-resolution graphics, along with lower prices and PC ease of use. Reducing the cost of GIS hardware and providing a familiar Windows interface will make GIS both more available and more usable.

Operating System

As previously mentioned, a key component of the personal workstation is the Microsoft Windows NT operating system. Microsoft has steadily improved its operating system products over the years. The Microsoft disk operating system (MS-DOS) was developed in the early 1980s for the first IBM PCs. The Windows operating system, introduced in the latter half of that same decade, was a milestone. Perhaps the most important new feature of Windows was the graphical user interface (GUI) that replaced the DOS command line and syntax, making it far easier to use.

Windows NT is targeted toward the data server and personal workstation market, and takes advantage of the advanced features of the Intel family of microprocessors. Windows NT was the first Microsoft operating system to natively support 32-bit applications, combining the power of UNIX with the user friendliness of Windows. It gave users access to the wealth of personal productivity tools already developed for Windows. Built-in networking capabilities provide access to information maintained on workstations, PCs, servers, and mainframes.

Windows NT supports *multitasking*. Initially, PCs contained a single microprocessor, and the DOS operating system ran only one application at a time. Changing applications required the user to close one application and open another. The Windows operating system introduced multitasking, which allowed a user to have two applications open at one time (though only one could be executing).

This feature increased productivity because it allowed users to easily switch between applications by clicking on a window or hitting a "hot key" combination. However, having only one active application or task at a time means waiting for one task to complete before the other can become active. For example, a user must wait for a print command to complete before continuing with word processing.

Windows NT also supports *symmetric multiprocessing*, which enables multiple microprocessors to be productive at all times, increasing overall system throughput and taking advantage of available processing power. Multiprocessing had traditionally been limited to large, relatively expensive computer systems. Multiprocessing enables multiple tasks or applications to run in parallel on different microprocessors. Performance is improved, in that two tasks running in parallel will finish more quickly than the same two tasks running serially. Personal workstations are designed to operate with more than one microprocessor in order to take advantage of the multiprocessing capabilities of Windows NT.

Microsoft also developed an important new standard with the introduction of Windows' Object Linking and Embedding (OLE). GIS users typically employ several other types of computer applications, such as word processing, spreadsheets, and

CADD. They may need to include information created using one of these other programs within a GIS file. For instance, the text from a word processing file may be placed within a GIS map.

Yet a more subtle problem for technical users is the fact that they must use several software packages, including both office and technical applications, to get the job done. Even if the technical and office programs were on one desktop computer, they would still have to work among several programs during the course of a typical job. Moreover, the final product may include several components created using different merged computer programs. OLE provides the means of integrating these applications. The following illustration depicts the integration of GIS and office automation software.

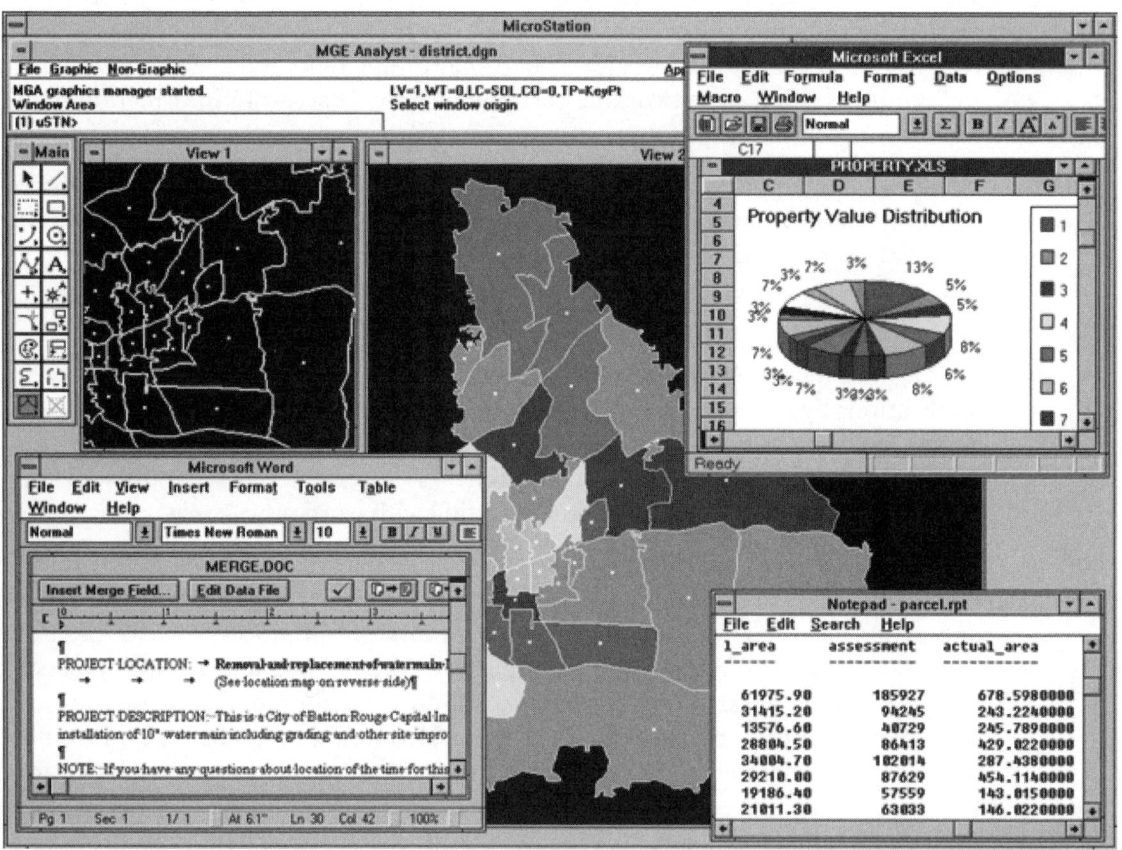

GIS integrated with office automation.

OLE enables the user to create documents consisting of multiple sources of information from different applications. These data "objects" can be almost any type of digital information, including vector CADD and GIS graphics, spreadsheets, text, raster images, digital audio, and digital video. The Windows standard for object technology is called the Component Object Model (COM). Each object will be created and maintained by its program, but through the use of OLE the functions of several different programs can be integrated. The user will feel as if a single application is being used, even though it has all of the functionality of several other applications.

End users of OLE applications do not need to be concerned with managing and switching between the various programs; they focus solely on the compound document and the task they are performing. Computer programs that take advantage of OLE will have greater ability to interact seamlessly, allowing users to focus more on creating and managing information rather than on remembering how to perform procedures.

OLE also offers the means of developing a new generation of applications that interoperate more effectively. It is an application programming interface (API) that defines a set of functions programmers can use to incorporate advanced visual editing and automation capabilities into their programs. Thus, OLE is the foundation of a new model of computing. With OLE, programmers can define a set of operations and make them accessible to other applications. The result of this capability is that applications can interact with one another without human intervention.

The extension of "object" technology across networks means that users will have access to almost unlimited information, regardless of its type or how it was created. Furthermore, this access will require almost no knowledge of where the information is located. Users will have more control over their computer environment and will be able to achieve more work with less effort by manipulating objects, rather than application programs, on their own computer or across networks. UNIX operating system providers have initiated a similar standard, known as the Common Object Request Broker Architecture (CORBA).

Software

As previously mentioned, an important trend in computing systems architecture is the development of a common graphical user interface (GUI), or window, to all applications software. A GUI such as Windows or the UNIX "X-Windows" environment is now a standard feature for the leading GIS software packages. These *point-and-click* menus have made GIS software much easier to use than earlier *command-line* interfaces, which usually baffled and discouraged users because of their complicated syntax. The GUI also reduces training time by presenting the user with intuitive windows, icons, dialog boxes, and on-line help.

Related to the development of the GUI is the use of standard application programmer interfaces (APIs). GIS vendors recognized that their customers would want to tailor a core GIS product to suit their particular needs. They offered macro development languages for this purpose, but each GIS vendor had its own such API.

Today, GIS products may be customized using standard APIs, such as the Visual Basic and Visual C programming environments. This eliminates both the need for specialized training of the GIS customer's programmer and the GIS vendor's cost of developing and maintaining a proprietary API. Using graphical APIs, even GIS end users will be able to customize their applications with little training.

The advent of *view-only* software has made GIS data available to "casual" GIS users. These users include the public, as well as staff members that do not use GIS on a consistent basis. They rarely need sophisticated analytical tools. Rather, they primarily need to view and query graphic and attribute databases. No data entry or editing commands are available in a view-only GIS package, but the software is very inexpensive and permits nearly anyone to access the GIS database in order to make simple queries with only minimal instruction.

Likewise, products are also available that place GIS data on the Internet. Allowing the data provider to strictly manage what data is available, these GIS data servers also allow Internet "browsers" to explore both GIS spatial data and nongraphic attribute databases, or to link to other Internet sites. Some GIS products now support the Wireless Application Protocol (WAP). This is the standard that enables a personal digital assistant (PDA, or "notebook" computer) to communicate over a cell phone network with the Internet. GIS data viewer products and Internet web servers have multiplied an organization's potential number of GIS users a thousandfold. The following illustrations show a simple view-only GIS product, an Internet browser viewing GIS data, and a hand-held computer viewing GIS data over a wireless network.

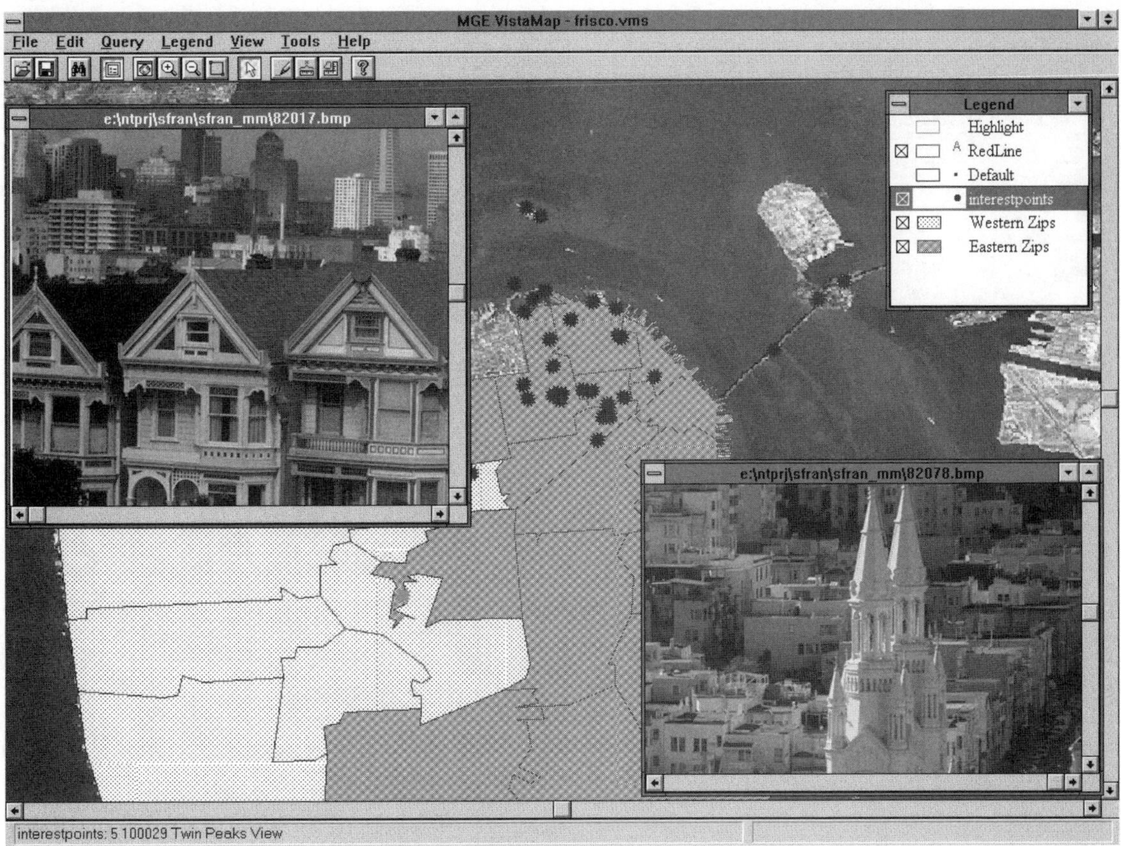

View-only GIS package.

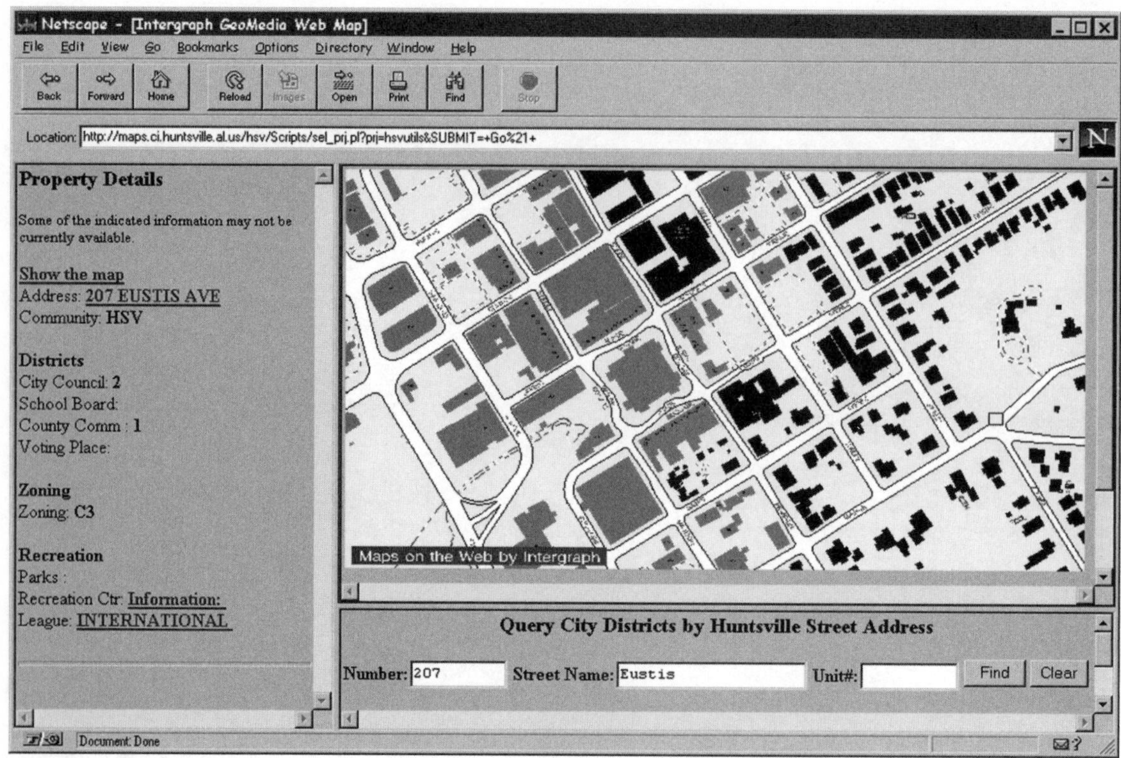

GIS data viewed on the Internet.

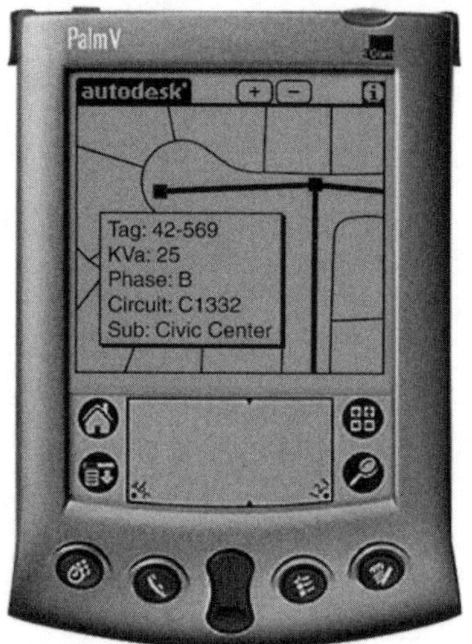

GIS data viewed on a hand-held computer over a wireless network.

GIS vendors are also offering the ability to link raster images and other multimedia data files to vector graphics for retrieval and display. For example, the user might point to a building outline in the vector file, and the software will retrieve a raster photograph of the front of the building, the building floor plan, a copy of the occupancy license for the structure, or a digital video clip showing a building "walkaround." Again, the improvements in the price/performance characteristics of computers previously discussed is a key factor in making the use of raster images in GIS databases feasible.

GIS vendors have also recognized the fact that many organizations use CADD for the engineering design of new facilities and utilities. As a result, CADD data may be imported into GIS database at several stages: planning, design, construction, and "as built." Because GIS is not as well suited for engineering design as CADD, many GIS software packages now support automatic links to CADD data to facilitate the import of these data elements into the GIS.

Most GIS software packages now make use of a relational database management system (RDBMS) to store and manage nongraphic (attribute) data. A number of powerful RDBMSs are supported by GIS packages today, including SQL Server, Oracle, and Sybase. Many of these GIS packages offer a single interface to several RDBMSs. Thus, the GIS can access and retrieve data from databases distributed throughout the network, even if they are supported by different RDBMS software. Using open systems architecture and an open relational interface, an organization can employ multiple computer platforms and multiple databases for different applications or departments. Users are not restricted to any particular proprietary database model, RDBMS, or hardware technology.

Some GIS software and database vendors have collaborated to make it possible to store spatial data in a RDBMS. This development makes data management easier and enables better integration between GIS and other types of data, including personnel, financial, and logistical information.

Summary

In summary then, here is George Korte's Top Ten list of the most important information technology trends affecting GIS (with apologies to David Letterman).

- No. 10: GIS support for the Wireless Application Protocol (WAP).
- No. 9: The integration of GIS and multimedia.
- No. 8: GIS support for links to and automated import of CADD data elements.

- No. 7: GIS that can be accessed through a web browser.
- No. 6: GIS spatial data managed by an RDBMS and integrated with financial, personnel, operational, and logistical data.
- No. 5: OLE/COM and CORBA integration of GIS “objects” with other computer applications.
- No. 4: GUIs providing greater ease of GIS use and customization.
- No. 3: High-powered, affordable personal workstations.
- No. 2: Low-cost, high-speed PCs with software for GIS data viewers and desktop mapping.

And the number one GIS technological trend is:

- No. 1: Open systems, high-speed data networks, distributed computing, distributed databases, and the Internet.

The major result of these trends will be a GIS on nearly every office desktop, spatial analysis tools commonly used throughout the organization, and further improved, decentralized decision-making.

GIS Data Sources, Collection, and Entry

Gathering the Information Needed to Make a GIS Useful

In This Chapter...

Roughly two-thirds of the total cost of implementing a GIS involves building the GIS database. This cost is often the biggest obstacle to justifying a GIS program. Therefore, GIS programs seek to build the GIS database at the lowest possible cost.

However, cost is not the only consideration in building a GIS database. Many other factors, most importantly data accuracy, can have a significant impact on the usefulness of the GIS—its users' return on the GIS investment. This chapter examines five primary methods of building a GIS database: using existing GIS data, digitizing, GPS surveying, using digital orthophotography, and using satellite imagery.

Existing GIS Data

One method of building a GIS database is to simply purchase GIS data another entity has collected. Existing digital map data can usually be purchased for much less than the cost of creating it. This is especially true if the seller has many buyers. The seller can recover the initial investment over a larger number of sales; therefore, the seller can charge a lower price on each purchase.

A typical county or large city may have to spend hundreds of thousands of dollars to obtain new digital topographic mapping for a GIS base map. On the other hand, it may be able to purchase data containing the same data themes (e.g., roads, drainage, topography, buildings, and vegetation), and covering the same area, for a few hundred dollars. What is the catch? Well, it is a big one. The horizontal and vertical accuracy of less expensive existing data is likely to be much lower than that of the new topographic mapping, and is likely to provide far fewer details of map features.

However, many GIS programs start with existing digital map data because of its relatively low cost. The plan may be to supplement and improve this data over time with more accurate and more detailed data, but in the long run this strategy might mean the total cost may be greater than simply buying the more accurate data to begin with. Nonetheless, this "incremental" approach to GIS database construction is popular. Moreover, the attraction of this approach is not just that it lowers the initial GIS investment threshold; it also helps the GIS program produce visible results more quickly.

> **NOTE:** *See Appendix C for federal and commercial sources of existing GIS data.*

Building a GIS database "from scratch" can take a very long time. Alternatively, when a GIS database is built with existing digital map data, users can get on to initial applications sooner. In addition, showing some benefits in short order is a good way of securing management support and additional funding for the GIS program.

State and local governments, as well as local and regional utilities, also collect and sell GIS data. They normally charge only a nominal fee for this data. Of course, the data will be limited to the geographic area for which the data supplier is responsible. The format of the data will depend on the GIS the agency or utility has chosen, which may not be compatible with your GIS. However, they should be able to provide the data in a popular data transfer format, such as the USGS DLG or Autodesk DXF formats. In the case of state agencies and utilities, the available data themes are usually only those related to their area of "business."

Local governments that have implemented GIS typically offer data related to land value, use, and ownership, as well as public works, topography, flood plains, and so forth. Many local governments are also responsible for utility services, including water, storm, and sanitary systems. It is far beyond the scope of this book to list all such local sources. You should need to make only a few phone calls to a state agency, utility, or local government to determine what GIS data is available and how to obtain it.

Digitizing

Digitizing is the process of tracing paper maps into a computer format. The term was coined to describe the fact that the maps are stored as digits in the GIS database. Most vector GIS data is collected by this method. From the late 1970s to the late

1980s, a digitizing table such as that shown in the following illustration was used almost exclusively.

Manual digitizing table.

When using a digitizing table, the paper map is carefully taped down on the table's surface. A grid of fine wires is embedded in this surface. This grid senses the position of the crosshair on a hand-held cursor. When the cursor button is depressed, the system records a point at that location in the GIS database. The operator also identifies the type of feature being digitized, or its attributes. In this way the map features can be traced into the system. A process more commonly used today is to first digitize the entire paper map using a scanner, such as that shown in the following illustration, capturing it as a raster image.

Raster map scanner.

Light sensors in the scanner encode the map as a large array of dots, much like a facsimile machine scans a letter. High-resolution scanners can capture data at resolutions as fine as 2,000 dots per inch (dpi), but maps and drawings are typically scanned at 100 to 400 dpi. The image of the map is then processed and displayed on the computer screen. This raster map is then registered to the coordinate system of the GIS, and the map features manually traced as vectors. This is often called "heads-up digitizing," as opposed to table digitizing, for obvious reasons.

Vectorizing software is available to help speed the process of converting raster map features to vector format. This software automatically recognizes most features on the raster image, such as line work and symbols, and traces vectors over them. Nonetheless, some occasional manual intervention and instruction is required to complete the task. Further manual digitizing on screen is usually needed to complete the data entry process.

Another digitizing method is "keyed data entry." If the map data contains coordinate values for points, or the bearings and distances of lines (e.g., tax parcel lines), the coordinates of the points, or the bearings and distances of the lines, can be entered on the computer keyboard. Because these types of data are compiled by exacting land surveying techniques, the resulting GIS data are more precise than data collected by tracing methods. Topographic mapping is complied by a photogrammetrist using a stereoplotter, such as that shown in the following illustration.

Stereoplotter.

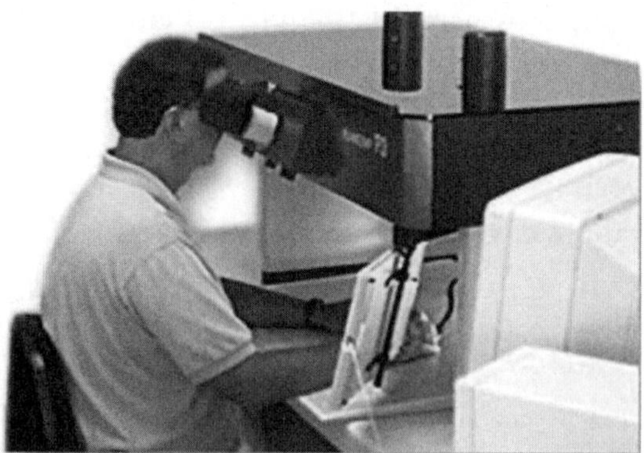

As its name suggests, a stereoplotter contains stereo pairs of aerial photographs. Through an exacting and rigorous technical process called aerotriangulation, the overlapping pairs of aerial photographs are registered to one another and viewed as a 3D image in the stereoplotter.

The photogrammetrist views the stereo model and carefully traces building outlines, streams, roads, topographic contours, and other features, which are encoded directly into the GIS database. This process is called stereocompilation. Photogrammetric mapping is therefore, in a sense, also a manual digitizing process. Additional vector data is typically added from other data sources, such as paper maps and drawings.

GPS Surveying

One alternative to traditional manual digitizing to be explored uses Global Positioning System (GPS) survey techniques. The GPS is a constellation of 24 satellites orbiting 11,000 miles above the earth, traveling on six separate paths. The GPS is depicted in the following illustration.

Global Positioning System.

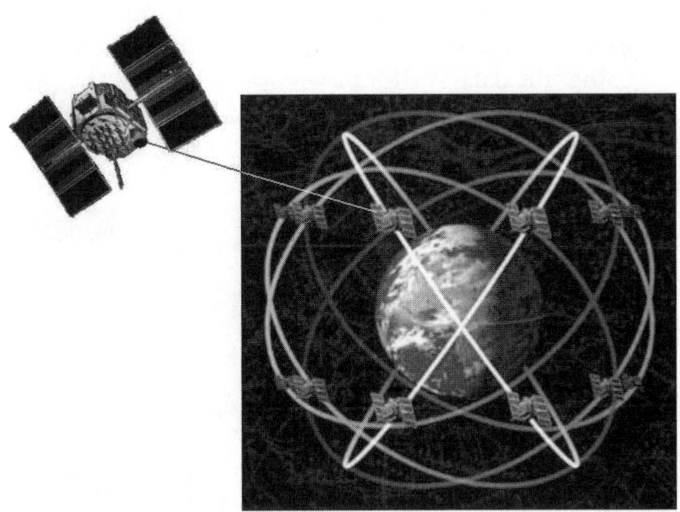

GPS satellites are equipped with atomic clocks, computers, and transmitters, each satellite broadcasting 24 hours a day. Reading the signal from at least four satellites, GPS receivers on earth are able to determine both their elevation and their location on the earth's surface. During the 1980s, mapping companies began using GPS survey techniques to set the ground control for aerial photography and photogrammetric mapping projects. The cost was much lower and the results more reliable.

GPS surveying is also being used to collect detailed mapping data directly in the field. The properly equipped surveyor, environmentalist, forester, or other scientist becomes a "human digitizer," collecting both locational and attribute data as he moves about. This data often supplements the information manually digitized from hard-copy maps and drawings, as previously described.

This process is used to locate utility lines, wetlands boundaries, park improvements, and so forth. It can also be used to verify and edit GIS data in the field, greatly reducing the time and cost required to "ground truth" data derived from satellite imagery and aerial photographs. Vehicles equipped with this type of system can as they travel road networks digitize their location and alignment while simultaneously recording video images of such elements as the pavement, traffic signs, and roadside buildings.

The method is not only two to five times faster, but 30 to 50 percent less expensive than conventional methods of utility inventory. Based on the GPS survey method and equipment used, the locational accuracy of such data ranges from +1 cm to +5 meters. Moreover, no line of sight between survey points is required for GPS data collection. Obstructions between points, such as trees and buildings, do not affect GPS data collection, which is also virtually immune to bad weather.

The process does require a line of sight from the GPS receiver to the satellites, however. In areas with tall buildings or trees, a total station survey may be required to gather the data. A GPS survey system such as that shown in the following illustration is based on a "ruggedized" portable or pen-based PC integrated with a GPS receiver and a computer-aided drafting (CAD) program.

Field data collection using GPS.

To perform a utility survey, electromagnetic radio detection and audio location devices are first used to locate underground utility lines. These locations are then usually spray painted on the ground, color-coded according to the type of utility. The GPS surveyor then walks the lines, recording their locations and attributes as she moves along. The CAD program plots the location derived by the GPS receiver. Predefined symbols for utility appurtenances and lines are placed at their proper map coordinates according to the type of utility. The system also records the details of these objects' attributes.

Digital Orthophotography

An orthophoto is a rectified aerial photographic image; that is, the relief displacement or radial distortions, which are both inherent in aerial photos, have been eliminated. The orthophoto is geometrically equivalent to a conventional line map, which represents planimetric features on the ground in their true orthographic positions. Because of this, orthophotos possess the advantages of line maps, such as the ability to make measurements of distances, angles, and areas. However, orthophotos, unlike line maps, also contain the images of an infinite number of ground objects.

In the past, the idea of creating digital orthophotos was out of the question, largely due to the difficulty of storing large amounts of data, and lack of a technology that could provide enough power to produce the end product quickly and at a reasonable cost. Today, computer power and storage have reached a level in speed and cost that allow digital orthophotography to be a commercial reality. Digital orthophotos such as that shown in the following illustration provide all the information of a photograph, but at the same time allow the registration of vector maps used in GIS.

Digital orthophoto.

The compilation of a digital orthophoto involves the following basic steps. First, establish ground control for the aerial photography and digital elevation models (DEMs) that follow. Second, obtain the aerial photography. Third, scan the aerial photographs to produce continuous-tone digital raster images of the photography. Fourth, compile the DEMs, a series of x,y,z coordinate triplets that defines the ground surface. The density of this grid is related to the variation of the terrain.

Thus, a smaller grid spacing is used for rough areas than for relatively flat and even areas. Fifth, overlay and rectify the raster images with the DEMs, based on ground coordinates, removing image displacements in the process.

Digital orthophotos are particularly well suited for updates to traditional planimetric mapping. They are less expensive than photogrammetric mapping, show more detail, and could perhaps enable users to digitize map changes in-house instead of using an outside service bureau.

Satellite Imagery

Satellite imagery can also be used as a raster backdrop to vector GIS data. Satellite images such as that shown in the following illustration have supported numerous GIS applications, including environmental impact analysis, site evaluation for large facilities, highway planning, the development and monitoring of environmental baselines, emergency and disaster response, agriculture, and forestry.

Satellite image.

Satellite images are especially useful for urban planning and management, where they are used to detect areas of change, monitor traffic conditions, measure water levels in reservoirs and building heights, and many other applications. They are also particularly helpful for agriculture and forestry, providing information for crop and forest identification and inventory, growth and health monitoring, and even measuring tree heights.

In addition to image analysis, satellite images can also be used to create "image maps"; that is, maps that combine the raster satellite image with vector line work and text that delineate special features, such as boundaries, roads, and transmission lines. There are four important aspects of satellite imagery: spatial resolution, spectral resolution, temporal resolution, and extent.

Spatial resolution refers to the smallest spatial element that can be sensed, or "resolved," by the satellite. These elements are commonly referred to as pixels (short for "picture elements"). Imagery from current satellites typically has a resolution of 10 to 80 meters. A new generation of satellites, many of which will be launched this year and next, will provide resolutions of 1 to 4 meters. The following illustration shows the spatial resolution of selected satellite images.

Spatial resolution of selected satellite images.

Landsat MSS (80m)

Landsat TM (30m)

SPOT Multispectral (20m)

SPOT Panchromatic (10m)

New Satellites (1 to 4m)

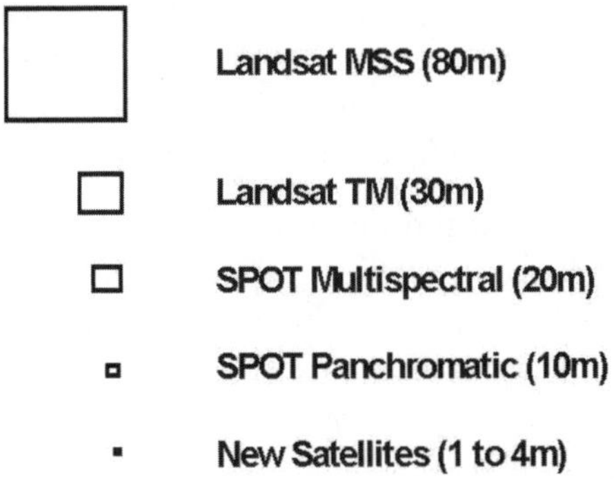

High-resolution, 1-meter satellite imagery will be useful for many more GIS applications, especially in remote areas of the globe, where the acquisition of aerial photography is both politically complicated and expensive. Even domestic applications of satellite imagery will expand. In urban areas, municipal public works, transportation, and zoning officials will be able to detect home and lot improvements, count parking spaces, or examine pavement conditions and markings. Telecommunications systems, disaster response teams, and flood risk assessment companies will also enjoy new benefits and substantial cost savings from higher-resolution satellite imagery.

Digital sensors aboard satellites divide the electromagnetic spectrum into "bands." The number and width of the bands determines the spectral resolution of the imagery. The following illustration shows the spectral resolution of selected satellite systems.

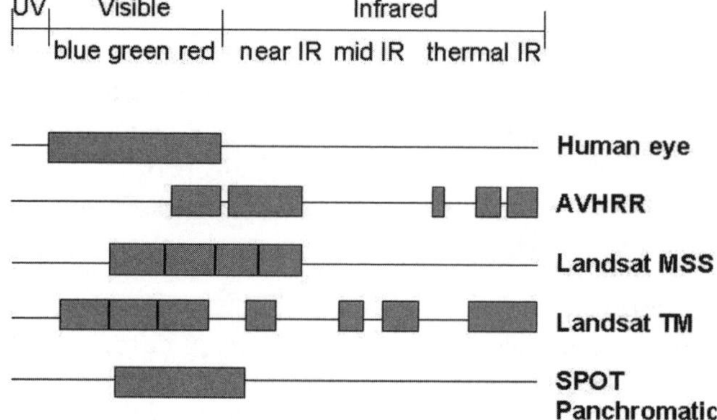

Spectral resolution of selected satellite systems.

The visible portion of the electromagnetic spectrum is the portion humans are capable of seeing; that is, red, green, and blue light. Whereas photographic film is capable of capturing only the visible and near-infrared portions of this spectrum, satellite data from the near- and mid-infrared portions of the electromagnetic spectrum allow GIS users to finely distinguish such things as vegetation types.

Temporal resolution defines when and how often an image of a particular area is captured. Existing satellite systems typically pass over an area every 3 to 16 days. The new satellite systems will be able to obtain images of a given area from as often as twice every day to once every three days, thereby increasing the temporal resolution of available satellite imagery.

Extent is the amount of area captured in one satellite image. Depending on the system, these images cover little more than a thousand to over 2 million square miles. The following illustration shows the extent of several existing satellite systems.

Extent of selected satellite systems.

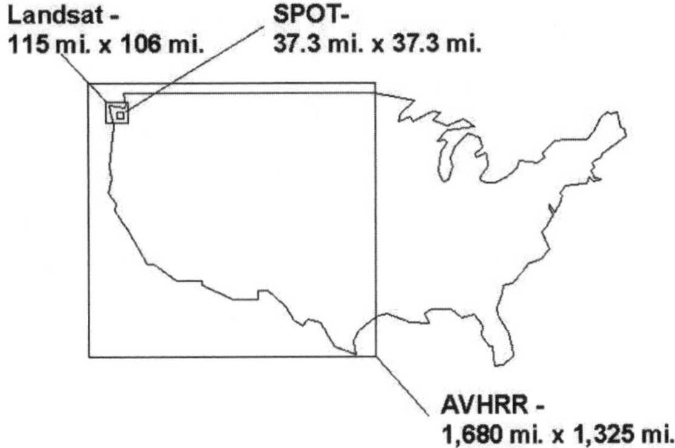

The large extent of a satellite image, combined with the fact that it is taken from hundreds of miles in space, makes the terrain correction and georeferencing of satellite imagery easier than that required for aerial photography. Owing in part to their higher spatial resolution, the next generation of satellites will generally have a smaller extent than existing imagery. To cover the same area, the new imagery will have to be combined in a mosaic. However, some of these new systems will offer forward-aft stereo images, making it possible to create DEMs and measure heights.

Despite their importance and usefulness, satellite images will not replace aerial photography. Airborne cameras can be sent aloft almost as often and as quickly as needed. They also offer the high image resolutions needed for large-scale (1inch = 1,000 feet or larger) mapping projects, such as those required by civil engineers and urban utility systems.

Summary

One more way to reduce the cost of the GIS database is to use a hybrid vector/raster database. The hybrid database includes vector GIS data and raster images. The raster images could be scanned paper maps and photographs, digital orthophotography, and/or satellite imagery.

The cost of obtaining or entering this data can be significantly less than that of creating comparable vector data. The raster image serves as the reference map, and only selected map features need to be digitized into vector format to have a useful GIS database. More features can be vectorized later as funding permits. For instance, a municipality might start with a raster photographic image covering its entire area as a base map, creating vector shapes of only building outlines. Tax assessment data can be attached to these building outlines, making the system a productive tool for the tax assessment department. Other features would be vectorized later to support applications in other departments.

Although the hybrid raster/vector database can be less costly than a totally vector base map covering the same area, its chief disadvantage is the tremendous data storage requirements of raster files. The recent and ongoing improvements in the price/performance characteristics of computer systems are key factors in making the use of raster images in GIS databases feasible.

GIS Data Formats and Standards

The Principal Types of GIS Data and Important GIS Data Standards

In This Chapter...

This chapter describes vector and raster GIS data, the two principal data models for GIS, as well as several important standards for storing, transferring, and integrating GIS data.

GIS Data Formats

In general, the two most widely used types of GIS data structure are vector format and raster format. The vector and raster models for storing geographic data have, respectively, unique advantages and disadvantages, both of which models can be handled by a full-function GIS. The following illustration shows the various ways these two data models would represent the same map features.

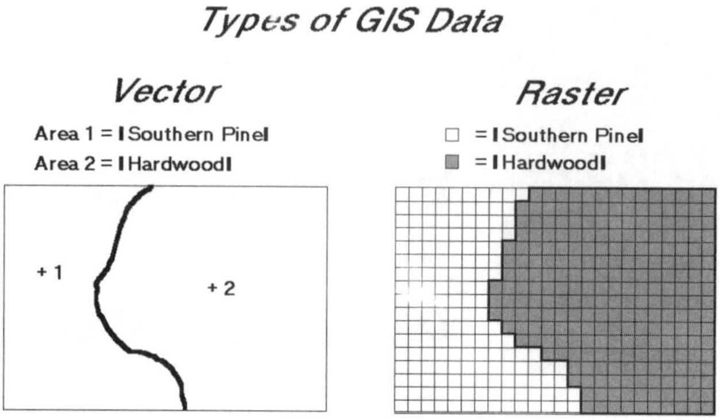

Vector and raster GIS data models.

Types of GIS Data

Vector

Area 1 = "Southern Pine"
Area 2 = "Hardwood"

+ 1
+ 2

Raster

□ = "Southern Pine"
■ = "Hardwood"

Vector Data

Chapter 2 discussed the differences between CADD, AM/FM, and GIS. All three systems use a vector model for representing drawing and map features. Vector digital map data is recorded as distinct points, lines (a series of point coordinates), or areas (shapes bounded by lines). In the vector model, information about points, lines, and polygons is encoded and stored as a collection of x,y coordinates.

The location of a point feature, such as a manhole, is described by a single x,y coordinate. Linear features, such as roads and rivers, are stored as a string of point coordinates. Polygonal features, such as sales territories and lakes, are stored as a closed loop of coordinates. The vector model is useful for describing discrete features, such as buildings, but less useful for describing continuously varying features, such as soil types or vegetative land cover. The vector format also provides for a more accurate description of the location of map features.

Raster Data

Raster data files consist of rows of uniform cells coded according to data values. An example would be land cover classification. The computer can manipulate raster data files quickly, but they are generally less detailed than vector data. As you can see from the previous illustration, the raster cells approximate the boundary line between the areas of hardwood and soft pine. The degree of approximation is related to the size of the cells. A grid consisting of smaller cells will follow the true location of the boundary line more closely; however, the size of the raster file will likewise increase.

Halving the grid spacing will quadruple the number of cells to be manipulated. Maps plotted from raster data may be less visually appealing than vector data files, which have the appearance of more traditional hand-drafted maps. For these reasons, the raster data model is generally used to model continuous map features. Like vector data, raster data can have attribute data attached to individual cells. This data can include map feature attributes such as types, measurements, names, values, dates, and classes.

Raster Images

A different type of raster data should also be mentioned. Photographs, drawings, paper maps, and other documents can be scanned into a raster digital format and attached to spatial data elements as attributes. For example, Chapter 3 described how the parcel data in a municipal GIS database could be linked to a tax assessment database. Scanned raster images of related documents could likewise be linked to the tax parcels. These might include deeds, easements, licenses, inspection reports, permits, and photographs of lot improvements, such as buildings. Click on the display of the tax map, and the GIS retrieves and displays the attached raster image.

The cells of these raster images are usually referred to as pixels (short for "picture elements") and are generally defined in one of three ways. Black-and-white drawings are typically scanned into a binary format in which each pixel is coded as either black or white. Continuous tone photographs are typically scanned and stored as shades of gray. Color photographs and maps are typically scanned and stored as color pixels. Scanners are available for each type of task.

Binary raster files are relatively small in comparison to grayscale and color raster files. Each grayscale or color pixel requires a much longer data record to describe it than a pixel defined simply as black or white. Moreover, binary raster data can be greatly compressed. Data compression is a process in which continuous cells of the same color value are coded together as a group. Obviously, the binary raster file of a scanned black-line engineering drawing has many white pixels that can be compressed in this manner, greatly reducing its size.

Data Standards

The federal government has developed two important GIS data standards. Although it is obvious that federal GIS users will be interested in these standards, others may be interested as well, especially if they provide data to or obtain data from the federal government. Moreover, many state, regional, and local government agencies are also adopting these GIS standards.

The material that follows describes the Tri-Service Spatial Data Standard (TSSDS), as well as the Spatial Data Transfer Standard (SDTS) and its Topological Vector Profile (TVP). Another emerging standard that promises to be significant to almost all GIS users is the Open Geodata Interoperability Specification (OGIS) being developed by the Open GIS Consortium (OGC).

DoD Spatial Data Standards (SDS) for Facilities, Infrastructure, and Environment

CADD/GIS Technology Center logo.

During the 1980s, Congress formed a bi-partisan commission to make recommendations on military base realignment and closure (BRAC). Naturally, this BRAC commission asked the Department of Defense (DoD) for summary information on the current status of its installations. The DoD in turn asked the service branches for summary reports, who turned to their major commands, who then asked the bases under their command to report.

The process of analyzing and summarizing installation reports highlighted an important problem. Although many DoD installations were moving toward digital mapping, there was no standard for storing the data. This made it nearly impossible to correlate and summarize mapping data from the various installations regarding their land holdings and real property, including buildings, utilities, and range facilities. As a result, the DoD has adopted the Spatial Data Standards (SDS) for Facilities, Infrastructure, and Environment for storing GIS data.

This standard was developed by the CADD/GIS Technology Center in Vicksburg, MS, which annually updates and expands the SDS. Established in 1992, the Center promotes GIS technology in the DoD. The Center evaluates technological developments, sets standards, coordinates activities, promotes system integration, accomplishes centralized procurement, and provides assistance for the installation, training, operation, and maintenance of these systems. Several field groups consisting of representatives from the Army, Navy, and Air Force assist the Center.

The SDS provides standardized groups and names for GIS features, as well as standardized feature attribute tables containing data about the features. The SDS are designed for use with the major commercial off-the-shelf (COTS) GIS and CADD software products, including ESRI's ArcInfo and ArcView; Intergraph's MGE and GeoMedia; AutoDesk's AutoCAD, Map, and World; and Bentley's MicroStation and GeoGraphics. SDS is also designed for use with relational database software, specifically Oracle and Microsoft Access.

The SDS is distributed via CD-ROM and the Internet *(http://tsc.wes.army.mil)*. A user-friendly interactive Microsoft Windows-based software application installs the SDS Browser and Generator applications on desktop computers and networks. The Browser application provides viewing and printing capability. The Generator application generates Structured Query Language (SQL) code for construction of the GIS database.

The SDS is a carefully developed, comprehensive, and excellent GIS standard. It is delivered in a manner that makes it very easy to install and use. Although the SDS is the standard for GIS implementations throughout the DoD, it is also the de facto standard of many federal, state, and local government organizations, as well as public utilities and private industry. State and local government agencies that adopt compatible GIS standards will not only find it easier to exchange data with neighboring local and state government agencies, but with neighboring DoD installations as well. You can obtain more information about SDS at the following address.

CADD/GIS Technology Center
U.S, Army Corps of Engineers
Waterways Experiment Station
Vicksburg, MS
www.tsc.wes.army.mil

Spatial Data Transfer Standard (SDTS)

GIS users often want to share or exchange data for a variety of reasons, usually having to do with overlapping geographic interests. A municipality may want to share data with the utilities that serve its citizens, and vice versa. Likewise, several state and federal agencies with regulatory or administrative responsibilities in the same geographic area may want to share or exchange spatial data. GIS software developed by commercial companies store spatial data in a proprietary format protected by copyright law. (Software developed by or under contract to government agencies are typically regarded as in the public domain and thus available to all.)

In general, one company's GIS software products may not read the data stored in another vendor's proprietary data format. Some GIS software companies have opened their data formats, meaning they have published their proprietary GIS data format and have committed to likewise publishing any future changes to it. This enables both users and competitors to write software that will directly read and write to their format.

When it is not possible for one GIS user's software to read another's data directly, it is still possible to exchange data through the use of a neutral data transfer format. This is two-step process: the source data is exported by the first GIS to the transfer file format, and then the second GIS imports the transfer file, creating a target data set in its own file format. Several such neutral transfer file formats are discussed in Chapter 24. However, most of these formats do not address all of the considerations involving GIS data.

The Spatial Data Transfer Standard (SDTS), the logo of which is shown in the following illustration, is a comprehensive standard for transferring GIS data. SDTS is maintained by the Federal Geodetic Control Subcommittee (FGCS) of the Federal Geographic Data Committee (FGDC). President Clinton's Executive Order 12906, dated April 13, 1994, charged the FGDC with developing standards for implementing the federal government's National Spatial Data Infrastructure (NSDI).

SDTS logo.

SDTS was originally published in 1992 as a Federal Information Procession Standard (FIPS Publication 173), and federal agencies are required to make their GIS data available to the public and other agencies in SDTS format. ANSI NCITS 320-1998, ratified by the American National Standards Institute (ANSI) in 1998, has superceded the FIPS version.

SDTS is intended to be the spatial data transfer mechanism for all federal agencies, but is available to state and local government entities, the private sector, and research organizations. SDTS is intended to facilitate the transfer of spatial data between different GIS, and specifies data exchange formats for both vector and raster GIS data. The SDTS is 300 pages long, a complex standard that specifies a feature coding system with thousands of feature codes, a model for defining GIS data quality, a complete attribute data dictionary, and an annex containing a glossary.

The SDTS specification is organized into the base specification (Parts 1 through 3) and multiple profiles (Parts 4 through 6). Parts 1 through 3 are related but relatively independent, each dealing with its own piece of the spatial data transfer problem. Parts 4 through 6 each define specific rules and formats for applying SDTS for the exchange of particular types of data in SDTS. The following are the six parts of the SDTS.

- Part 1: Logical Specifications
- Part 2: Spatial Features
- Part 3: ISO 8211 Encoding
- Part 4: Topological Vector Profile
- Part 5: Raster Profile
- Part 6: Point Profile

Part 1 of the SDTS defines a model for spatial data, spatial data quality, and spatial data transfer specifications. The conceptual model of spatial data defines 32 types of simple 0-, 1-, and 2D vector and raster spatial objects. (A point feature is said to be a 0D object, whereas a line has one dimension, length, and an area has two dimensions.)

The spatial data quality report is intended to allow a data user to evaluate the usefulness of the data. Five aspects of data quality are specified: lineage, positional accuracy, attribute accuracy, logical consistency, and completeness. An SDTS data set, or "transfer," consists of 34 types of modules. These modules carry information about the transfer's data quality, feature and attribute data dictionary, coordinate reference, spatial objects, and associated attribute and graphic symbology.

Part 2 addresses data content. It contains a catalog of spatial features and associated attributes, and describes a model for the definition of spatial features, attributes, and attribute values. It includes a standard list of terms, with definitions of more than 200 topographic and hydrographic features and their attributes, as well as more than 1,200 related terms. The current version of Part 2 is limited to small- and medium-scale spatial features commonly used on topographic quadrangle maps and hydrographic charts.

Part 3 specifies the physical format of the data transfer. This exchange format is independent of both magnetic media and computer system. This part explains the use of a general-purpose file exchange standard, ISO 8211, for creating SDTS file sets (i.e., transfers).

Part 4, the Topological Vector Profile (TVP), is the first of a potential series of SDTS profiles, each of which defines how the SDTS base specification (Parts 1, 2, and 3) must be implemented for a particular type of data. The TVP limits options and identifies specific requirements for SDTS transfers of data sets consisting of topologically structured area and linear spatial features.

Part 5 deals with the raster profile and extensions for 2D-image and gridded raster data. It permits alternate image file formats using the ISO Basic Image Interchange Format (BIIF) or Georeferenced Tagged Information File Format (GeoTIFF).

Part 6 describes the point profile, and contains specifications for use with geographic point data only, with the option to carry high-precision coordinates such as those required for geodetic network control points. This profile is a modification of Part 4, on the TVP, and follows many of the conventions of that profile.

Compliance with SDTS is now mandatory for federal agencies. SDTS is available for use by state and local governments, the private sector, and research and academic organizations. Information about SDTS is available from the following address.

SDTS Task Force
U.S. Geological Survey
1400 Independence Road
Rolla, MO 65401

Open Geodata Interoperability Specification (OGIS)

The Open GIS Consortium, Inc. (OGC), the logo of which is shown in the following illustration, is a not-for-profit trade association created to organize resources and industry support for the development of an Open Geodata Interoperability Specification (OGIS). OGC was founded in 1994 in response to widespread recognition of the problems previously discussed. OGC is an open-membership organization, currently supported by more than 100 public and private organizations.

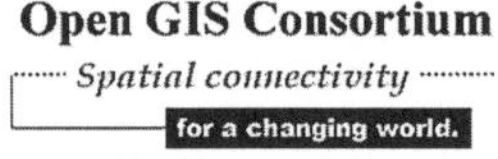

Open GIS Consortium logo.

GIS software vendors, database software vendors, system integrators, computer vendors, telecommunications companies, universities, information providers, and federal agencies have joined the Consortium to participate in creating a software specification and new business strategies that will help solve these problems. OGC's goal is "the interoperability of geospatial data and geoprocessing resources, [and] the sharing of resources between different GIS packages and organizations."

OGIS is to be a software architecture that will provide "transparent access to heterogeneous geoprocessing resources in a distributed environment." In other words, GIS users will be able to access one another's spatial data across a network even if they are using different GIS software programs. The OGC vision is a national and global information infrastructure in which GIS data and geoprocessing resources move freely, fully integrated with the latest distributed computing technologies, accessible to everyone, "geo-enabling" a wide variety of activities currently outside the domain of proprietary GIS products, opening new markets, and giving rise to new types of businesses and new benefits to the public.

OGIS is a comprehensive specification for software that provides access to both GIS data and software distributed among users of a computer network. OGIS gives software developers a detailed, common interface template for writing software that will interoperate with other OGIS-compliant software written by other software developers. The OGIS framework includes the following three parts.

- *Open Geodata Model (OGM):* A common method of digitally representing the earth and its related phenomena and attributes, both mathematically and conceptually.
- *OGIS Services Model:* A common specification model for implementing software utilities for GIS data access, management, manipulation, representation, and sharing between GIS users.
- *Information Communities Model:* A framework for using the Open Geodata Model and the OGIS Services Model to solve both the technical and institutional problems of GIS interoperability.

Information about the OGIS is available at the following address.

Open GIS Consortium, Inc.
35 Main Street, Suite 5
Wayland, MA 01778
www.opengis.org

Summary

Imagine that Microsoft Word could not read WordPerfect documents and vice versa, only ASCII text. Imagine that the only way people using these two popular products could share letters, reports, and the like would be to first save them as ASCII files. Then, after opening the ACSII file they would see the text of the document, but it would have no page layout, no text font, style, or formatting, no graphics, just…text. Although this would be bad enough for people in different organizations, imagine the hassles for people in the same organization.

In addition to the time lost performing file translations and reformatting documents, these files would have to be duplicated, greatly increasing the amount of hard drive storage required. It would not be possible to simply allow other people on the network to open a file in a shared directory. "Well, I guess they'd just have to standardize on a single word processing package, Korte," you say to yourself. Well, then, imagine some folks saying, "They'll get me to learn Word (or WordPerfect) when they pry my cold, dead fingers from the keyboard," or something to that effect.

As difficult as this may be to imagine about word processing software and documents, this is exactly the situation GIS users have had to deal with. It is common for different organizations in a region or metropolitan area to be using different GIS products, and storing their GIS data in different formats. This may be true of municipalities and utilities, two different utilities, or two different municipalities. There are also many organizations that use two or more GIS internally, usually in different departments.

Whatever the case, these organizations or departments often have common data needs, such as the need to have a base map of the same geographic area. They may also need to view one another's data. Obviously, the fact that the GIS data is in different proprietary formats can be a major obstacle to sharing this data. This chapter examined two important federal standards for spatial data storage and transfer, as well as OGIS, a standard that promises to make spatial data stored in differing proprietary formats accessible to all GIS users.

Types of GIS Analysis

How GIS Can Be Used to Analyze Spatial and Attribute Data

In This Chapter...

This chapter describes three fundamental types of questions (most commonly referred to as "queries") that a GIS can answer and gives examples of each: attribute data queries, spatial queries, and set queries. It also looks at other specialized types of analysis, including buffers, feature merging, network analysis, digital terrain modeling, and grid modeling.

GIS Queries and Terminology

In general, the three fundamental types of GIS analysis functions, attribute data queries, spatial queries, and set queries, can be performed on all types of spatial data: points, lines, areas, nodes, faces, and edges. The result of a query, called a "query set," is a group of features or elements of features that meet a given criterion. The basic distinction is that features are used when entire geographic entities are desired (such as whole parcels), and elements are used when only portions that meet specific spatial criteria are desired (such as the portions of parcels inside a flood zone).

Map Topology and GIS Analysis Terms

The following terms are commonly used in map topology and GIS analysis.

Map Features

The following terms are associated with map features, as distinguished from map feature elements, defined in the section that follows.

- *Area features:* Features delineated by closed boundaries. Counties, land parcels, and zones are examples of area features.
- *Linear features:* Sets of connected points that represent a feature that either has no width or that has width but is shown by a single line at the scale of the map being used. Rivers, railroad tracks, utility lines, and roads are examples of linear features.
- *Point features:* Features that either represent the location of a feature that has no dimensions or a feature that has width and length but whose perimeter cannot be mapped at the defined map scale. Elevation control points, stream gaging stations, oil wells, and small buildings are examples of point features.

Map Feature Elements

The following terms are associated with map feature elements, which are the component physical characteristics of map features.

- *Faces:* Two-dimensional topological entities, they are the smallest units of area. In a single map theme, before it is overlaid with other themes, a face corresponds to an area feature (for example, residential land use or commercial land use). After being overlaid with other area themes, these original faces are usually broken down into many faces because of intersections formed between the overlaid themes. In a topological file, faces store all attribute linkages of the features from which the face was made, as well as storing a calculated X,Y centroid.
- *Edges:* Nonintersecting lines between points. They are one-dimensional topological entities that represent part of the boundary of a face (or possibly the complete boundary in the case of islands) or the position of linear features. In terms of features, edges represent part of the boundary of an area feature or part of an original linear feature. In a topological file, records that contain edge information store the ID number of the faces on each side of the edge, the ID number of the start and end nodes of the edge, and the X,Y coordinates defining the edge.
- *Nodes:* Zero-dimensional topological entities. Nodes represent the start or end points of edges, the position of point features, or both (a node can be on an edge and be associated with a point feature). In a topological map file, records containing node information store their X,Y coordinates. For nodes representing point features, the records store the ID number of the face that the nodes are in and all attribute linkages.

Queries

The following are terms associated with queries.

- *Query set:* The result of a question or query posed to the topological file. The results from this query can be used to generate reports, graphic displays, and new map files. Query sets are groups of elements and/or features in a topological file that meet a given criterion.
- *Query string:* The actual queries posed to a topological file are specified by a query string. A query string is an expression of spatial and/or attribute criteria. For example, the query, "Show all parcel features that are zoned residential and that are within 500 meters of a major highway" might have been answered using the following query string.

```
from parcel where zone='residential'; within 500 m; from feature where
fname='major_highway'
```

- *Operators:* An operator in a query string performs an operation on the resultant content of an operand (a group of features and/or elements). An example might be the proximity operator "within," which keeps only the members of the first set that are within the specified distance of a member of the second set.
- *Operand:* The query set operated on.

Attribute Data Queries

In addition to the geometry of map features ("spatial data"), the GIS can store attributes of the map features. For instance, the GIS database may store a map of tax parcel boundaries. Attached to each parcel will be a database record containing its attributes. These might include the name of the owner, the street address, the assessed value of the property, and so on.

The GIS typically stores these spatial and attribute data in two different files, one containing the spatial data and the other containing attribute data. GIS software vendors typically provide interfaces to several database management systems (DBMS) that can store, manage, and analyze the attribute data. Corresponding records in the two files are linked by an identification number, which in this case might be the tax parcel number.

The GIS can search and display spatial data based on attribute criteria, and vice versa. For instance, the user could point to a map parcel and ask the GIS to retrieve and display its attributes. Conversely, the user could identify a record in the attribute database and ask the GIS to locate the corresponding parcel on the tax map.

But the GIS also provides for much more sophisticated queries. Building a "query statement," the user can define complex search parameters involving arithmetic and logical expressions. "Operators" in the query statement (+, -, =, not=, >, >=, <, <=, between, is null, and so on) specify the value to be found and can be combined in a single query. Thus, the GIS could answer the question "Which tax parcels are vacant, zoned for commercial use, and larger than one acre, and where are they located on the parcel map?"

The GIS would first build and send a query statement to the DBMS, which would search the attribute database for those parcels meeting the criteria. The DBMS would return their parcel numbers to the GIS. The GIS would then search its spatial data file to find these parcels and their coordinates. Finally, it would highlight those parcels on the display of the tax map.

Conversely, the user could describe an area on the parcel map and ask the GIS to retrieve the attribute records of all parcels within the area. This area could be a circle, a rectangle, or an irregularly shaped polygon. Logical operators could again be used to narrow the search or to specify which attributes to retrieve. The resulting query set could be used to generate reports, graphic displays, and new map files. The following illustration presents the result of an attribute data search.

GIS attribute data query result.

Spatial Queries

The GIS can also create a query set based on the spatial relationships of map features. Spatial operators in a query string define the spatial relationships that exist between map features. The following illustration shows the function of these spatial operators. Most of these can be combined to answer complex spatial queries. Six, however, are proximity operators (*between*, *within*, *beyond*, *entirely_within*, *entirely_between*, and *entirely_beyond*) that cannot be combined with each other.

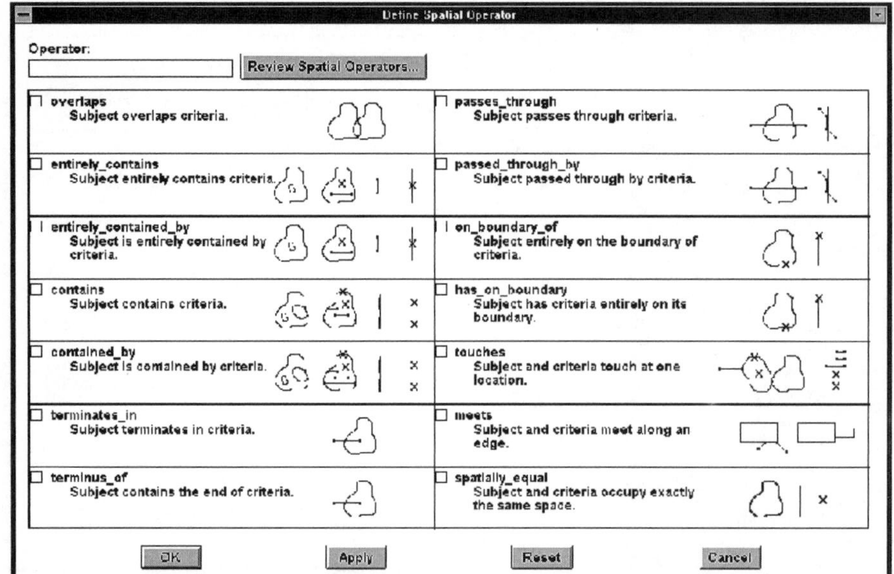

GIS spatial query operators.

Distance Operators:

between
within
beyond
entirely_within
entirely_between
entirely_beyond

For instance, a city's emergency management agency might want to compile a database of all property addresses that indicated whether or not the building was flood-prone. It could use GIS to compare a map of the 100-year flood plain to the tax parcel map. The user would first query for all parcels that are within the flood plain area (use the *entirely_contains* operator). These addresses would be classified as "flood prone." The GIS would next be used to find all parcels crossed by the flood plain boundary (use the *passes_through* operator).

These addresses would be classified as "indeterminate" and would receive further visual inspection to determine if the structure itself was flood prone. The remaining parcels would be entirely outside the flood plain (as a check, use the *disjoint* operator and make sure all parcels have been identified.). These addresses would be classified as "not flood prone." The following illustration shows the result of a spatial query asking for all tax parcels within a specified distance of a certain street address.

GIS spatial query result.

Set Queries

The GIS can compare or combine query sets in several logical ways. Queries that involve two or more query sets are called "set queries." Set query operators typically consist of INTERSECT, UNION, MINUS, and DIFFERENCE (these are commonly referred to as "Boolean operators," named after the nineteenth-century English mathematician George Boole). They behave in the following ways.

- UNION: Uniquely combines the content of both operands (query sets).
- INTERSECT: Keeps only items present in both operands.
- MINUS: Keeps only items from the first operand that are not in the second.
- DIFFERENCE: Keeps only items present in one, but not both, operands.

For instance, a military base planner might be studying possible sites for a new small-arms firing range. He might use a GIS to create several query sets containing areas not suitable for such a range. One set might contain buffered areas around endangered species habitats; another might contain all wetland areas; another might show all areas having greater than 5 percent ground slope; another, all developed areas; and so forth. He could then combine these query sets (using the UNION set operator) to delineate all areas on the installation that are not suitable for the range. The remaining areas would then be examined for potential sites. The following illustration shows the result of a set query.

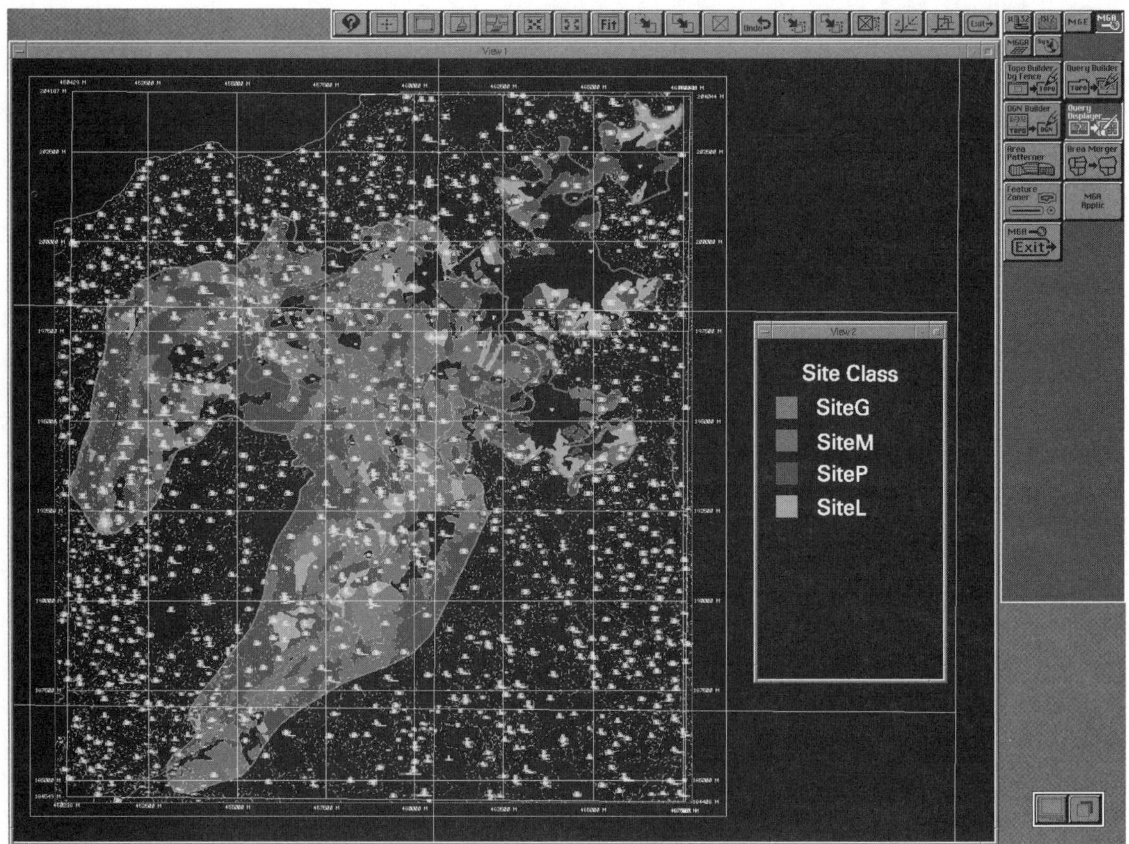

GIS set query result.

A GIS can perform several other important spatial functions that deserve mention. It can generate "buffers" of varying distances around point, line, and area features. The buffer becomes an area feature in the spatial data file that can be manipulated, analyzed, or queried, like any other feature.

There are also several types of "feature merging." For instance, very small area features can be eliminated by merging them with larger neighbors, a function sometimes referred to as a polygon "eliminate." Similarly, area polygons having the same attributes can be combined into a single, larger area, often referred to as a polygon "dissolve." The first of the following illustrations shows buffers and feature merging. The second illustration shows buffers drawn around bus stops.

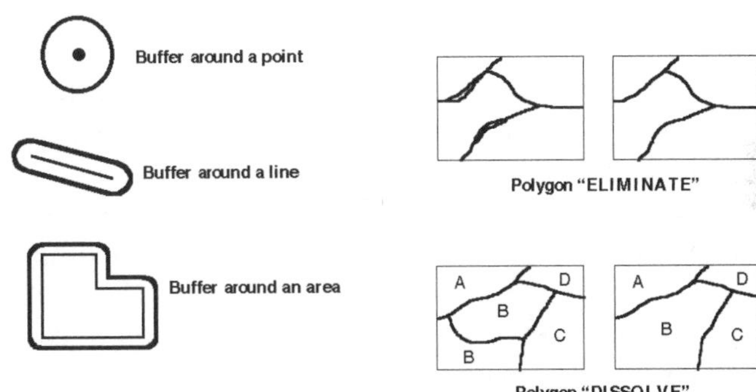

Buffers and polygon feature merging.

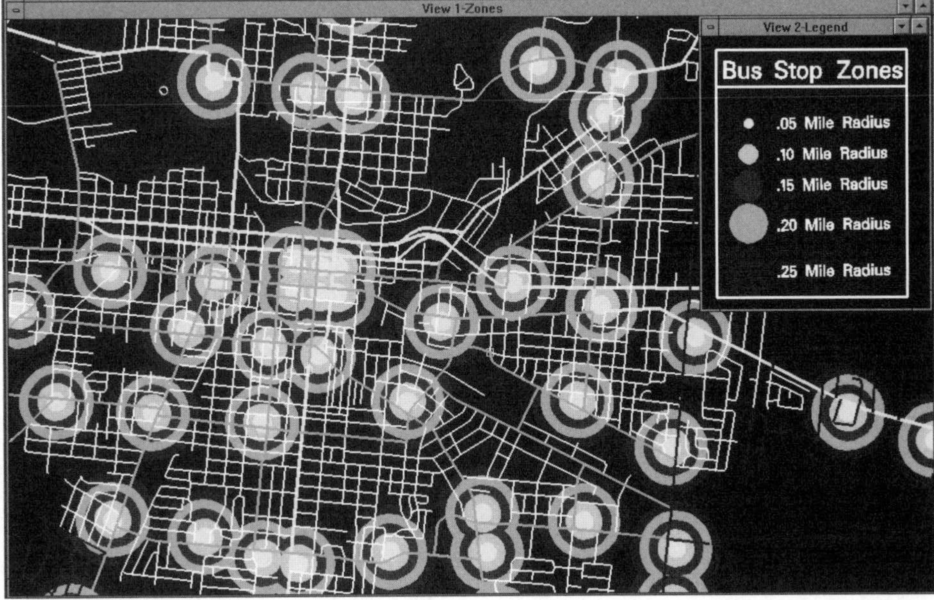

Buffers drawn around bus stops.

Other Types of GIS Analysis

The analysis functions previously described are typically provided as core capabilities of a GIS. Additional types of analyses may be available in separate modules so that the user can purchase them as required. These include network analysis, digital terrain modeling and analysis, and grid cell modeling and analysis.

Network Analysis

Network analysis is useful for organizations that manage or use networked facilities, such as utility, transmission, and transportation systems. Utilities employ network models to model and analyze their distribution systems and meter-reading routes. Municipal public works departments use networks to analyze bus and trash routes, whereas businesses use them to plan and optimize the delivery of goods and services. Network analysis can also be applied to retail store planning. For instance, analyzing driving times can aid in the determination of retail store trade areas. Three principal types of network analysis are network tracing, network routing, and network allocation.

Network tracing determines a particular path through the network. This path is based on criteria provided by the user. For example, an electric utility manager could ask the GIS to trace the path between an electric substation and the transformer nearest a proposed building site to examine the power distribution system's capacity along that route. Network analysis can also extract data for a separate engineering analysis program. A water system network could be traced and its geometry extracted for use in an engineering program that analyzed water pressures throughout the system.

Network routing determines the optimal path along a linear network. The selection of the path can be based on numerous criteria, such as "shortest distance," "fastest route," "no left turns," and "minimum cost." The path can pass between two points or through several selected points. These might be delivery points, for instance. The routing analysis checks for route restrictions and impacts in order to avoid conflicts. Depending on the network being modeled, these route restrictions and impacts can include one-way streets, no turns, blockages, impedance, costs, factors, and demands.

"Point-to-point" routing calculates a route from a source to a destination, whereas "multiple-point" routing calculates a route with multiple intermediate stops. Network routing supports delivery and service-oriented routing applications, such as government and utility vehicles, as well as maintenance and delivery services. It supports vehicle routing, transit planning, emergency services, transportation system planning and maintenance, timber haul routing, commercial business delivery, and utility maintenance. The following illustration presents the results of a network routing.

Network routing analysis.

Network allocation assigns portions of the network to "supply centers" or "destination points." If the supply center were a school with the capacity to handle a given number of students, the GIS could determine what portion of the community the school could serve. The system would allocate the school's capacity on a block-by-block basis, using census data on the school-aged population.

If the destination point were a downtown subway stop, network allocation could be used to determine all points on the city street network within a five-minute walk. Allocation routing can be used to determine the coverage areas of bus stops, service distances, retail trade areas, and times for emergency or utility vehicles. Allocation analysis can calculate and display the likelihood of a customer patronizing one store over another based on distance, cost, or time.

A dramatic example of network analysis was reported by Prince William County, Virginia, the subject of the case study presented in Chapter 27. The Prince William fire and rescue department used the county GIS to analyze emergency response service levels throughout the county. The GIS was used to compute and display emergency-travel times from the nearest fire station to each segment of the county road network. The network model was then used to play "what if" games to review the location of planned fire stations.

A major portion of the southwestern corner of the county was found to have unacceptable response times from the two nearest fire stations. This led to the decision to plan an additional fire station to serve that area of the county. On the other hand, the analysis also showed how the county could avoid building two of three previously planned new fire stations. By simply changing the proposed location of one station, the need for one of the other stations was eliminated. Moreover, the analysis showed that adding a new roadway linking two existing county roads would eliminate the need for the third station.

The new link brought a large area of the county within the minimum acceptable travel time from an existing fire station. The cost of building the new roadway, which was only a few miles long, was less than the cost of acquiring the land for, building, and equipping the planned new fire station. When the large annual cost of manning and maintaining a new station was taken into consideration, the cost saving was tremendous. Without the GIS, these analyses would have been very tedious to perform by hand and possibly would not have been conducted because of the amount of work required. The following illustration presents the result of a network analysis.

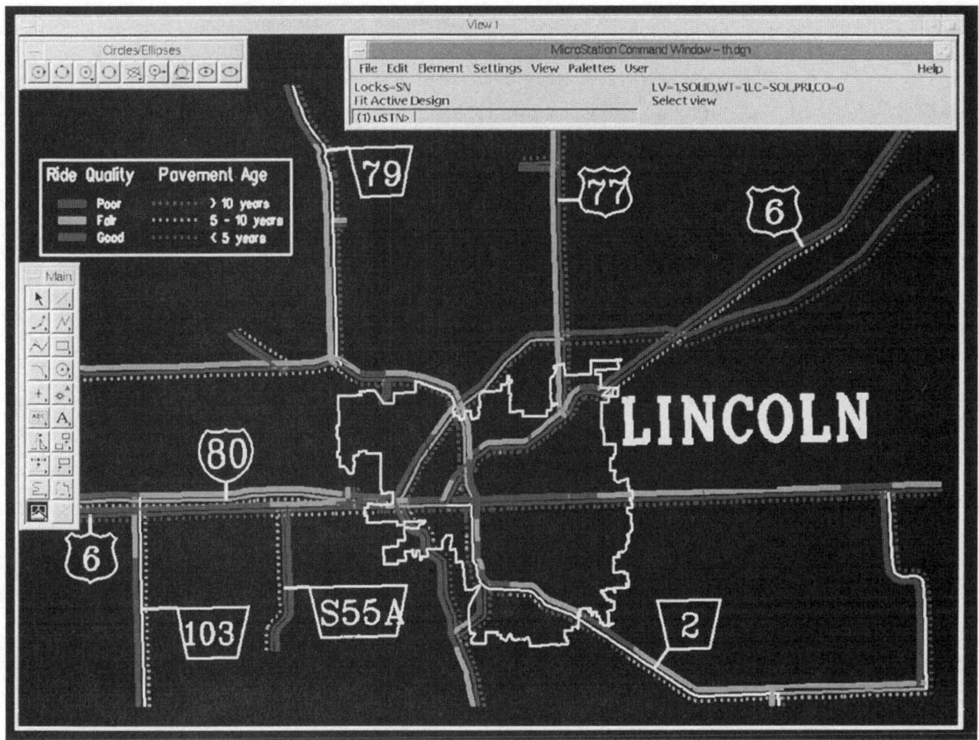

Network analysis.

Digital Terrain Modeling and Analysis

A digital terrain model (DTM) is generally used to represent the surface of the earth, although other surfaces can be represented as well. There are two primary types of DTM data models, the triangulated irregular network (TIN) and a grid-based data model. The following illustrations show these two types of DTM models. A DTM can be created from a series of data points that include the XY coordinate of each point, as well as its elevation (Z value). It can also be created from 3D topographic contour lines or 2D contours that have been tagged with their elevation value.

Grid-based digital terrain model.

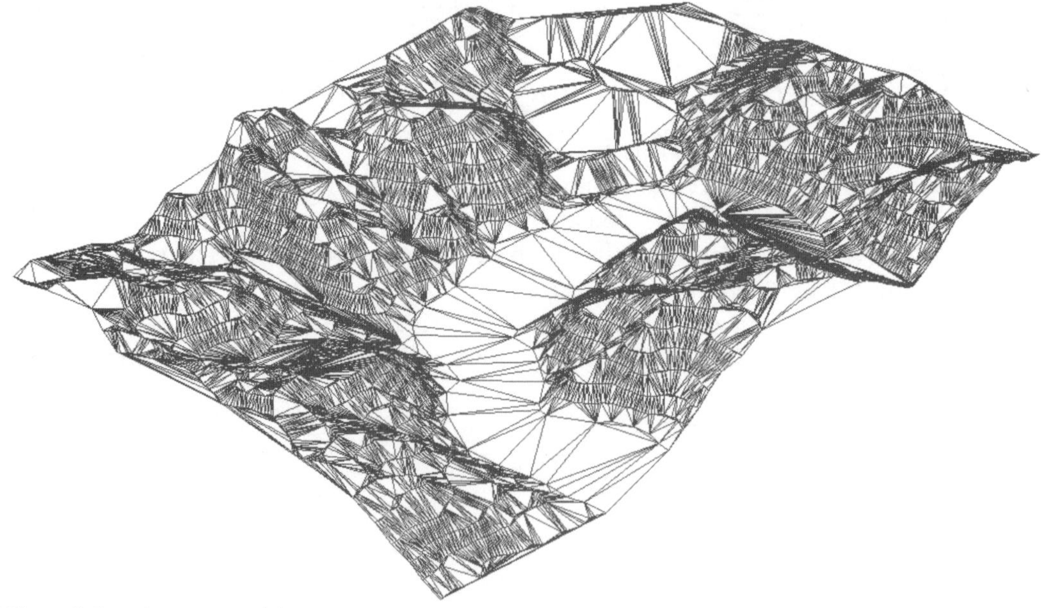

TIN-based digital terrain model.

The DTM module of a GIS will enable the user to edit and manipulate the data model. This would include editing Z-values in the TIN or grid model, as well as the addition, deletion, or movement of the data points in a TIN model. Two or more DTMs can be matched along adjoining edges and merged to create a single, larger data model.

The program can apply arithmetic and logical operations to DTM grids (e.g., comparison, difference, add, subtract, multiply, or divide) in order to modify existing models or to create a new one. The DTM can also be used to generate topographic contours using a variety of options. The options use different algorithms for plotting, smoothing, and simplification of the contour lines.

Once a DTM has been created, it can be displayed or plotted as TIN triangles, as a grided surface mesh, or as elevation contours. The view of the DTM can be presented or enhanced in a variety of ways, including plan (from directly overhead) or perspective (from an oblique angle) views, or color-coded by elevations. Multiple displays can be combined for a more complete visual analysis. A perspective view can be continuously repositioned and redisplayed to present the effect of "flying over" the terrain.

The DTM module is integrated with the core GIS program. This enables users to integrate 3D terrain data with 2D GIS data. For instance, 2D map features from the GIS can be "draped" over the 3D DTM surface. The following illustration shows a 2D GIS data draped over a DTM. Note the viewshed analysis shown in view 3 of the image.

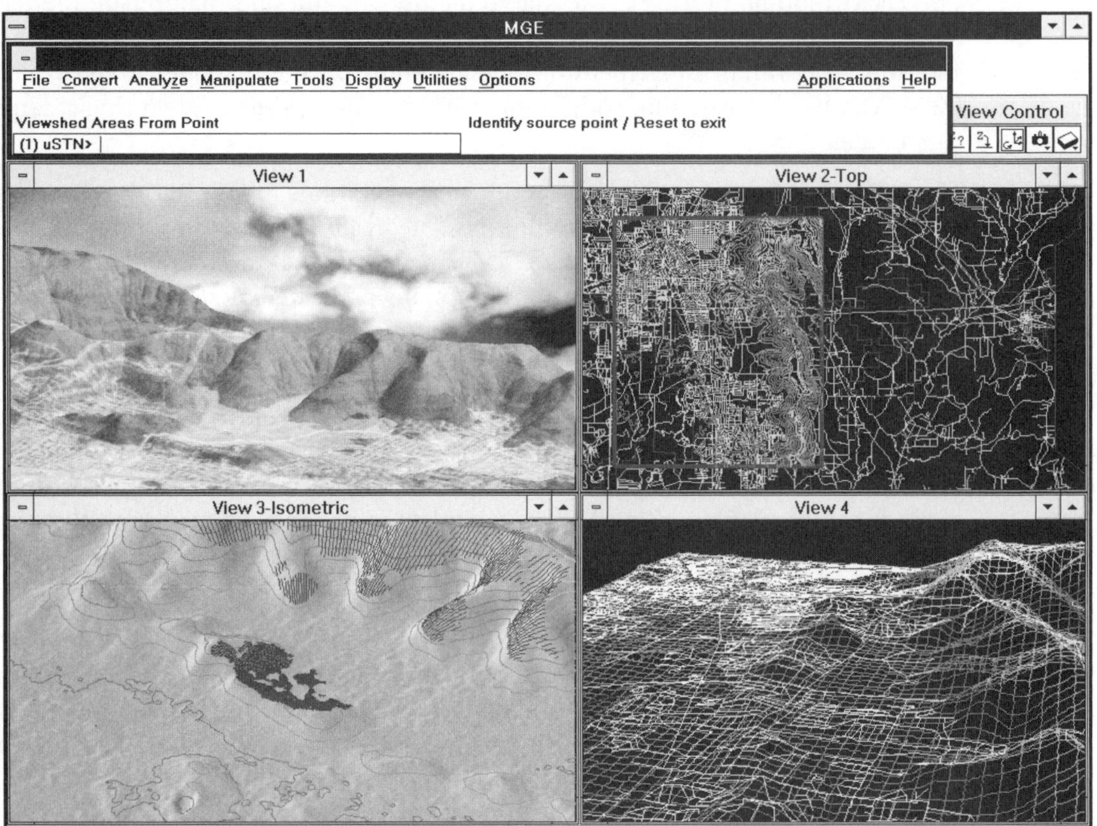

2D GIS data draped over a DTM.

DTMs are used for a variety of applications, including topographic mapping, land planning, civil engineering, and environmental and natural resource management. The U.S. Geological Survey offers Digital Elevation Models (DEMs) for many of its popular 1:24,000 scale, 7.5-minute by 7.5-minute maps. USGS DEMs use a grid-based DTM data model, with elevation points spaced every 30 meters (approximately 20 per inch in both directions on the map). DEMs have been used by communications engineers throughout the United States to plan the locations of cellular telephone and microwave transmitting stations.

The DEM is used by specialized a GIS to analyze transmission patterns and coverage areas. The U.S. National Imagery and Mapping Agency (NIMA) produces DTM data for areas of strategic military importance throughout the world. This data, known as Digital Terrain Elevation Data (DTED) has a number of military applications, from generating terrain perspectives in flight simulators to providing the cruise missile with a path to its target.

Grid Cell GIS Modeling and Analysis

Chapter 8 described the grid cell GIS data model. This type of GIS data model is often referred to as "raster" GIS data and is not to be confused with the grid-based DTM discussed previously. Grid cell GIS data can derived from a variety of sources, including digital satellite images, as well as paper maps, and photographs or similar hard-copy images that have been scanned into a raster format. Many GIS applications are better suited for a grid cell GIS model than a vector GIS model. These include natural resources management, urban and regional planning, land assessment, environmental modeling, and corridor analysis. The following illustration shows a grid cell GIS model.

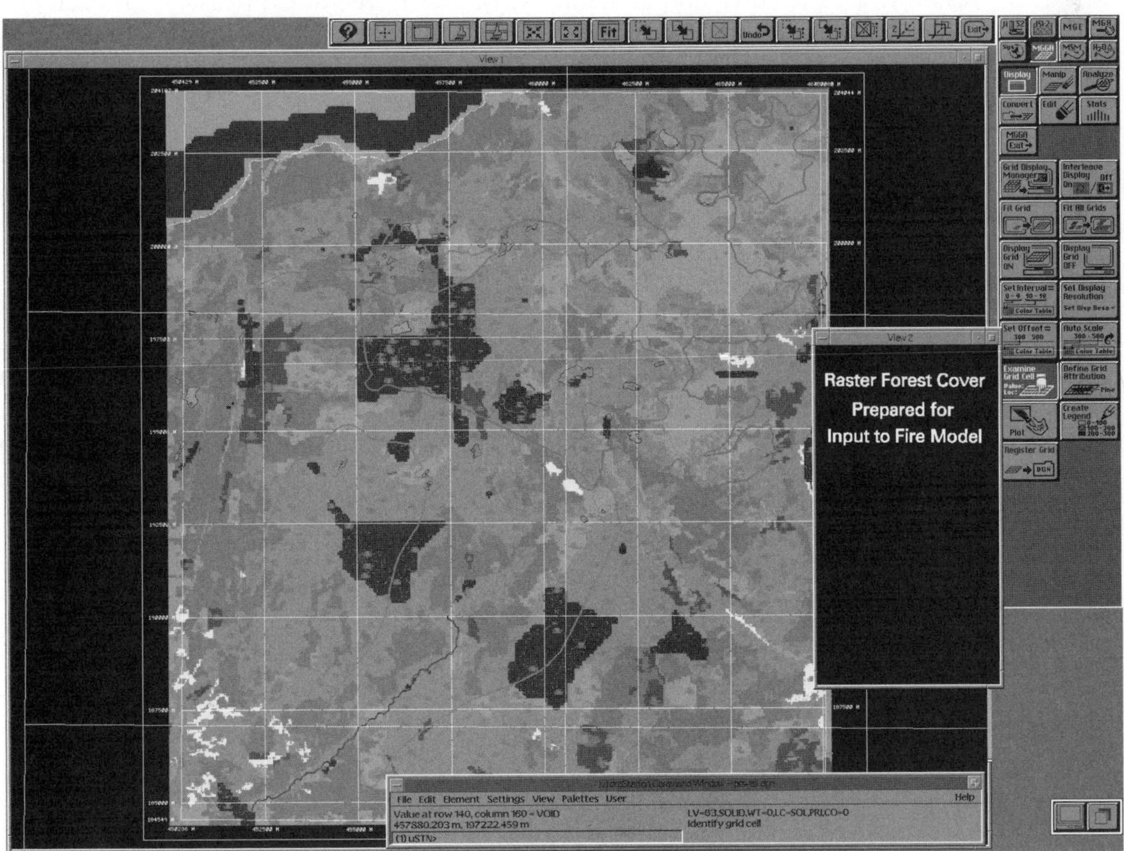

Grid cell based GIS data.

Grid-based analysis frequently involves the overlay of two or more grid cell GIS files. These comparisons can include multiple variables and constants, and multiple regression statistical functions. Several examples of these analyses are generating statistical models for prediction studies, modeling changes in the environment, and demographic analysis.

A grid cell GIS offers several tools for spatial analysis. It can create a buffer zone around a specified attribute value, or range of values, making it easy to identify areas of special interest. It can also conduct "proximity analysis"; that is, finding all grid cells having specified values and lying within a certain distance of a given value. It can also rank, reclassify, extract, merge, resample, thin, and densify the grid cells.

Grid cell data can be translated to and from vector GIS data. The tools for creating vector GIS data from grid cell GIS data offer options for filtering and smoothing the resulting area features and attaching attributes to them. Conversely, grid cell GIS data layers can be created from vector GIS features. The process will read the attributes of the vector map features contained in an attached database and apply them to the resulting grid cell model.

Summary

This chapter briefly explored the three fundamental types of GIS queries: attribute data queries, spatial queries, and set queries. It also discussed other types of modeling and analysis: network analysis, routing, and allocation; digital terrain modeling and analysis; and grid cell modeling and analysis. Many of the benefits of GIS stem just from the fact that map data are made available in electronic form; therefore, they are more easily stored, managed, edited, and transmitted. However, GIS also provides the ability to quickly analyze that information in a variety of ways, often in ways that were not feasible with paper maps. The analytical functions of GIS are at least as important as the map storage and distribution functions.

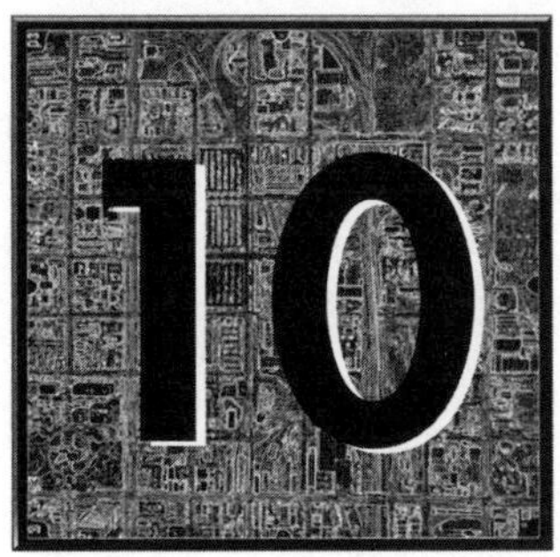

GIS on the Web

Using the Internet (or an Intranet) to Distribute GIS Data

In This Chapter...

The need for serving GIS data to customers is tremendous. Yet during the 1980s and much of the 1990s, most GIS data remained in the hands of a relatively few GIS specialists and analysts. The primary means of distributing this data to those outside this inner circle of full-time GIS users was still paper maps, drawings, or charts plotted by the GIS. Web-based GIS has changed all that.

Now all that valuable GIS data can be made available to literally thousands of GIS viewers at a relatively marginal cost. This chapter briefly introduces the concept of web-based GIS and explains the important difference between serving static maps and live GIS data on the Internet. It also describes several web sites you can visit today to see examples of GIS on the Internet.

Web-based GIS

Over the past several years the Internet has exploded in both popularity and usefulness, and the World Wide Web ("WWW" or simply "the Web") is the most commonly used aspect of the Internet. In its simplest form, the Web consists of a web browser on the user's PC interacting with a web server over the Internet. The web server sends hypertext markup language (HTML) or similar code to the browser software, which uses it to create the computer user's display.

HTML is basically ASCII text the browser interprets to paint the screen. The user clicks on hypertext links to retrieve different pages. Thus, a web site can be thought of as a book. The book's table of contents resides in the home page of the web site. The user selects pages of the book to be displayed by clicking on the hypertext links displayed on the home page.

Government organizations want to add digital maps to their public service web sites, whereas commercial organizations want to provide customers with geographic information that helps sell products, such as the location of their nearest retail outlet. Mapping departments want to provide GIS capabilities to non-GIS specialists responsible for corporate operations, such as routing deliveries, answering customer queries, or managing property.

Many organizations are also using web servers and web browsers to distribute corporate information on their internal LANs. These so-called intranets can pass GIS data as well. Consider a military installation. Personnel all over the base may already have web browsers on their PCs. The base only needs to add web capabilities to its GIS to distribute data about the facility. This greatly reduces, or even eliminates, the need for paper maps. Moreover, personnel all around the base have instant access to the latest map and facility information.

Maps on the Internet

It was only a matter of time before maps and map-related information would be available to Internet users. Many useful sites are already on line. For instance, MapQuest *(http:\\www.mapquest.com)* is a very popular web site that not only presents maps of countries and cities throughout the world, but also searches for a specific street address or business name and presents a street map of its locality. It even assists with trip planning, providing both directions and a route map between two designated locations. The following, in alphabetical order, are examples of some interesting web sites that serve maps to Internet users.

- *Cartographic Images:* This collection, found at the following address, is dedicated to the cartographer's art through the ages. It includes maps from the ancient world, the Dark Ages, and the Renaissance.

http:\\www.iag.net/~jsiebold/carto.html

- *The CIA's World Factbook:* Even the Central Intelligence Agency has placed its collection of political maps on-line, available at the following address. This site is also loaded with information on each country.

http:\\www.odci.gov/cia/publications/nsolo/wfb-all.htm

- *Exploring the West:* The University of Virginia's library has a tremendous collection of colonial American works. The library's web site Exploring the West, found at the following address, includes the maps compiled by Lewis and Clark during their explorations.

http:\\w.lib.virginia.edu/exhibits/lewis_clark/home.html

- *National Geographic Society:* Of course, you would expect National Geographic magazine to offer a web site, found at the following address, devoted to mapping. This site includes links to other Internet map sites.

http:\\www.nationalgeographic.com/ngs/maps/cartographic.html

- *The Perry-Castañeda Library Map Collection:* Housed at the University of Texas at Austin, this library includes hundreds of maps from the past to the present. These include more than 300 maps of the United States that chronicle its rise from a small colony to a world power. This information is available at the following address.

http:\\www.lib.utexas.edu/Libs/PCL/Map_collection/Map_collection.html

GIS on the Internet

Although the Internet map sites described in the previous section are interesting, informative, and useful, these are static maps. Typically, a paper map has been scanned and the resulting raster image stored on the web server. For historical and other types of maps that do not change, serving static maps over the Web will suffice. However, corporations, utilities, localities, the military, and many other organizations maintain maps and related data that change frequently, and they typically use a GIS to do so. More than simply viewing static maps, their employees and customers must browse, explore, and query dynamic maps derived from active GIS databases.

The following web sites provide "live" access to real data that is actively maintained or periodically updated by the web site owner. These are divided into U.S. Web Sites and Foreign Web Sites, with entries listed alphabetically under each section. These lists are by no means exhaustive. The intent is to provide just enough of a description of the application that you can determine whether or not you would find that web site interesting. These lists also attempt to give you a feel for the wide variety of GIS data available on the Web.

U.S. Web Sites

The following are examples of U.S. web sites. They include sites developed by city, county, and state entities, as well as federal government and national private sector organizations.

- *Arkansas Road Conditions:* The following web site presents information on the latest road conditions in the state.

http://www.ahtd.state.ar.us/roadconditions/map2.asp

- *Bell County, Texas:* The following web site presents the county's real estate appraisal data.

http://www.texastax.com/

- *City of Huntsville, AL:* The data on the following web site comes directly from the city's GIS database. It includes the following: Street Names and Addresses, Street Centerlines and Structures, Recreational Zones, Political Boundaries, City Council Districts, Zoning Boundaries, and County Commission Boundaries.

http://maps.ci.huntsville.al.us/hsv/

- *City of Ontario, CA:* The web site at the following address includes the following features.
    - *Parcel Search Utility:* Find parcel locations by address or parcel number. The user can also zoom in on an area and then identify a parcel. The utility provides a list of information on the chosen parcel.
    - *Development Status Utility:* Find the status of current developments within the city.
    - *Site Selector Utility:* Locate potential sites by defining desired parameters. This aids the city's Redevelopment Agency in their promotion of prime business locations.
    - *Infrastructure Utility:* Identify various utilities in the city.
    - *Election Polling Place Locator Utility:* This application enables citizen users to find their assigned polling location.
    - *El Niño Information & Mapping Utility:* Allows the user to see if a property is within a flood zone. Storm drain and catch basin information is also available.

http://www.ci.ontario.ca.us/gis/index.asp

- *Environmental Protection Agency Maps on Demand:* The EPA's Envirofacts Warehouse retrieves environmental information on Superfund sites, drinking water, toxic and air releases, hazardous waste, water discharge permits, and grants information. Maps of environmental information for the entire United States are available through the EPA's Maps on Demand (MOD) and are published through the following web site. MOD accesses data available through the Envirofacts Warehouse.

http://www.epa.gov/enviro/html/mod/

- *Louisiana Department of Transportation and Development (DOTD):* The following web site shows state-maintained roads and current road improvement projects, including budgets and schedules. The web site is constructed so that the user can view the entire state subdivided into three different sets of political districts: the U.S. House of Representatives, the state Senate, and the state House of Representatives.

http://192.234.241.12/legislature/

- *Louisiana Oyster Leases:* The following web site presents information about oyster leases in Louisiana. Click on the Map Page link.

http://www.govtech.net/gtmag/1999/june/geoinfo/geoinfo.shtm

- *Polk County Iowa Tax Assessor's Office:* The following web site offers information on residential, commercial, industrial, and agricultural properties.

http://www.co.polk.ia.us/departments/assessor/assessor.htm

- *San Diego Geographic Information Source (SanGIS):* SanGIS is a joint effort of the city and county of San Diego. Its mission is to maintain and promote the use of a 14-gigabyte GIS "data warehouse" for the San Diego region. The following web site offers public access to SanGIS data. The accuracy of the data varies between themes and within themes, depending on the source documents used to create the data. Most of the source documents used for SanGIS data were at 1 inch = 200 feet or 1 inch = 400 feet, making it more accurate than many other GIS data sources. In general, the SanGIS data has an overall accuracy of +/– 10 feet. SanGIS mapping themes include the following.

    - Agriculture Areas
    - Appraisal Areas
    - City Council Districts
    - Community Plan Areas
    - Enterprise Zones
    - FEMA Flood Hazard Areas (potential to flood during a 100-year storm)
    - Fire Department Engine and Battalion Districts
    - Generalized Vegetation
    - Habitat Evaluation Model Map
    - Library Branches
    - Neighborhoods

- Redevelopment Project Areas
- Regional Trails and Pathways Map
- Seismic Safety Study/Geologic Hazards and Faults
- Sheriff's Facilities and Command Areas
- Solid Waste Collection Areas
- Steep Slope Areas
- USDA Soil Survey Data
- Zoning, General Plan, Floodplain, and Assessor Parcels

http://www.ruis.org

- *Vermont AOT:* The following web site presents information on all airports in the state of Vermont.

http://www.aot.state.vt.us/vaotgis/vaotgis.asp

Foreign Web Sites

The following are examples of web sites outside the United States. They include local and regional sites, as well as national federal and private-sector entities.

- *British Borehole Catalogue:* The following web site includes over 600,000 borehole records from the British Geological Survey, covering the entire United Kingdom.

http://www.geomediaservices.co.uk/geomediaserv/htm/maps.htm

- *British Sites of Special Scientific Interest (SSSI):* A Site of Special Scientific Interest (SSSI) is the term used to denote an area of land notified under the British Wildlife and Countryside Act of 1981 as being of special interest for nature conservation. The SSSI data set, available at the following address, was developed by English Nature, the statutory body for nature conservation in England, and is intended to assist in the conservation of natural heritage and resources.

http://www.geomediaservices.co.uk/sssi/

- *Helsinki.net:* This is a guide to events and services in the Helsinki area, including traffic information, found at the first of the following addresses. See the second of the following addresses for maps.

http://www.helsinki.net/

http://www.helsinki.net/gwm/infomap/infomap.asp

- *PropertyLive:* The British National Association of Estate Agents (NAEA) and Intergraph UK developed PropertyLive for both the public and real estate agents. The following web site finds properties for sale or rent, as well as realtors, in any town in England, Scotland, or Wales.

http://www.propertylive.co.uk/PL/html/browse/default.htm

- *Quebec Ministry of Natural Resources (MNR):* The Quebec MNR is divided into four sectors: Energy, Mines, Forest, and Land. The following web site is published by the Mine sector and is in French.

http://tm.mrn.gouv.qc.ca/4041/intro.asp

Summary

The major GIS software vendors have developed products designed to publish GIS data on the Internet. These products are described in Chapter 14. Their corporate web sites include web pages that provide demonstrations of these products. Of course, although these demonstrations may use data that has been derived from actual GIS programs, they typically are not using "live" GIS databases. Rather, the data resides on one of the vendor's corporate servers. These demonstrations, although interesting and illustrative of the usefulness of web-based GIS, are nonetheless "canned" demos.

Selecting and Implementing a GIS

Part Two examines the planning, selection, and implementation of a GIS, including the use of a GIS consultant and GIS staffing. It also describes the products offered by four leading GIS software developers. The chapters in Part Two are:

- *Chapter 11, Planning a GIS,* describes a through planning process as the first stage of implementing a GIS. It discusses the major issues to focus on in the planning stage.
- *Chapter 12, Implementing a GIS,* presents a seventeen-step process for assessing the need for a GIS, setting up procurement criteria, and establishing goals toward successful implementation.
- *Chapter 13, Selecting a GIS,* discusses the primary considerations in selecting a GIS, including system functions, vendor support, the user environment, and costs.
- *Chapter 14, Four Leading GIS Vendors' Products in Review,* looks at GIS products from Environmental Systems Research Institute, Intergraph, MapInfo, and Autodesk.

- *Chapter 15, Keys to Successful GIS Implementation,* reviews critical issues regarding successful GIS implementation. These include management support, data conversion, training, and others.
- *Chapter 16, The Pitfalls of a GIS,* is a discussion of the most common problems encountered in implementing and managing a GIS, with suggestions for avoiding and overcoming them.
- *Chapter 17, Using a GIS Consultant,* looks at the reasons for using a GIS consultant, as well as the functions of a consultant and criteria for selecting one.
- *Chapter 18, Managing and Staffing a GIS,* examines GIS system management and describes the various staff positions and functions needed to run a GIS.
- *Chapter 19, Managing and Maintaining a GIS Database,* discusses the various basic configurations of a GIS and explains data maintenance issues.

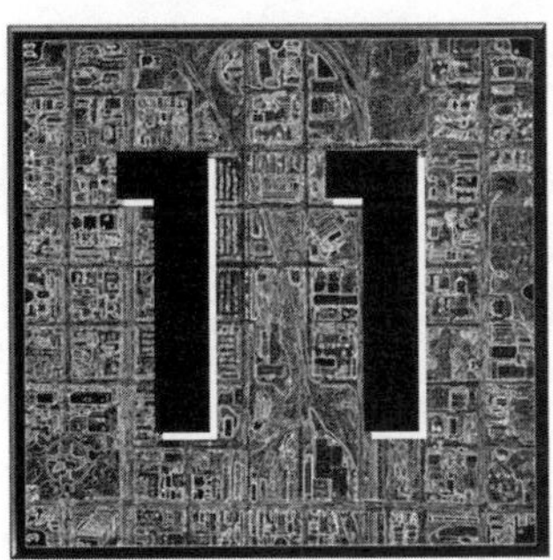

Planning a GIS

How Good Planning Can Promote Successful GIS Implementation

In This Chapter...

The need for planning can be summed up by the saying, "If you don't know where you're going, any road will get you there!" The benefits of GIS for government agencies and utilities are substantial and well documented. Nonetheless, adopting this complex new technology requires thorough planning and vigilant control.

The best way to begin a GIS program is to carefully review your organization's needs, and then develop a strategic plan that will systematically guide the selection and implementation of the system. This chapter describes seven ways a carefully prepared implementation plan can help ensure the success of a GIS program. The components of a typical GIS plan are then presented.

A GIS Plan

The sections that follow discuss seven aspects of a GIS plan that need to be a part of any successful plan.

Sort Out the Issues

The choices that must be faced in the initial stages of a GIS program are complicated. The strategies selected will have a tremendous impact on the design of the entire project, and will significantly affect the time, energy, and cost required to implement it. Moreover, if the GIS is to be used to its fullest potential, it will be integrated throughout the organization. Thus, the entire company can be affected. The following are the components of a successful GIS plan.

Examine Goals and Strategies

For these reasons, even the smallest GIS application is no simple undertaking. GIS is a significant venture for any organization, and therefore demands the careful planning that any significant investment requires. Planning of this scope forces any organization to examine its basic goals and strategies.

Guide the Implementation

A GIS plan provides the guidelines for an organized, systematic, and efficient implementation of this new technology. It should document the steps to be taken, their schedule, and the persons responsible for accomplishing them. Such a plan can effectively coordinate the various components of a complex program.

Forecast Requirements

A GIS plan serves as the basis for developing budget requests and staff requirements. It is the best means of ensuring that the present and future needs of all users will be met by the GIS being investigated or proposed.

Justify the Program

An effective GIS plan can also help obtain prompt and full funding of the GIS program. GIS projects are characterized by large up-front costs, a significant time period before break-even, high risk, and often profound cultural effects of job changes for users. These factors combine to cause great concern among top-level managers and decision makers who must approve GIS funding. A good GIS plan can provide the level of understanding and confidence they need to approve the program.

Set Goals

A GIS plan can define goals and thus lend a sense of direction and purpose to the GIS program. Goals help maintain the morale of GIS personnel as they encounter the problems and setbacks that can occur. Clearly defined goals also provide a means of measuring success. Demonstrated achievement can help justify further funding and continuation of the program, and give employees a sense of satisfaction and accomplishment.

Involve Users

Users need to play a responsible role in the planning, selection, and implementation of a GIS. If they do not take part in this process, the organization loses the opportunity to benefit from their experience. Employees who have no say in the way they do their work may resent radical changes, such as those associated with a GIS. Thus, failure to involve the users can also create a sense of indifference, or even hostility, toward the new system. A formal GIS planning process will interview potential users for their needs, problems, and suggestions. This gives them a sense of "ownership" of the program.

The Components of a GIS Plan

A GIS plan should typically address all or most of the following topics.

- Introduction and background to GIS
- Summary of existing operations
- Summary of existing needs and problems
- General description of a GIS
- GIS hardware and software
- GIS database
- GIS data maintenance
- Data communication
- Staffing and organization
- Training
- Implementation phases and schedule
- Financial analysis

Such a plan describes the proposed GIS program, presents the justification for a GIS, and guides its implementation. It also describes the current operations and potential GIS uses, and contains a schematic description of the content of the GIS database and describes the sources of the data. It presents a schematic configuration for GIS hardware, and a general description of the GIS software and communications functions required.

New staff positions needed to support the GIS should also be included, as well as proposed user training and support programs. The plan also presents the procedure, schedule, and budget for implementing GIS hardware and software, and for converting existing data to the format of the GIS. It relates the benefits that can be expected from a GIS, both quantitative and qualitative, and provides a cost/benefit analysis.

Summary

Because it has the potential for widespread application throughout an organization, a GIS program is neither a small nor simple undertaking. Due to the size and complexity of the project, it offers the potential for both great success and great failure. Careful project planning and execution is the best way to minimize risks and obtain the greatest rewards.

This chapter contained descriptions of seven benefits of preparing a GIS plan. Even after a GIS has been installed, this plan can be of great value to the operation. Updating it periodically is a good way of reviewing the results and benefits of the GIS program. Such a review can also aid the process of obtaining further funding for the program.

Implementing a GIS

A Step-by-Step Guide to Selecting and Installing a GIS

In This Chapter...

What's wrong with this picture?

"How is your GIS program going?"

"Well, we're off to a good start, but I'm not sure where we go from here. We can't get money to upgrade the system because management says it isn't sure GIS is such a good investment. They say they want to see results first. But, how can we show results unless we have the right tools? They just don't understand the technology. But, man, you should see the way we set up the database to handle temporal analysis..."

The lack of an adequate implementation plan and failure to sell such a plan to management is one of the principal reasons many GIS programs never get off the ground. For the same reason, others have stalled when the initial efforts at using GIS failed to produce convincing results.

Part I of this book explained what a GIS is, how it can be used, and some of its important features. GIS produces many benefits, but it is an information system that is complex to initiate and maintain. Implementing a GIS is far more complicated than making a simple computer system purchase.

Having a well thought out implementation plan is critical to the success of the system. This chapter outlines a seventeen-step approach to implementing a GIS. Depending on the size of the organization and the extent of its GIS applications, the level of detail and staff committed to this process may vary. However, the sequence of steps and the logical process involved remain essentially the same. (See the following illustration.)

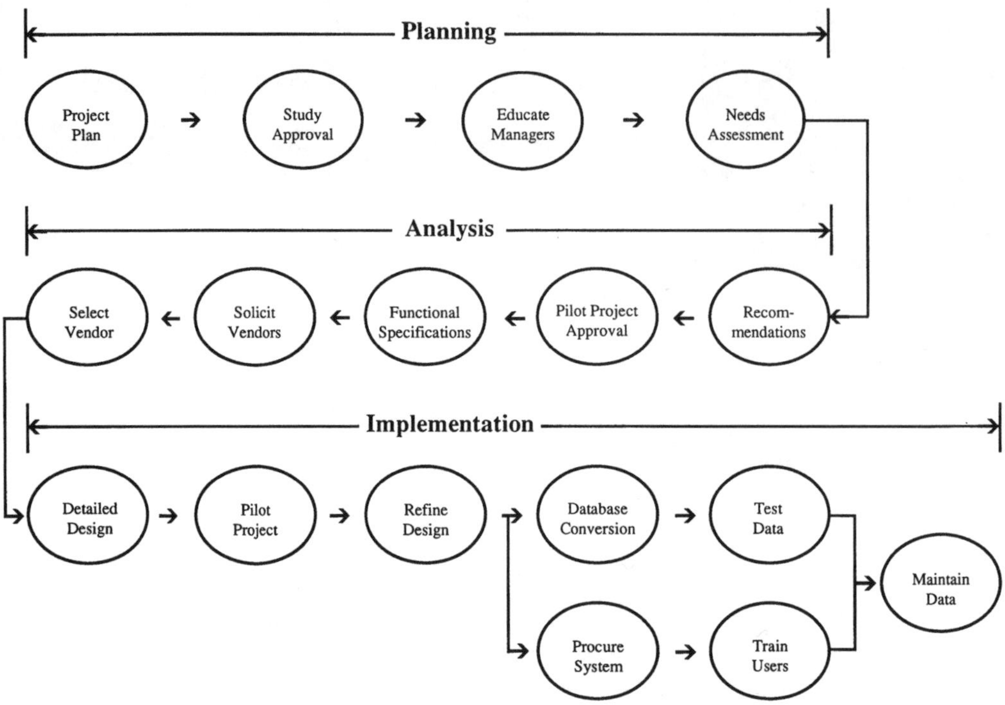

The GIS implementation process.

Implementation Overview

The process should be directed by a project manager, ideally the GIS manager. Moreover, the process should be overseen by a GIS steering committee. This steering committee could be chaired by the GIS project manager and consist of representatives from several of the key departments that will be using the GIS. The GIS project manager provides overall guidance for the process and seeks direction and approval from the steering committee. The project manager also serves as the chief coordinator for the distribution of reports and for the gathering of review comments.

Note that there are a number of key decision points in the process. The full commitment to the GIS (including all hardware, software, and data conversion) is not made until near the end of the assessment process. Management is given the opportunity to postpone or kill the project at several stages. It is far better to spend several thousand dollars on a feasibility study and pilot project, yet decide to postpone full implementation of a GIS, than to spend several hundred thousand dollars on a system that does not produce the expected benefits.

The process presented here has been used, with some variations, for a large number of successful CAM, AM/FM, and GIS installations. It has been used by utilities, cities, corporations, consulting firms, and state and federal agencies. It parallels the AM/FM Life Cycle approach advocated by AM/FM International. The seventeen steps outlined in this chapter are divided into three phases: planning, analysis, and implementation.

Phase One: Planning

The first four steps in this approach constitute the planning phase of a GIS implementation.

Step One: Develop a Project Plan

The obvious but often overlooked first step is to plan out the process of justifying, evaluating, selecting, and implementing a GIS. The plan may simply be to follow the seventeen steps outlined in this chapter, but it should also include an assignment of responsibilities for action, a timetable, and a budget. The plan should make a reasonable business case for GIS that demonstrates the financial justification for the investment. Moreover, it should be apparent that the GIS plan could and would become a road map to successful implementation of this complex technology.

Step Two: Obtain Study Approval

This is the time to involve top management, if they have not already been involved. Top management support is *critical* to the success of a GIS program. The top decision maker who will approve the GIS should be apprised of the plan and approve it. He or she should recognize that simply approving this plan does not authorize full implementation of the GIS, but that the process involves periodic opportunities to approve or disapprove the concept as the assessment proceeds.

Step Three: Educate Managers

It is important that the GIS study process involve the managers of those departments that might eventually use the system. "Top management support" includes their support.

The best way to obtain this support is through education, which can take the form of a half-day technology seminar. The seminar should give them an overview of GIS technology, applications, costs, benefits, and pitfalls. The seminar should also describe the GIS implementation process. Alternatives are to invite a vendor to bring in a system for a demonstration of GIS capabilities, or to arrange for managers to attend a GIS conference or class.

Step Four: Review Existing Operations and Needs

The next step is to look at how the organization is currently using geographic data, including its collection, analysis, storage, presentation, and distribution. (In this context, the term *geographic data* refers not only to maps but to related information such as tax rolls, utility data cards, permit records, police reports, and so on.) This review should examine the operations of all potential GIS users for reference during the planning process. The planning process must ultimately demonstrate to management that a true need for the new technology exists. The term *requirements analysis* or *needs assessment* is often used to describe this phase of the process.

A needs assessment can be developed through interviews with key personnel in each of the departments that will likely use the GIS. Other organizations may use or produce GIS data that would pertain to clients' GIS operations. These organizations can be interviewed as well. This would typically include all of the municipalities, utilities, and regional agencies in the region. Interviews can be facilitated using a survey questionnaire designed to make sure that all topics are addressed.

The process should define the mission, operation, and organization of each department; that is, how it collects, uses, analyzes, and distributes geographic data, as well as the needs and problems it has using this data. The process should also collect data on the costs of existing operations that could benefit from the use of a GIS. This cost data will be used for the financial justification of the investment. The results of this research should be documented in writing. The departments included in such a report should have the opportunity to review and comment on it.

Phase Two: Analysis

Steps five through nine constitute the analysis phase of a GIS implementation.

Step Five: Analysis and Recommendations

The next step is to analyze the collected data, examine the feasibility of a GIS, and develop an implementation plan. A GIS implementation planning workshop is a powerful tool to use in developing this plan. A personal anecdote might serve to highlight the significance of such a workshop.

I did not employ a planning workshop in my first two GIS planning assignments. I interviewed the client's key personnel and submitted a needs assessment. The client circulated the needs assessment among the persons I interviewed and relayed a few clarifying comments to me. These were incorporated into a final needs assessment. I then returned to my office, drafted a reasonable GIS implementation plan, and delivered it to the client. I received a few minor comments, incorporated them, and submitted the final plan, satisfied that I must have done a good job if there were so few comments on the plan I proposed.

However, as I reflected on the work I had done, I could not help suspect that the implementation plan would simply sit on the client's shelf. I knew intuitively that something was missing. Then I realized the client probably viewed the document as my plan, not his plan. The staff had reviewed my draft plan but had never really "bought into it." They had no sense of ownership of the plan.

I then recalled the technique that architects use to develop the initial parameters of a new design project. They meet with the client over an agenda that includes all of the principal issues that need to be addressed. As the participants tackle each issue, the results are listed on a "flip chart." Pages from the chart, filled with the decisions that have been reached, are taped to the walls around the room. By the end of the meeting, the preliminary design parameters are up on the walls for all to see.

This process does more than simply define the initial design parameters; it also cements the working relationship between the architect and the client. Moreover, it gives the client a valuable sense of involvement in and "ownership" of the design.

A GIS planning workshop can be conducted in much the same way. The participants should include, at the very least, all of the persons interviewed for the needs assessment. At the end of the workshop, the participants sit back, look around the room, nod their heads, and say, "Yeah, we can do this." Topics for such a workshop might include the following.

- Review of workshop objectives
- Potential GIS applications
- Existing equipment, software, data, and operations
- Base mapping requirements
- GIS data requirements
- GIS software requirements
- GIS hardware requirements
- Data network communications
- GIS data maintenance

- Staffing and organization
- Training
- Costs and benefits
- Implementation phases and responsibilities

Decisions reached in the planning workshop should be documented in a report. This report should be distributed to the workshop participants for comment and revised accordingly. This report contains the strategic plan for the GIS implementation.

Step Six: Obtain Pilot Project Approval

This is the first key decision point. The needs assessment report and implementation plan are presented to top management for review and approval. If management approves a pilot project, it is committing to a significant expenditure. Nonetheless, management should also be aware that there will be another opportunity to decide on the full implementation of the system after completion of the pilot project.

When completed, the final GIS plan is distributed to top management. The GIS project manager and steering committee may need several working sessions with top management to explain the program. A formal presentation of the plan may be in order. Such a presentation should be aimed at the level of technical knowledge of the audience. Usually a high-level presentation is appropriate.

The presentation should focus on making the business case for the investment. It should stress benefits, costs, and organizational impact. Basic examples of GIS applications should be presented clearly and in simple terms. The proposed system should be described using straightforward language. Given management approval, the process proceeds to the next step.

Step Seven: Prepare Functional Specifications and Standards

In this step, the results of the previous analysis are reformatted for presentation to vendors. There should be two sets of specifications prepared. The first set should describe hardware and software requirements, and the second set should describe the requirements for converting existing data to the format of a GIS database.

The hardware and software specifications should describe the functions required and provide a general description of the hardware configuration. These specifications should also present the number and types of hardware devices to be provided.

Any requirements for networking and for interfacing with existing computer systems or databases must also be described. It is usually desirable to communicate minimum requirements as well as optional but useful features.

The specifications for data conversion should clearly identify the source material to be converted, the format to which it is to be converted, and the standards that will be used to test the data. It should also clearly define the minimum criteria for acceptance of the data.

Step Eight: Solicit Vendors

The functional specifications for hardware and software and the specifications for data conversion are next incorporated within the organization's standard request for bid (RFB) and request for proposal (RFP) formats.The RFB and RFP should clearly state how the bids and proposals will be evaluated, the contract terms and conditions, the criteria for award of the contract, the project schedule, insurance requirements, penalties, the formats in which bids and proposals must be presented, and any other instructions to the offerers. The solicitation should be sent to several hardware and software vendors, as well as to data conversion vendors. Some vendors will provide both hardware/software and data conversion.

Step Nine: Evaluate Bids and Proposals, and Select Vendors

There are a variety of techniques for evaluating bids and proposals, but some type of point rating system is most often used to achieve a ranking of offerers. Many ranking schemes emphasize price, but few organizations select a GIS vendor solely on this basis. Other important considerations are corporate experience, project management approach, staff qualifications, and technical approach. Many evaluation systems also require a capabilities demonstration by offerers who meet the minimum selection criteria. When the evaluation is complete, a vendor is selected.

Phase Three: Implementation

Steps ten through seventeen constitute the implementation phase of a GIS implementation.

Step Ten: Detailed Database Design

Once a vendor for hardware and software has been selected, it is possible to refine the schematic database design into a detailed design for that vendor's system.

Detailed database design involves defining how graphics will be symbolized (i.e., color, weight, size, symbols, and so on), how graphics files will be structured, how nongraphic attribute files will be structured, how file directories will be organized, how files will be named, how the project area will be subdivided geographically, how GIS products will be presented (e.g., map sheet layouts and report formats), and what security restrictions will be imposed on file access. The result is a detailed specification that will govern the creation of the database.

Step Eleven: Conduct a Pilot Project

A pilot project involves constructing the GIS database for a small, representative portion of the project area. The key goals of the pilot project are to test the detailed database design and the cost estimates for data conversion. Management should review the results of the pilot project and an updated cost/benefit analysis, and then make a final decision regarding the implementation of the GIS. This is the last opportunity to approve, cancel, or delay the GIS project before major expenditures are made.

Step Twelve: Refine the Detailed Database Design

The creation of the digital database usually represents the largest portion of the total investment in GIS. Therefore, it is wise to take every opportunity to fine-tune the database design before undertaking the conversion of the entire project area. Changes in the database design can be made even after the data has been converted; however, *it is much less expensive to build the database correctly the first time than it is to edit the data later.*

Step Thirteen: Database Conversion

Quite often, the first phase of this database conversion process is a surprisingly large effort to collect, coordinate, and clean up the data before it can be converted to a digital format. This is because few organizations have all of their existing manual maps and files totally complete and up to date.

Therefore, the first step in data conversion is to compile all existing manual records and check them for completeness and homogeneity, collecting missing information and correcting conflicting information. When the data has been compiled and all source records are complete, the mapping data is digitized and the attribute data is entered and "attached" to (associated with) the graphic data.

Step Fourteen: Procure GIS Hardware and Software

Obviously, if database conversion is done in-house, at least part of the system must be purchased and installed, and users trained, before this work can begin. If a contractor does the data conversion, the GIS hardware and software purchase can be delayed, but it must be installed and in place well before data conversion is completed. There must be enough time to thoroughly train users.

Step Fifteen: Train Users

GIS is a relatively new technology and represents a radically different way of conducting everyday business for most employees in organizations that use it. For many, it may be the first time they have been required to use a computer system on a daily basis. Others may be familiar with computer commands and keyboard instructions, but they may be very unfamiliar with computer mapping concepts. They may know how to find information on existing paper maps, but are unfamiliar with coordinate grid systems, map layers, and map accuracy concepts. They will almost certainly be unfamiliar with the powerful tools for spatial analysis and attribute query a GIS offers.

However, the ultimate benefit and effectiveness of a GIS depends in large part on the ability of these users to master the system. For these reasons, a thorough training program for all personnel who will be using the GIS is important. Moreover, the training must be completed in time for users to take over the maintenance of the database following data conversion.

Step Sixteen: Test and Correct the Data

Whether data conversion is performed in-house or by a service bureau contractor, the quality of data must be checked. The process of data conversion is labor intensive and the information is usually complicated. Source materials can also be difficult to interpret. This makes them subject to the judgment of the person entering the data. All of these factors combine to make the conversion process prone to error. Moreover, the reliability of data will affect the success of the GIS. If the database is known to have errors, it will lose credibility. This erodes users' confidence in and commitment to the GIS.

Step Seventeen: Maintain Data and Support GIS Users

Because the physical world is constantly changing, the GIS database must be updated to reflect these changes. Once again, the credibility of the GIS database is at stake if data is not current. Usually, *the effort required to maintain the database is a fraction of that required to create it.* This ongoing maintenance work is normally assigned to in-house personnel as opposed to a contractor. The entire process should be planned well in advance. Equipment and personnel must be ready to take over maintenance of the database when the data conversion effort is complete.

Database maintenance requires two supporting efforts: ongoing user training and user support. Ongoing user training is needed to replace departing users with newly trained personnel. This enables data maintenance to be carried out on a continuous and timely basis. It is also important to offer advanced training to existing users, which provides them with the opportunity to improve their skills and to make better use of the system.

GIS is a complicated technology, making operating problems inevitable. User support will help users solve these problems quickly. It will also customize the GIS software to enable them to execute processing tasks more quickly and efficiently. User support is usually provided by in-house or contract programmers. In addition to troubleshooting common command and file problems, user support requires a knowledge of the operating system and macro programming language.

Summary

The seventeen steps outlined in this chapter represent a road map to a successful GIS project implementation. There are many GIS programs that have disappointed or left a management group dissatisfied because the results did not meet its expectations. In almost all of these cases, the GIS project did not follow a systematic process of planning, analysis, and implementation as described in this chapter. Instead, these projects invariably skipped one or more of these steps. The following chapters discuss some important considerations while proceeding with the process of assessing and implementing a GIS.

Selecting a GIS

Key Considerations That Influence the Selection of a GIS

In This Chapter...

This chapter looks at some of the key areas to consider when selecting a GIS, including system functions, vendor support, user environment, and cost. Frequently an organization that is planning a large GIS operation, such as a municipality or utility, will develop detailed specifications for many or all of these areas. Often a GIS consultant is hired to do this work. These specifications then become a key component of a Request for Proposal (RFP) that is issued to potential hardware and software vendors. Vendor proposals are evaluated on how well they meet the requirements of the specifications.

This chapter does not attempt to provide detailed product specifications and requirements. Rather, it is intended to provide a list of general areas that must be considered when choosing a GIS or specifying its requirements.

Many organizations fall into a sort of trap when choosing a GIS. They focus solely on comparing the features and functions of the various systems. They read a lot of product literature, attend product demos, talk to other GIS users, and become well educated on the state of the art. They usually find that the differences among GIS products are many but sometimes subtle. Moreover, there are a large number of features and functions to compare, which can make the task daunting and the evaluation process tedious and confusing.

The trap lies in failing to start with the most important step, which is to define the needs for a GIS. Given this definition of their requirements, the organization can focus on how well the various GIS products will satisfy their needs, not on what one system offers relative to another. Simply put, the best GIS may be the one that meets their minimum requirements at the lowest total cost. That GIS may not necessarily be the cutting edge system with the most features and functions.

Functions

The process for GIS planning described in the previous chapter included a requirements analysis. This analysis is used to determine what functions will be required of the GIS. The most important question is, Will this system meet my needs in these four areas: input, manipulation, analysis, and presentation? Functions that may be needed are outlined in the following sections.

Data Input

Data input requirements may include the following.

- Manual digitizing
- Scanning
- Keyed bulk data entry
- Automatic checking and corrections for digitizing errors
- Acceptance of existing raster and vector data

Data Manipulation

Data manipulation requirements may include the following.

- Data revisions
- Thinning and weeding digital line work
- Sliver polygon removal
- Transformation between map projections
- Edge matching of adjoining map files
- Transformation of data to fit specified control points (i.e., rubber sheeting)
- Merging data from a variety of digital and hard-copy sources into a common digital database
- Raster to vector conversion
- Merging polygons with common attributes
- Computing distance buffers
- Aggregating data within specified parameters

Data Analysis

Data analysis functions may include the following.

- Point, line, and polygon overlay analysis
- Geometric measurements and calculations
- Analysis of proximity and contiguity
- Spatial data queries
- Attribute data queries
- Coordinate geometry calculations
- Digital terrain modeling and analysis
- Network analysis

Data Presentation

Data presentation requirements may include the following.

- Display and plot of raster and/or vector data
- Display and plot of data at user-defined scales
- Display and plot of digital terrain models
- Automatic plot of attribute data as map text
- Automatic generation of map symbols based on attribute data
- Automatic dimensioning
- Specific printer and plotter capabilities
- Specific report and map output formats

Vendor Support

In that GIS usually requires significant after-the-sale support from the vendor, the next consideration is what support will this vendor provide? This key question pertains to hardware and software support, as well as to future developments.

Hardware Support

Hardware support considerations include the following.

- Location of field service technicians
- Size of the field service staff
- Size of the available parts inventory
- Schedule for maintenance services
- The cost of basic and "extra" services

Software Support

Software support considerations include the following.

- Procedures for logging and resolving customer problems
- Availability of an "800" telephone number
- Size of the customer support staff
- Vendor policy for responding to customer requests for changes
- The cost of software support

Future Developments

Future developments that may be critical include the following.

- The likelihood that the vendor will remain in business
- The vendor's program of product enhancements
- Vendor support for old technology following product updates
- The ability to move data files from one hardware platform to another, especially to new generations of computers

Keep in mind that the cost of creating the GIS database and training users will be several times larger than the cost of the hardware and software. Therefore, preserving this investment in the database and user training is most important.

User Environment

You must next consider how well the vendor's system will support the user. The following are key areas to consider.

- User training aids
- Ease of use
- Product documentation
- On-line user help screens and error messages
- Computer resource accounting
- Ease of access to separate application packages
- Support for multiple users and multiple tasks
- Transferring files among users
- User access privileges and file security features
- Sharing peripheral devices (e.g., file servers, plotters, and printers)
- Electrical power and air conditioning requirements

Cost

Cost considerations can be broken down into two major categories: initial and ongoing expenditures, and user productivity.

Initial and Ongoing Expenditure

Initial outlays and ongoing costs must both be considered. These costs include the following.

- Hardware and software purchase
- Data conversion
- Hardware and software maintenance
- User training
- System management
- User support

- Software development and customization
- Database maintenance
- Supplies
- Hardware and software upgrades

User Productivity

Chapter 20 discusses these costs in greater detail and presents typical cost ranges for several items. User productivity is another cost concern. Several criteria to examine are the time required to perform the following.

- Input new data
- Edit and maintain existing data
- Analyze data and generate reports
- Prepare plots, especially for large map files or a large number of plots
- Perform daily file backups
- Display information at the workstation, especially large map files

Summary

The lists of requirements and potential requirements in this chapter demonstrate that the number of significant considerations for selecting a GIS is large. In addition, there are several dozen GIS software packages available today. Therefore, the most manageable selection of a GIS requires foremost a careful analysis of the needs for a GIS and the proposed applications of the system, and then an analysis of the available GIS products.

A final note: People often ask experienced GIS users, "What is the best GIS?" I have found that in many cases "the best GIS is the one you know best." Be wary of the fact that an "experienced GIS user" may only have experience in the use of one GIS, or one application of a GIS. He or she may not be able to give you an unbiased opinion based on an objective evaluation of all GIS alternatives based on application.

Four Leading GIS Vendors' Products in Review

In This Chapter...

In order to give you a better idea of the types of features found in a GIS system, this chapter reviews the core products of the four leading GIS vendors according to the GIS industry: Environmental Systems Research Institute, Inc. (ESRI), Intergraph, MapInfo, and Autodesk. Each vendor's technical and sales literature was reviewed to develop these product descriptions. Product claims and comparisons were eliminated, toward presenting a straightforward, no-nonsense explanation of each product's capabilities. In addition, there is no intention to offer any comparisons or evaluations of the capabilities of these products.

Environmental Systems Research Institute, Inc. (ESRI)

ESRI's corporate headquarters is in Redlands, California, with eleven U.S. regional offices and ten international subsidiaries. The company develops and maintains GIS software, and provides software installation and support, database design application, and programming and database automation. Its core GIS products are ArcInfo, ArcView, and ArcIMS.

ArcInfo

ArcInfo 8 is accessed through three applications: ArcMap, ArcCatalog, and ArcToolbox. Users will typically interact with the system by having two, or all three, of these applications open.

- ArcMap is the map-centric application for editing, displaying, querying, and analyzing map data. In addition, ArcMap contains a scientific charting and graphic system, an object-oriented editor, and a report writer. ArcMap is the environment for working with map data and creating production-quality cartographic output.

- ArcCatalog is the data-centric application that locates, browses, and manages spatial data. ArcCatalog can create and manage spatial databases. ArcCatalog is used to lay out data schema in the database and specify and use metadata.
- ArcToolbox is an environment for performing geoprocessing operations, such as data conversion, overlay processing, buffer creation, and map transformation.

ArcInfo 8 supports two primary geographic data models: the georelational model (i.e., coverages and shapes with attributes) and an object-oriented model called a geodatabase. The geodatabase allows users to add behavior, properties, rules, and relationships to their data, and supports topologically integrated feature classes. It supports complex networks, relationships among feature classes, and other object-oriented features. The geodatabase model supports an object-oriented vector data model. In this model, entities are represented as objects with properties, behavior, and relationships.

Support for a variety of geographic object types is built into the system. These object types include simple objects, geographic features (objects with location), network features (objects with geometric integration with other features), annotation features, and other more specialized feature types. The model defines relationships between objects, together with rules for maintaining the referential integrity between objects.

The ArcInfo geodatabase model is implemented on standard relational databases with the ArcSDE application server. ArcSDE defines an open interface to database systems. It allows ArcInfo to manage geographic information on a variety of database platforms, including Oracle, SQL Server, and others. In addition to providing the bridge between ArcInfo and spatial database implementations, the ArcSDE application server defines an open C application programming interface (API). This API defines a relational (simple feature) view of the geodatabase.

The georelational view provides a simpler, nonobject view of the data model. The relational model corresponds to the "simple feature" model used in ArcView GIS, MapObjects, Oracle Spatial, and other systems and standards (such as OGIS simple features and SQL 3 spatial). Software and applications that understand georelational data (i.e., ArcView GIS 3.x, ArcPlot, MapObjects, and so on) can view and make use of geodatabases.

ArcInfo 8 provides a series of out-of-the-box GIS applications, as well as a customization capability. ArcInfo 8 can be customized using drag-and-drop and menu-driven tools. Industry-standard Microsoft Visual Basic for Applications (VBA) is provided for all scripting and application customization jobs. Any component object model (COM)-compliant programming language can also be used to customize and extend ArcInfo 8.

In ArcInfo 8, display, query, and analysis of map data are supported by the ArcMap application. ArcMap provides a direct data read capability, which allows data sets to be used on the fly without translation or use of an intermediate format. ArcMap supports ArcInfo coverages, ESRI shapefiles, Spatial Database Engine (SDE) layers, map libraries, ArcStorm layers, DXF and DWG, DGN, and many image types. In addition, ArcMap supports on-the-fly projection at the individual layer level.

ArcMap is a menu-driven application for working with map data. It is an integrated application for creating and editing spatial databases, displaying and querying geographic data, performing complex analysis, generating quality reports and charts, and making high-quality maps. Features include the following.

- Integrated map display, editing, and production environment
- Windows user interface
- Data visualization for interpretation/analysis
- Out-of-the-box usability
- Creation of charts and reports
- Mapping capabilities
- Computer-aided design (CAD) editing on intelligent GIS databases (ArcSDE databases, coverages, and shapefiles)
- Drag-and-drop customization

ArcMap includes an integrated Object Editor capable of multiuser geographic and attribute entry and update. The Object Editor can work with coverages, shapefiles, and geodatabases stored in a database management system (DBMS) using ArcSDE. The Object Editor supports the following.

- Creation and updating of shapefiles, coverages, and geodatabases
- Editing of features according to rules/behavior (i.e., network connectivity, attribute consistency, and so on)
- Display of raster and vector data

- Snapping to vector data (including CAD files)
- Versioning and conflict resolution across work groups
- CAD/sketching function directly on the GIS database
- WYSIWYG editing
- Editing within magnify windows
- Integrated tracing
- Customization for user-defined tasks/tools

The Object Editor includes many of the graphic editing functions of CAD editing packages for editing map features, using rule-based tools for creating and maintaining spatial databases. The Object Editor also enables users to directly edit data in a DBMS via ArcSDE.

ArcCatalog is used for locating, browsing, and managing spatial data. It resembles the Windows Explorer, but can view geographic data and metadata. It hosts all data management tasks and helps GIS database administrators maintain spatial and tabular GIS data. Users can select data sets with ArcCatalog and then drag and drop them into ArcMap for query and analysis. Data handled by ArcCatalog includes ArcInfo coverages, ESRI shapefiles, geodatabases, SDE layers, ArcStorm layers, INFO tables, images, GRIDs, triangulated irregular networks (TINs), ArcSDE, CAD files, address tables, dynamic segmentation events, and other ESRI data types and files.

ArcCatalog also supports editing and viewing of metadata. ArcInfo 8 has been designed to create metadata for any data set supported/created by ArcInfo, as well as any other data set identified and cataloged by the user (e.g., text, CAD files, and scripts). ArcToolbox is used for performing geoprocessing operations such as conversion, overlay processing, buffering, and map transformation. More than 120 tools are organized in a tree view, and each tool has a menu-driven interface with wizards or dialogs.

Previous versions of ArcInfo were usable across a range of UNIX and Windows NT platforms. ArcMap, ArcCatalog, and ArcToolbox are designed for use solely on Windows NT (and in the future, Windows 95, 98, and 2000). They are also designed to integrate and work directly with ArcInfo UNIX environments. Alternatively, all of ArcInfo can operate on a single Windows platform in a client/server mode. Users can implement an integrated UNIX/Windows network to perform tasks. A single license manager process on a network can be used for both UNIX and Windows NT seats.

ArcInfo on Windows can work with data stored and maintained on UNIX systems. Furthermore, the ArcToolbox application can execute its geoprocessing functions on remote UNIX/Windows servers. In this way, users can compose and execute geoprocessing jobs in a Windows environment and run them on UNIX hardware elsewhere on the network.

ArcView

ArcView is a desktop mapping and GIS tool that enables the user to select and display various combinations of data. ArcView is integrated with other applications, creating an environment for analysis and desktop publishing, and spreadsheets, databases, word processing, publication graphics, and other software applications that extend the functionality of ArcView. ArcView works directly with ArcInfo, ArcCAD, PC ArcInfo, and SDE databases. It can import a variety of data sources, including CAD files, spreadsheets, raster data, and other databases. The key features of ArcView include the following.

- “Windows” graphical interface
- Integrated charts, maps, tables, graphics, and multimedia
- Visual mapping
- Wizards for map composition
- Labeling and text tools
- Industry/application-specific symbols
- Geographic hot links to supported data formats
- Analysis wizards for geoprocessing operations
- Address matching and geocoding
- Utility for projection and datum transformation
- Geographic and tabular data editing
- Integration of images and CAD data
- Client/server access to data warehouses
- Five CDs of data included
- Report writing using Crystal Reports
- Self-paced tutorial

- User manual
- Online help
- Customizable with developer environment
- Expanded analysis capabilities using optional extensions

ArcView runs on Microsoft Windows and UNIX computers. ArcView can map tabular data residing in Microsoft Access, dBASE, FoxPro, ASCII, INFO, SQL, Open Database Connectivity (ODBC), and/or SAP R/3 databases. ArcView supports the following GIS data formats. You can read data from ESRI shapefiles, ArcInfo coverages, ArcInfo dynamic segmentation coverages, route systems, PC ArcInfo coverages, AutoCAD (DXF and DWG), MicroStation (DGN and MSG), TIFF 6.0 (including GeoTIFF), VPF, ADRG, CADRG, CIB, NITF, MrSID, JPEG (JFIF), ERDAS IMAGINE, ERDAS LAN and GIS, BSQ, BIL, BIP, SunRaster files, BMP, GRID (as image data), and DIGEST (ASRP and USRP) ArcSDE with Spatial Database Engine (SDE), or Oracle Spatial Data Option (SDO) with SDE.

You can import data from ArcInfo, MapInfo (MIF), S-57, SDTS (raster, point), and ASCII. ArcView also supports two extensions: ArcView Network Analyst for network solutions and ArcView Spatial Analyst for combined raster-vector analysis.

ArcIMS

ArcIMS is a platform for the exchange of Web-enabled GIS data and services. Internet maps can be created using shapefiles, ArcSDE data sets, and images. ArcIMS operates in a distributed environment and consists of both client-side and server-side components. The ArcIMS server-side components include the following.

- ArcIMS Spatial Server, which processes requests for maps and related information
- ArcIMS Application Server, which handles the load balancing of incoming requests and tracks what MapServices are running on which ArcIMS spatial servers
- ArcIMS Application Server Connectors, which are used to connect the web server to the ArcIMS Application Server
- ArcIMS Manager, which is a suite of web pages that provides access to all ArcIMS server-side functions and tools needed to set up and administer Internet services

ArcIMS client-side components include HTML and Java clients, as well as support for a full suite of other ESRI clients. The ArcIMS client viewers process data on the client machine, performing some tasks without further interaction with the server. Users can also develop custom desktop applications using the ArcIMS connector interface. Users can build custom Visual Basic and Visual C++ applications that use ArcIMS services.

The ArcIMS Viewers determine the functionality and graphical look of the web site. Users can customize templates for adding logos, graphics, colors, and functions. The ArcIMS Viewers offer tools for viewing and querying spatial and attribute data, performing spatial analysis tasks such as selecting and buffering features, measuring distances, geocoding, and sharing ideas about data with others, using such tools as Edit Notes and Map Notes. The ArcIMS Viewers also feature legends, overview maps, saving and retrieving projects, and map printing. The ArcIMS Manager consists of the following three stand-alone components.

- ArcIMS Author, which allows users to define the mapping application content
- ArcIMS Designer, which generates the web service the end user will view
- ArcIMS Administrator, which controls the operation of the web mapping site

The ArcIMS authoring and site administration tools provide a web site management environment. In addition, tools can be distributed to various administrators and can be run in a combined browser interface or as stand-alone Java applications. The ArcIMS Manager resides on the web server computer and can be accessed remotely for administering a site.

Generating an online map involves adding data content and setting other map properties that create a MapService. A MapService allows the content of a map configuration file to be published on the Internet, and sets the framework for the web site functionality. The output from the ArcIMS Author is a map configuration. This file can also be edited in a text editor independent of the ArcIMS Author environment.

The ArcIMS Author is a menu-driven applet that steps users through the map content definition process. The ArcIMS Author allows users to define connections to databases, symbology, and other mapping parameters. The ArcIMS Author includes features for setting scale dependencies so that a map renders the appropriate amount of detail at a given scale.

The ArcIMS Designer leads its users through a series of panels, including selection of MapServices, templates, and the operations and functions that will be available to the client web browser. The ArcIMS Designer lets the user define whether to allow query, editing, Map Notes, Edit Notes, or data integration functions. The ArcIMS

Designer takes the user through the steps of creating a web site and MapService, defining the page elements, map extent, visible layers, and overview map, as well as setting features such as the scale bar. With the ArcIMS Designer, users can also select toolbar functionality from a predefined menu of options.

The ArcIMS Administrator manages MapServices, servers, and folders. With the ArcIMS Administrator, users can perform the following.

- Add and reconfigure ArcIMS sites
- Load balancing
- Manage ArcIMS Spatial servers
- Assign tasks to servers
- Monitor client and server communication
- Automatically update web site configuration
- Compile statistical information

NOTE: *For further information on ESRI, call 800-447-9778 or visit* www.esri.com *on the Internet.*

Intergraph

Intergraph Corporation of Huntsville, Alabama, is a software developer, hardware manufacturer, and services provider, specializing in computer graphics systems for technical applications, including GIS, automated cartography, photogrammetric mapping, and AM/FM. Its core GIS products are MGE, GeoMedia Professional, GeoMedia, and GeoMedia Web Map.

MGE: Modular GIS Environment

Modular GIS Environment (MGE) is a suite of GIS products. The foundation of the MGE product line includes the following five software components.

- MGE Basic Nucleus for data query and review.
- MGE Base Mapper for data collection.
- MGE Basic Administrator for database setup and maintenance.
- MGE Spatial Analyst for the creation, query, spatial analysis, and display of topologically structured geographic data. Analysis capabilities include Boolean and spatial operations. Spatial operations include buffer zones, spatial

relationships, calculation of area and length, and merging new themes into existing files.

- MGE Base Imager for multispectral digital image enhancement and analysis. Data sources can include satellite, video camera, scanned black-and-white or color photography, digital terrain model (DTM) files, and coordinate digitizers. Raster and vector data can be combined and overlaid.

MGE uses the MicroStation CAD package for most graphic data creation, editing, and output. To handle nongraphic attribute data, MGE supports six relational databases through Intergraph's Relational Interface System (RIS), which allows direct access to Informix, Oracle, Ingres, DB2, SYBASE, and Microsoft's SQLServer. RIS also supports the Open Database Connectivity (ODBC) standard, enabling access to other industry-standard databases. GIS data translators are included with all MGE modules, including translators for ASCII data, U.S. Census Bureau TIGER data, USGS DLG data, and data stored in SDTS format. The key capabilities and features of MGE's core modules enable users to perform the following.

- Set up and manage GIS projects
- Perform data entry, digitizing, data cleanup, and data management
- Import other GIS data formats into MGE projects
- Integrate engineering design files in the GIS database
- Perform image display and analysis
- Process scanned drawings, aerial photographs, and satellite imagery
- Manipulate both image and vector data for heads-up digitizing
- Query, analyze, and display spatial data
- Cut and paste database reports into spreadsheets and word processing documents
- Compose thematic maps interactively

In addition to the five core MGE modules, Intergraph offers other specialized MGE modules. Of these, several key products are MGE Advanced Imager, MGE Grid Analyst, MGE Network, MGE Projection Manager, MGE Terrain Analyst, and MGE Map Publisher.

GeoMedia Professional

GeoMedia Professional (Pro) provides the following general capabilities.

- Data capture, automation, and maintenance
- Enterprise data management
- Spatial analysis
- Map production
- Applications development

GeoMedia Pro provides data capture and editing tools in a CAD-like, Windows-based drafting environment designed to reduce the number of steps required for each task. Integrated vector and raster "SmartSnaps" support the capture of vector data from raster images. As the cursor is moved, snap points are identified automatically. Table digitizing and vector transformation are also provided for data that requires geometry transformation.

GeoMedia Pro includes tools designed to capture data with minimal editing. Automatic vector breaking and coincident geometry digitizing reduce problems such as slivers and intersections without nodes. Coincident geometry digitizing captures a new polygon by selecting an existing polygon to digitize the common boundary, reducing the possibility of gaps or slivers. Automated error detection capabilities locate problems in existing data sources or that are created during digitization.

Errors are corrected using intelligent feature placement and editing tools, such as dynamic queued editing and coincident geometry editing. Dynamic queued editing presents each problem in sequence. As each problem is fixed, the queue is updated. When a point on a shared boundary between two features is added, moved, or deleted, the coincident geometry editing capability automatically changes both features. Data is annotated using labeling and text placement tools.

GeoMedia Pro provides access to popular scanned image formats, including Intergraph formats, Geotiff, TIFF, MrSid, ESRI World, CALS, JPEG, Hitachi, IGS, and BMP. GeoMedia Pro's data server interface accesses CAD and vector GIS formats on the fly, directly reading MicroStation, AutoCAD, ArcInfo, ArcView, MapInfo, MGE, and FRAMME files. GeoMedia Pro's editing tools support the edge matching of files derived from data for multiple map sheets.

GeoMedia Pro makes live connections to native GIS data in multiple data warehouses simultaneously. It performs coordinate system and projection transformation on the fly. With real-time access to native data, queries and thematic maps are automatically updated to reflect changes in the data warehouse. GeoMedia Pro supports a variety of industry-standard relational databases, including Access, SQL Server, and Oracle8i Spatial.

GeoMedia Pro includes tools that support spatial analyses, including buffer zones, spatial and linear segmented data queries, and thematic maps. GeoMedia Pro can access multiple databases in their native formats and create topology on the fly. Satellite imagery can be integrated with other geographical information to add new features. GeoMedia Pro creates dynamic maps that integrate multimedia and GIS using hyperlinks to files containing sound, images, or text.

GeoMedia Pro provides tools for map layout using a CAD-like drafting environment, providing tools for the placement of map information, as well as the typical graphics found on printed maps and charts. With "SmartPlot" tools, marginalia such as legends, scale bars, and North arrows are automatically generated and dynamically linked to the map information. Any changes to the map are reflected in the layout. GeoMedia provides multiple selection methods for controlling the map data to plot: map window, existing area feature, rectangle, polygon, paper size, geographic frame, or projected frame.

Multiple map frames can be placed to present overviews and insets along with the primary map, or multiple maps can be placed on a single layout. GeoMedia Pro's map layout tools are OLE-compliant, so that images, spreadsheets, and other Windows-compatible information can be associated with map features to prepare reports, annotate drawings, and create presentations. GeoMedia Pro can print and plot to scale with WYSIWYG feature symbology.

GeoMedia Pro supports customization by storing scripts, styles, projections, and other information. GeoMedia Pro is compatible with standard Windows development tools, including Powersoft's Powerbuilder, Microsoft's Excel (with VBA), Visual Basic, and Visual C++. The toolbar, menus, and shortcuts can be customized using a drag-and-drop action.

GeoMedia

GeoMedia is a GIS visualization and analysis tool, as well as a platform for custom GIS solutions, providing the following tools.

- Data integration with major GIS vendor formats
- Spatial analysis tools
- Map layout
- An open development platform for creating custom applications
- Microsoft Windows-standard user interface

GeoMedia provides access to leading GIS product formats through "live" data connections to "native" GIS data repositories. GeoMedia can integrate data using on-the-fly coordinate transformation and feature definition. GeoMedia data server technology supports views of multiple GIS data sets in various formats, and analyzes this information by running queries, buffer zones, and thematics across multiple GIS formats. Accessible GIS formats include MGE, FRAMME, MGE Segment Manager, MapInfo, AutoCAD, Oracle SC Relational, MicroStation, ArcInfo, ArcView Shapefiles, Microsoft SQL Server, and Oracle8i Spatial. GeoMedia supports Oracle, including Oracle8i Spatial, as the geospatial warehouse.

GeoMedia's analysis tools include nine types of spatial functions, such as "entirely contained by" or "touches"; a suite of arithmetic operations; and the tools necessary to create "what if" queries. It can create buffer zones and spatial overlays, and query spatial data within a specified area. GeoMedia's dynamic segmentation capabilities can query and segment linear and point data from multiple MGE databases.

GeoMedia's thematic mapping features create maps with color-coded and patterned attribute data, and integrate multimedia with GIS, including hyperlinks to files that contain sounds, images, text, and satellite imagery. GeoMedia also supports geocoding, which translates tabular data, such as street addresses, into spatial data.

GeoMedia can create maps and presentations with tools for map layout. This includes tools for the placement of map information, as well as the typical graphics found on printed maps and charts. Marginalia such as legends, scale bars, and North arrows are automatically generated and dynamically linked to the map information. Changes to the map are automatically reflected in the layout. GeoMedia provides multiple selection methods for controlling the map data to plot: map window, existing area feature, rectangle, polygon, paper size, geographic frame, or projected frame.

Multiple map frames can be used to present overviews and insets along with the primary map, or multiple maps can be placed on a single layout. GeoMedia is OLE-compliant, so that images, spreadsheets, and other Windows-compatible information can be associated with map features for preparing reports, annotating drawings, and creating presentations. GeoMedia is compatible with standard Windows development tools and can be customized using OLE automation tools such as Powersoft's Powerbuilder, Microsoft's Excel (with VBA), Visual Basic, and Visual C++.

GeoMedia Web Map

GeoMedia Web Map is Microsoft Windows-based technology that combines and distributes GIS information from multiple sources over an intranet or the Internet to clients running Windows and an industry-standard web browser. GeoMedia Web

Map enables distribution of GIS information with vector-based "smartmaps" that contain hyperlinks on individual features within the map. Using templates delivered with the product, the user can define map definition files (MDFs) to serve various end-user needs.

For example, data from different sources can be combined on a single map, allowing users to retrieve data from a utilities database and a land-use database, and display them together in a map. Each feature in a smartmap will know which database to search for its attribute information. Once defined, those intelligent maps are published on the Web in response to user queries. Users interact with the intelligent maps using the ActiveCGM open data format.

When a user clicks on a map feature, GeoMedia Web Map provides associated information about that feature from the data available in the respective local or distributed databases. Maps are recreated every time geographical information is updated on the server, so that users always view current information. Smartmaps also deliver reduced Internet packet sizes and improved response times for typical applications.

GeoMedia Web Map combines raster and vector information within the same map, and descriptive information can be linked to map features through standard relational database techniques to create hyper-linked map features. GeoMedia Web Map reads MGE, ArcInfo, ArcView, MicroStation, MapInfo, AutoCAD, Oracle Spatial Cartridge/Spatial Data Option (SC/SDO), and Microsoft Access data directly, without translation.

GeoMedia Web Map can be customized using standard web development tools such as Java, Visual Basic, or Microsoft FrontPage. With the addition of GeoMedia Web Enterprise, web clients can also analyze and manipulate GIS data, performing spatial queries, buffer zoning, geocoding, network analysis, and coordinate transformation.

> **NOTE:** *For more information on Intergraph GIS products, call 1-800-345-4586, or visit* www.intergraph.com *on the Web.*

MapInfo

MapInfo Corporation is headquartered in Troy, New York. MapInfo sells its products principally through its network of VARs (value-added resellers). These VARs offer services to implement, provide training for, and integrate MapInfo's products. The company's core GIS product is MapInfo Professional.

MapInfo Professional

MapInfo Professional provides data visualization, including step-by-step thematic mapping, and three linked views of data: maps, graphs, and tables. MapInfo Professional supports raster map registration and imports a variety of standard raster formats and vector formats. Maps can be digitized to create vector images, and edited using several functions: "reshape objects," "snap-to editing," "move objects," "select multiple nodes for deletion," "overlay nodes," "copy objects," "create polylines from regions," "create regions from polylines," "smooth/unsmooth," "revert table," and "clear map objects only."

MapInfo Professional data can be visualized to create thematic maps using ranged shading, bar charts, pie charts, dot densities, graduated symbol, and individual values. Range classification of maps can be based on equal count, equal ranges, inflection point, natural breaks, standard deviation, quartile, and user-defined. Thematic options can be combined to create maps that display multiple variables from many tables. Thematic joins can be based on count, sum, value, average, min/max, weighted average, proportional sum, proportional average, and proportional weighted average.

MapInfo Professional provides multiple data views with a zoom range of 55 feet to 100,000 miles. The user can set the layer display order and view hundreds of layers as seamless maps. MapInfo Professional supports 18 map projections, and performs map projection transformation/display on the fly. The user can overlay various projections and convert between projections.

MapInfo Professional will attach data to, and retrieve information about, any object. It also provides data analysis capabilities. Polygon overlay operations include data aggregation/disaggregation, erase, erase outside, split, combine, and overlay nodes. MapInfo Professional will create buffers around any object or group of objects-points, lines, polylines, and polygons. The user can perform geographic selection to find any object, points within a radius/polygon, lines/polylines within a radius/polygon, and polygons within a radius/polygon.

MapInfo Professional supports donut/island polygons, as well as object buffering to examine data within a specific proximity. MapInfo Professional also supports geocoding to street address, ZIP codes, census tract, and user-defined match options.

MapInfo Professional offers SQL capability with geographic extensions on expressions (contain, contains entirely, within, entirely within, intersection/union). Queries can be based on expressions, can display derived fields, and can perform aggregations, subqueries, multi-table joins, and geographic joins. MapInfo Professional provides relational query operators, as well as standard and geographic operators. MapInfo Professional supports standard mathematical, statistical, calendar, trigonometric, logical, measurement, conversion, sorting, and aggregation functions.

MapInfo Professional provides direct access to spreadsheet and database data. MapInfo Professional can directly read from and write to ODBC-compliant databases and issue geospatial queries to ODBC-compliant databases storing point data. MapInfo Professional supports geographic queries to remote database servers containing point data. The user can create, edit, or combine map features to test scenarios before committing resources.

New maps or tabular databases can be created based on new information or the results of queries and edits. MapInfo Professional provides tools for managing database fields. The field types supported include character, floating point, integer, decimal, logical, and date. The number of database records is limited only by hard disk storage. A record may contain up to 250 fields.

MapInfo Professional provides automatic or manual labeling, supports hundreds of symbols and user-created bitmaps, and offers an unlimited number of colors. Labels are scalable when zooming. MapInfo Professional supports any Windows-compatible printer or plotter and outputs to Bitmap (.BMP) or Windows Metafile (.WMF) formats.

MapInfo Professional offers a 3D viewing capability. Based on Microsoft's implementation of OpenGL software graphics interface, it allows freehand tilt and rotation of the image. The traditional pan, zoom, and Info tools also operate in the new window. Users of 3D can view any map that contains a continuos thematic grid layer; for example, by importing digital elevation model (DEM) files or interpolating from a layer of points that contains elevation values. The image of any other layers is draped over the 3D surface.

With MapInfo Professional's Internet connectivity, users' maps can be linked to the world. Any object in a map can contain a URL that is automatically launched in the default web browser when that object is selected. The same applies to other types of document files the Windows operating system can launch (*doc*, *xls*, *ppt*, *tab*, *wor*, *mdb*, and so on). MapInfo Professional can also generate HTML Image Maps that can be included in web pages.

Custom MapInfo Professional solutions can be created or integrated into other applications developed in programming languages such as Visual Basic, PowerBuilder, C++, and Delphi. MapInfo Professional will access and integrate global positioning system (GPS) data into applications to track data in real time or collect data from remote field locations.

MapBasic and MapInfo Professional

MapBasic is a BASIC-like programming language used to create customized mapping applications or integrate mapping functionality into existing applications. The MapBasic Development Environment includes a text editor for typing in programs, a compiler for creating an executable from a program, a linker used to link separately written modules of programs when creating large or complex applications, and on-line help, which provides reference information for each statement and function in the MapBasic language.

MapBasic can be used to modify the MapInfo user interface by adding customized toolbars, menus, and dialog boxes to suit the specific needs and technical sophistication of the end user. Unneeded functionality can be hidden. Wording can be changed to reflect the terminology appropriate to a specific application. Complex database queries can be reduced to a single MapBasic statement, and repetitive operations can be automated.

MapInfo Professional provides OLE 2.0 support. Users can drag and drop maps into other applications and activate MapInfo features from within other OLE 2.0-compatible applications. Over 300 megabytes of maps and additional market and demographic data are included with MapInfo Professional.

> **NOTE:** *For further information on MapInfo, call 518-285-6000 or visit* www.mapinfo.com *on the Internet.*

Autodesk

Autodesk, the developer of AutoCAD, has been a leader in the CAD software market since the 1980s. Autodesk introduced AutoCAD Map as its first GIS product, and has since then added Autodesk MapGuide as a second core GIS product.

AutoCAD Map

AutoCAD Map is software for mapping and GIS analysis in the AutoCAD environment. It provides GIS analysis tools and features for creating, maintaining, and producing maps and geographic information, along with the underlying functionality of AutoCAD. The following are some of its key features.

- *Data integration:* Import and export maps from standard CAD and GIS file formats. Provides database-linking capability to add "intelligence" to maps. Link internally and externally to Microsoft Access, dBase, Oracle, FoxPro, Paradox, and other ODBC-compliant databases with drag-and-drop configuration. Raster tools provide support for multiple raster file formats.

- *Multiple map access:* Work with large data sets and multiple maps, simultaneously sharing them with other users without version conflicts. Work with multiple maps, in different coordinate systems, in a single session. Save map projects, and the settings for the session are saved automatically for recall at a later time. Provide multiuser access to single or multiple maps with entity-level locking.
- *Map creation and cleanup:* Digitizing, coordinate transformation, and drawing cleanup. Digitizing tools attach object data or SQL links. Transform maps with different coordinate systems into one standard or customized system. Edit map data with rubber sheeting, edge and boundary trim, and so on. Prepare for topology, with automated drawing cleanup tools.
- *Querying:* Access and query data from different sources and combine the information into one map, alter object properties during queries, save routine queries for future use, and perform topology queries.
- *GIS spatial analysis:* Topological operations, path tracing, and other analytical capabilities. Create, edit, and save node, network, and polygon topologies. Perform flood analysis and buffer, shortest, and optimal path trace. Conduct polygon overlay and dissolve operations. Create thematic maps based on associated data.
- *Coordinate conversion:* Support for global and local coordinate systems. Combine multiple maps into a common coordinate system. Define a custom coordinate system.
- *Construction tools:* Build new maps using construction tools. Use polar construction tools to place field locations of objects the way they were measured. Switch between multiple editing alignments.
- *Presentation and plotting:* Thematic mapping and plotting. Flexible plot styles support changes to the visual properties of any object, including color, line type, line weight, and pen assignment at plot time. Predefine map's appearance and reapply the same settings.

Autodesk MapGuide

Autodesk MapGuide consists of three main components: Autodesk MapGuide Author, Autodesk MapGuide Server, and Autodesk MapGuide Viewer. Autodesk MapGuide Author is used to design intelligent maps for Internet, intranet, or extranet distribution or for inclusion in a live presentation. It can define a map's appearance, security, and GIS interactivity. Autodesk MapGuide Server is the administration software that brokers requests and delivers live, interactive maps over the Internet or intranet.

It can also be used to monitor and keep a record of requests, gathering information about patterns of use. Autodesk MapGuide Viewer enables nontechnical end users to access and interact with intelligent maps through their web browsers or through custom applications. Autodesk MapGuide Viewer enables users to pan and zoom, make queries, create dynamic buffering zones, measure distances, print to scale, and run custom reports based on selected objects.

> **NOTE:** *For further information on Autodesk, call 415-507-5000 or visit* www.autodesk.com.

Summary

These four GIS product lines illustrate the wide variety of functions GIS software can perform. It is difficult to compare them with one another because of the large number of features and options that each offers. Therefore, if you are trying to select a GIS, it is very important to begin with an assessment of what functions you need in a GIS and compare the various GIS programs to your needs, rather than to one another.

Keys to Successful GIS Implementation

Eight Important Issues That Influence GIS Implementation

In This Chapter...

This chapter discusses the eight key components of a successful GIS operation. These are:

1. Management support, leadership, and vision
2. Data conversion and maintenance
3. Hardware and software
4. User training
5. Data communications
6. Software customization
7. User support
8. Funding

Thriving GIS programs have adequate resources in all of these areas. Ineffective, faltering, or disappointing GIS programs are lacking in one or more of these key areas. One GIS has great hardware and software, and great people managing it, but is restricted to one project or department, rather than serving the entire organization, because top management does not support it.

In another organization, top management is really enthusiastic about GIS, but the program never achieves its potential because of turnover among the GIS staff. Yet another GIS program has all the right tools, personnel, and management support, but lacks the funds to move ahead. GIS programs succeed to varying degrees depending on their effectiveness in each of these areas.

Management Support, Leadership, and Vision

Establishing and maintaining a GIS requires a significant expenditure of time and money. Therefore, it most often needs approval by the top decision makers in an organization. Although this is true of many other capital improvement programs, a GIS is different in that the top-level decision makers are usually not familiar with GIS technology. Some may understand the basics of computer technology, but few are familiar with mapping concepts in general and computer mapping in particular.

In many cases, the initial support and enthusiasm for a GIS develops among mid-level managers, technicians, and professionals who hear about it from peers in other organizations, read about it in trade publications, or learn about it at industry conferences. They must then interest top management. If top management is not fully convinced of the justification for a GIS, they might grant only tentative or partial approval of the program. This poses a real danger. Without management's total support and full commitment to a GIS, sometimes, for example, only portions of land records are automated, or only portions of the GIS are implemented.

This generally means that the system will not reach its full potential. Moreover, if the project runs into technical problems or *cost overruns*, top-management support is critical to keep it afloat. Obtaining top management's full support may delay the start of a GIS project, but in view of the benefits that could result as the GIS project proceeds, it is well worth the effort and the delay.

A GIS program also needs a clear vision of objectives. This vision may serve as the foundation of the system's implementation plan. Some GIS programs are too ambitious, trying to do too much too fast. Others have a scope that falls short of the technology's potential, often creating a "stovepipe" operation that only serves one department's needs and neglects those of other departments. Likewise, good leadership is needed to keep the vision in full view of the GIS team, offer encouragement, organize and manage the GIS operation, make course corrections, and answer to top management.

Data Conversion and Maintenance

Typically the largest portion of the cost of a GIS program is converting existing land and facility records to a digital format. In fact, *the cost of data conversion can range between 60 and 80 percent of the total cost of implementing a GIS.* Most of this cost goes into the labor-intensive process of collecting and digitizing existing map information. Therefore, it is important to carefully consider what information really needs to be included in a GIS.

If money is tight (isn't it always?), one can save a significant amount by postponing conversion of map data that has the least amount of user interest. For example, topographic maps, soil maps, geologic maps, sanitary sewer laterals, and water system house connections represent a significant data-entry cost, but may be needed by only a few GIS users. It may be acceptable to delay their addition to the GIS database, spreading the cost of database construction over several phases.

A significant task in data conversion often overlooked or underestimated is preparation of data for digitizing. This process, often referred to as *scrubbing* data, is normally required when existing records, maps, and files are incomplete or not current. Moreover, the data scrub will need help from the persons most familiar with the records. This may be a big job for the staff. Similarly, when the data has been converted to a digital format, it must be thoroughly checked before being accepted as the GIS database. This may also demand an extra effort by in-house staff. These tasks can be particularly difficult to complete when they have been underestimated or overlooked entirely.

Another consideration is whether to perform the data conversion using in-house personnel or a contractor. A major advantage of using in-house staff is that the effort may at least *appear* to be less costly. This depends on the type of internal cost accounting system in use. In most municipalities and utilities, the only cost attributed to GIS data conversion is the cost of salaries (or salaries plus fringe benefits) of the employees doing the work.

On the other hand, a contractor's fee includes salaries, fringe benefits, overhead costs (i.e., insurance, office administration, utilities, rent, supplies, equipment depreciation, equipment maintenance, and marketing costs), and a profit. In fact, a contractor's fee will typically run more than twice the total salaries of the employees who do the work. Of course, a public agency also incurs most of these types of costs, but its accounting system typically does not attribute these costs to individual projects, such as a GIS data-conversion effort.

Doing the work in-house may give the organization better control of the work. It will also provide a better opportunity to modify the database design during the data-conversion process. However, in-house personnel are usually far less experienced at this type of job than contractor personnel. Contractors specialize in this work and have learned the most efficient techniques for getting the job done, with the fewest errors.

Contractors can also be held to a fixed fee for their work. This puts a firm cap on the cost of the conversion work. Moreover, in-house conversion will normally take longer than a contractor would require because a contractor can usually dedicate more equipment and personnel resources to the effort in order to get the job done faster.

Organizations commonly contract with a service bureau or consultant for the initial database creation. When this contractor has delivered the database, it becomes the municipality's responsibility to maintain it. Therefore, the GIS implementation plan should ensure that all of the resources needed to take over maintenance of the database are in place well in advance of database delivery. This includes employee training.

After land or facility records have been turned over to a contractor for digital conversion, changes are still occurring in the real world. The contractor usually creates a database from the information shown on the drawings at the time her work begins. If she is creating topographic base maps, she digitizes information shown on the aerial photography at the time the photos were taken.

Therefore, the organization must monitor changes in the real world that will not be included in the contractor's database. When the contractor finishes her effort and the database is delivered, the first maintenance task is to incorporate these interim changes. The interim period may be anywhere from several months to a year or more, depending on the size of the project area and the amount of data converted. The GIS implementation plan should anticipate this need.

One cost saving measure is data "cost sharing." In many regions there are several organizations that need topographic base maps, tax parcel maps, or other land data covering the same geographic area. Such groups normally include local government agencies, utilities (e.g., electric, gas, cable TV, telephone, water, and sewer systems), regional planning agencies, state agencies, and others. Physical proximity presents an excellent opportunity for these organizations to share the cost of creating the GIS database.

Although there is a real opportunity for cost savings, the political complexities can be difficult to surmount. Nonetheless, the idea of creating a consortium of agencies that share a common land database has been successfully implemented in a number of major metropolitan areas. Therefore, it is an idea worthy of serious consideration by any municipality or utility embarking on a GIS program.

Hardware and Software

Computer hardware and software are the tools employees will use to input, analyze, manipulate, and present GIS data. Obviously, the choice of tools will have a major impact on the success of the GIS program. Some guidelines for choosing GIS products were previously discussed. The principal consideration is selecting tools that will best satisfy the needs of the organization and its users, which is not always the most advanced or most popular system.

User Training

GIS introduces a great number of employees to a very different way of doing work. Typically, they have little exposure to computer systems in general, and none to GIS in particular. Most employees sincerely want to do well in their work. However, to do well in their jobs, they need to feel confident about what they are doing, and they need to have the knowledge and skills necessary. Trying to work with a new technology such as GIS can destroy employee confidence and morale. This can only be overcome through a comprehensive, planned training program.

Initial training following system installation can almost always be purchased from the vendor or from a local authorized training center. Subsequent training can also be purchased, but many organizations set up in-house training programs run by employees with previous experience. The important point here is that *some* ongoing program is needed to train new users. They will replace employees lost through attrition, and all users will need to learn to operate new equipment added as system usage expands and needs change.

The training program should recognize that the learning curve to full proficiency is three to six months. Moreover, the training program should make a realistic assessment of employee turnover. In fact, it should assume that turnover will be greater among GIS users because their new skills will be in greater demand than their previous ones. A basic GIS user training program will usually require about 40 hours of the employee's time. This may be spread over a number of sessions.

For instance, the employee could participate in two half-day training sessions per week over a five-week period. System managers may require one to two weeks of additional training to learn system management, network management, user account management, and more advanced GIS and database functions the average user will not need.

Data Communications

Few organizations that use a GIS will confine its use to one department. Most often there are numerous departments that need access to the database. Distributing the GIS data is the key to making this data available to them. A data communications network is the most commonly used vehicle for GIS data distribution, so the timing of the GIS implementation plan is often tied to the installation of the network.

A local area network (LAN) is used most often, but there are other options for distributing GIS data. Some organizations distribute vector GIS files over the network, but use CD-ROM disks to distribute more voluminous types of raster data, such as aerial photography and satellite imagery. Whereas vector data may be updated on a daily basis, raster data sets rarely change, often only once a year or so. GIS data can also be transmitted by modem over a telephone line, but because the data files are relatively large, this process can be too slow to be practical.

The Internet is also used to distribute GIS data. This is most often done to make the information available to the public. Obviously there will be security considerations, and some data may not be made available to Internet users. In some cases, access to certain types of data can be password-protected so that only authorized users can have access to it. A related technique is use of an intranet, which basically refers to the idea of accessing data using a web browser over an organization's internal LAN.

Software Customization

GIS vendors normally design their software to meet the needs of the largest possible number of users. They incorporate general capabilities suitable for many types of applications. For this reason, a GIS package will not provide optimal use of the technology unless it is customized for a particular user's needs.

The best way to get maximum use of a GIS is to customize the software. The GIS vendor typically offers a *macro* command language. This enables a user to string several commonly repeated operations together under one command. The vendor also usually offers a means of customizing the user interface; namely, the tablet menus and screen menus that provide access to the GIS commands. These two features can save a tremendous amount of time. This software development work is usually done by a GIS programmer.

It is important to find a programmer familiar with both the GIS programming language and the application. It is also important to encourage good communication between the programmer and users. This will ensure that the programming meets their needs.

User Support

GIS is a complex technology, making it relatively difficult to implement and manage. One way to greatly increase the chances of a successful GIS implementation is to involve an experienced GIS manager and/or system support technician. Even if an experienced person is not hired into the organization, it is possible to obtain this experience from a contractor. This may be the GIS vendor or a qualified consultant. The length of the support contract may vary, but two months would seem to be a reasonable minimum.

In situations in which an organization is unable to ensure that an experienced employee will always be available, GIS support can be contracted for on a long-term basis. Although GIS vendors invariably offer extensive telephone support, having experienced help on site is far more effective. Personal experience suggests that, without exception, *the most successful GIS programs either hired an experienced GIS manager or system engineer, or obtained this support through a long-term contract.*

Funding

A complete GIS program for a municipality, utility, or military installation can cost hundreds of thousands to millions of dollars to implement, and tens to hundreds of thousands to maintain each year. Obviously, adequate funding for the program is vitally important. Although there are many factors that affect the availability of funding, the lack of financial justification need not be one of them. In a later chapter you will look at the process of estimating the costs and possible savings a GIS can produce, and how the business case for GIS can be made.

I once helped a very large suburban water and sewer utility develop a financial analysis for their proposed GIS program. We showed that the GIS would produce substantial cost savings. This analysis was reviewed by internal financial officers. A phased implementation program was presented to the utility commission, which approved it.

During the following year's budget process, every information technology (IT) budget was cut back except that of the GIS program. In fact, when the utility commission remembered the cost savings that had been documented and convincingly supported, and after receiving assurance that the program was proceeding according to plan, they decided to *increase* its budget above the requested amount. "The sooner we implement this system, the sooner these savings will begin," was the essence of their reasoning. Needless to say, the GIS program manager was pleasantly surprised.

Summary

Many people investigating GIS technology concentrate on the most obvious components of a GIS; namely, the software and hardware. This chapter has pointed out some other key considerations beyond the system itself. These issues are vital to the success of the GIS program. They will demand significant corporate resources, including money, time, and organizational energy.

In regard to the "energy" and commitment factors, a personal anecdote highlights a salient point. I participated in my first GIS selection process in 1981. I was with a large, private civil engineering firm. Computer mapping technology was still very new in those days. Our selection committee sought the advice of one of the few experienced users at that time, Mid States Engineering of Indianapolis, Indiana. We met with several of their key people, including the president of Mid States, Sol Miller. We thoroughly discussed the relative advantages and disadvantages of the two leading systems of the day, but we could not conclude which system would be best for our firm.

Finally, Mr. Miller looked at us and said: "Let's put it this way. We may not know which system is the best for you to *succeed* with, but we do know that you can *fail* equally well with either system. The most important thing is how you are going to put it to work." Today, the reliability and capabilities of GIS technology have progressed to the point that problems in GIS programs are nearly always nontechnical in nature. Instead, they are typically related to financial, organizational, planning, and human resource issues.

The Pitfalls of a GIS

Common Problems in GIS Implementation and Management

In This Chapter...

Numerous GIS programs have produced less than satisfactory results. The blame is often placed on the equipment, the software, or a vendor. However, Sol Miller's wise observation at the end of the last chapter leads one to suspect that there may be other reasons. Close investigation usually finds that the root of the problem lies in organizational and operational issues.

It is easy to find software commands, workstations, and plotters that do not function as planned, but these problems can usually be fixed quickly and with relatively little expense. The truly significant problems (those that can derail an entire program) are far beyond the control of vendors. Employees of software and hardware vendors sometimes shrug in frustration over a customer's faltering GIS program; not because the system does not work well, but because the customer is unable to make it work well.

Following are fifteen common problems encountered during the implementation and management of a GIS program. One can enhance the chances of success for a GIS program by simply being aware of these potential issues and taking appropriate steps to avoid them and to deal with them effectively when they do occur.

Failure to Consider Problems

Many organizations fail to grasp the full scope of a GIS project in terms of time, money, organizational energy, and management credibility. Problems are bound to occur. When an organization fails to anticipate those problems and their potential magnitude, it is not prepared to deal with them. This delays resolution of these problems and makes them more expensive to solve. Moreover, the project can lose credibility if these problems were not foreseen by the original project plan and justification. It is very difficult to find continued funding for a discredited project.

Overstating Benefits

There is a great temptation to emphasize only the benefits and the "gee whiz" of GIS technology during the initial justification stage. Potential difficulties, such as personnel changes, are often downplayed. The total costs and the fact that the payback period can last several years are often suppressed due to fear that the project may not get off the ground. However, this approach sets the stage for tremendous dissatisfaction if the results do not meet or exceed management's expectations, or if during the course of the assessment process the true costs and payback period gradually accrue in decision makers' minds.

Using a GIS as an Experiment

Organizations sometimes find the new GIS technology interesting or even exciting, and on this basis they decide to give it a try. However, they fail to make a true commitment to fully integrating the technology into their operation.

As a result, the GIS program receives too little funding, too few support personnel, and too low a priority to have a chance at real success. Moreover, the results of such "experiments" cannot be extrapolated to a fully operational and fully integrated system. The organization may eventually become convinced that the technology really *does* work, but it must still start from scratch to develop plans for a truly effective GIS program.

Failure to Define Goals

Goals give a sense of direction and purpose to any program. They are vital for maintaining the morale of GIS personnel as they encounter the problems, setbacks, and disappointments that inevitably occur. But to be effective, these goals must be defined clearly, and in understandable terms. Moreover, without goals there are no measures of failure or success, and users often lack a sense of satisfaction and accomplishment unless they can *see* that they have reached their goals. It can become difficult to justify continuation of the program without demonstrated achievement.

No Long-term Planning

Given the GIS program's goals, the next critical element is a long-term plan for achieving them. The plan should include the sequence of key events, their schedule, and the assignment of responsibilities for completing them. This plan also serves as the basis for developing staff requirements, equipment purchases, and budget requests.

Many organizations get into GIS when one department manages to obtain funds for the purchase of a workstation and GIS software. This group applies the technology to its operations and most often meets with some success. Then another department finds out about it, gets its own funding, and initiates its own GIS effort. This goes on until the day comes that someone asks, "Gee, can we get all of these GIS operations to work together?"

Then comes the bad news: "Well, we can do it, but it will cost this much, and this group will have to change their standards, and we will have to do some retraining, and..." Finally, someone realizes that the workings of the entire organization should have been considered before the first department set up a GIS. Plans should have been drawn up for the eventual implementation of a fully integrated GIS, even if the plans called for department-by-department growth.

Lack of Management Support

The need for management support was discussed in the previous chapter. It is such an important issue that it bears repeating. A GIS program most often requires the approval of the top decision makers in an organization. These decision makers are usually less familiar with GIS technology, which makes it difficult for them to support the concept without careful project justification.

Mid-level managers, technicians, and professionals must convey their support and enthusiasm to top management. Without top management's total support and full commitment to the GIS, the system implementation can be dangerously shortchanged. In that much of a GIS's power comes from its ability to integrate data, a partial implementation generally means that the system will fall far short of its potential benefits.

Lack of User Involvement

Users need to play a responsible role in the planning, selection, and implementation of a GIS. If they do not take part in this process, two negative things happen. First, the organization loses the opportunity to benefit from the experience of the users when planning and selecting the system. After all, who is better qualified to speak out regarding the needs for a GIS than the people who will use it? Because they know their jobs, they know how the GIS must perform to get the work done.

Second, failure to involve users also creates a sense of indifference, or even hostility, toward the new system. Employees who have no say in the way they do their work resent radical changes such as those associated with a GIS. As a result of these two negative impacts, the GIS is crippled and its chances for success are greatly reduced.

Failure to Specify Requirements

The pitfall of failing to define goals is clear, but even if goals are well defined, it is possible to fail at translating them into specific requirements. I am often confronted with a question that goes something like "The Alpha GIS does this real well and that pretty well, while the Beta GIS does this plus the other thing, but does not do that at all. Which is the better GIS?"

When I hear this type of question, it seems that the person has been swayed by fancy technology and lost sight of his true requirements. He has forgotten to ask the basic question "What do I need the GIS to do for me?" A clear statement of requirements is necessary to effectively select GIS hardware and software, or to build a GIS database. Not having a list of specific requirements fosters mistakes in the selection and purchase processes.

Computerizing Existing Problems

There is a temptation to simply incorporate new GIS technology into an existing organization and operations. I once heard a specialist in business process reengineering refer to this as "paving the cow path." However, *the implementation of a GIS presents an excellent opportunity to reevaluate existing workflow processes and organizational structure.* A GIS can increase productivity, integrate departmental operations, and better organize land data. However, it will not solve most management, operational, and organizational problems. These problems must be resolved *before* the GIS can reach its full potential.

Continuation of Existing Manual Systems

Some organizations choose not to terminate traditional manual processes when an automated process is installed. They want to ensure that they have a backup. Or, they may want to make sure that the new process is really working before they let go of the existing process. It is advisable in many cases to continue both systems in parallel for a short time until all problems are ironed out of the new system. However, there must be a definite commitment to stake everything on the new system as soon as it is operating satisfactorily.

Continuing the old manual process undermines the new system in two ways. First, it sends a mixed message to users: Is management fully committed to making the GIS work or not? Second, it enables users to postpone training on the new system. The old process is always available, so they can avoid using the new system as long as they want.

Lack of User Training

I was once told by the manager of a GIS that had been in operation for several years that the greatest obstacle to further use of the system was user training. I was surprised, because I had expected it would be for lack of funding for new equipment. Instead, he said he could get more than enough equipment; his problem was not having enough qualified users for it. GIS users who are not fully trained must rely on a specialist to do anything beyond the most routine tasks.

This creates the need for a central staff of GIS experts to support users. This central support staff often becomes swamped by user requests. The GIS group then increases in size until it becomes a major department in the organization. Moreover, users cannot possibly make the best use of the system, even if they have specialists to support them, because they do not understand its full potential.

On the other hand, fully trained users need only occasional support from GIS specialists. The central GIS group remains small and efficient, and users know how to get the most out of the system.

Lack of User R&D Support

Although a large GIS support group may not be desirable, or even possible, some user support is certainly necessary. GIS vendors develop products that meet the needs of a maximum of customers. They cannot tailor each system to the needs of a particular buyer. Therefore, no GIS can be used to its full potential right out of the box. Some programming is required to make the GIS meet each user's particular requirements. This calls for the development of graphic and database standards, custom menus, and macro commands. In addition, an expert must be available to troubleshoot occasional problems users are not normally equipped to handle.

Systems That Cannot Be Expanded or Modified

There is only one pitfall that is primarily a technical issue. It is common knowledge that computer technology has been changing rapidly over the past three decades. As a rule of thumb, *it is recommended that an organization plan to replace its computer hardware approximately every four years.* Peripheral devices, such as plotters, may be expected to last longer. Moreover, an organization's requirements change over time. Some changes can be expected; many are totally unexpected.

When hardware is obsolete or when user requirements change, it is important that new hardware can be incorporated with the least number of changes in the operation. It is most important that the GIS database can be transported to the new hardware without major modifications. The data represents the largest portion of the investment in GIS, and it should have a useful life of 20 years or more (allowing for continuous updating to incorporate technological and other changes). The best way to avoid this pitfall is to ask prospective vendors to explain how they have handled changes in hardware and operating system technology in the past, and what their policy is for the future.

Budget Overrun or Underestimation

Any program is usually in serious trouble when it has a budget overrun. Budget overruns usually result from one of two things: either the project was poorly managed and controlled, or the budget was underestimated to begin with. In the case of GIS, the latter occurs more often than the former.

One of my GIS clients was a county in Virginia for whom I conducted a GIS needs assessment and helped prepare a GIS implementation plan. At the outset of the study, I was told by the county manager that the one thing he most wanted me to do was to *not underfund budget estimates.* He said the worst thing he has to do as county manager is return to the board of supervisors and ask for additional funding of a capital project. Of course, whatever the reason for the budget overrun, it tends to put the program in serious jeopardy with management. *A significant budget overrun is one of the most likely reasons for the cancellation of a GIS effort.*

Failure to Report Results

A GIS project is also undermined when the GIS team fails to report the results of their efforts to top management, regardless of whether those results are good or bad. When top management loses track of the project's status, it tends to lose its sense of involvement and its commitment to the project as well. When top management is unaware of the status of the program, the GIS becomes susceptible to unfounded criticism and rumors.

Management has no information with which to counter such misinformation or make good decisions. It also lacks the incentive to support the program. Moreover, bad results appear worse when they have not been reported openly and promptly. Even if the results are poor, reporting them in a timely fashion fosters an attitude of cooperation in resolving problems.

Summary

Anticipating problems is half the battle in solving them. It is like the story of the country-boy pitcher who threw a two-hitter in the first game of a doubleheader. His brother was the team's other pitcher and threw a no-hitter in the second game. Someone asked the boy how he felt about having his brother outdo him. He replied, "Shucks, if I'da knowed he was gonna throw a no-hitter, I'da throwed one too!" Being aware of the potential pitfalls of GIS is the best way to avoid them.

Using a GIS Consultant

Reasons for Using an Outside Consultant to Assist in GIS Planning and Management

In This Chapter...

The most important decisions about a GIS program are the pivotal, strategic decisions made during its planning process. Seasoned, practical experience is needed to make these decisions effectively and with confidence. Unfortunately, this is also the time when an organization knows the very least about GIS. To solve this problem, many organizations have sought the assistance of a GIS consultant, the topic of this chapter.

Benefits of Hiring a GIS Consultant

Even GIS programs that are well underway still need assistance from time to time. I once helped a large water-sewer utility prepare the financial justification for its GIS program. The analysis showed a strong return on investment and resulted in full funding for the program. However, some time later I received a call from one of their top managers asking me to meet with him. "George, our GIS has made a lot of progress in building the database but we seem to be bogged down right now and unable to deploy real applications," he told me. "Do you have a silver bullet for us?"

I suggested the idea of a simple strategic planning workshop. It would focus on three questions: "Where are we now in using GIS?" "Where do we need to go?" "What do we need to do to get there?" He liked the idea, so we convened a one-day session with representatives of all departments that could potentially use GIS. We worked through an agenda of a dozen topics and I wrote up a brief report summarizing the discussion and the decisions they reached. A few months later he told me it was just the thing they needed. Borrowing the expertise of a GIS consultant offers many benefits, including those discussed in the sections that follow.

Specialized Expertise and Education in GIS Planning

A GIS consultant has more than just a formal education in systems analysis for computer hardware and GIS software. She also has expertise in all of the technical aspects involved with GIS implementation. These include land survey, aerial photography, photogrammetry, GIS database construction, quality control, and project management.

Practical Experience in a Variety of GIS Applications

Although experience in GIS implementation is by far the most important qualification, a GIS consultant should also have a background in many areas of GIS applications. These include engineering, planning, public works, utility systems, land records, tax assessment, and public safety.

Knowledge of GIS Pitfalls

A GIS consultant has seen an assortment of successful and less-than-successful GIS programs. He has learned from his own and others' mistakes. Therefore, he is in an excellent position to keep his client from making the same mistakes.

Short-term Contract Labor

Hiring a GIS consultant delays the commitment to hiring a full-time GIS staff until the GIS plan has been developed that will justify them. Moreover, the GIS staff hired later need not have the senior experience and broad background required of a GIS consultant.

Quick Response and a Shorter Planning Process

A GIS consultant can hit the ground running. This reduces the time required to get the planning process underway. Moreover, the consultant can direct the client along the shortest route to project justification and funding. Simply having this expertise available gives the client the confidence to proceed more quickly.

More Cost Effective than Self-education

An organization can attempt to educate itself in all aspects of GIS to develop its own plan. However, the staff time required to do so, the long delay, and the lower quality of the resulting plan can far outweigh the cost of having an expert available.

Coordinated, Unbiased Planning for All GIS Users

A GIS plan developed exclusively by internal staff may be slanted toward the needs and interests of the departments responsible for its preparation. On the other hand, a GIS consultant is a neutral party, having no bias toward the operations and use of the system. Therefore, she is in the best position to make an impartial assessment of the needs of all potential GIS users.

Independent, Neutral Analysis of a Major Capital Program

The total cost of a GIS program for a typical municipality can range from a few hundred thousand to several million dollars. Elected officials and top managers frequently demand that major capital programs such as a GIS receive an independent, expert review for practicality, feasibility, and cost justification. They are looking for a sound business case that will prove the merits of the project. Moreover, they need ample reasons to support this effort, as opposed to the many other capital programs competing for scarce funds.

Outside Responsibility

In many cases, management wishes to transfer certain responsibilities to outside parties or acquire the objective recommendations of outsiders. *The GIS consultant can assume responsibility for making difficult or unpopular recommendations and for estimates of GIS costs and savings.*

The GIS Consultant's Mission

A GIS consultant specializes in helping organizations through the critical processes of first reviewing their needs, then planning, selecting, implementing, and managing a GIS. He helps them determine the best GIS program, justify its cost, and avoid the common pitfalls of GIS implementation. The investment in a consultant's services benefits the GIS program throughout its lifetime. Typically, the objectives of a GIS consultant's services are the following.

- Review and document the client's current business process
- Document the client's current problems and limitations
- Document the client's requirements for the GIS
- Document potential GIS applications
- Develop a conceptual design for the GIS
- Develop a strategic plan for implementing GIS
- Estimate the costs, savings, and financial justification for a GIS
- Describe the qualitative benefits a GIS offers
- Develop a specification for purchasing GIS hardware and software
- Develop a scope of work for developing the GIS database
- Assist in the testing and acceptance of the GIS database
- Assist in the testing and acceptance of GIS hardware, software, and applications
- Assist the client with the process of selecting, purchasing, and implementing a GIS

What to Look for in a GIS Consultant

When seeking a GIS consultant, look for the qualifications discussed in the sections that follow.

Experience

Obviously, a qualified GIS consultant has an appropriate education and practical experience in GIS applications. However, it is also important to seek a consultant who has worked with organizations that have a mission similar to yours and that are comparable in size.

Neutrality

Many firms that offer GIS planning and consulting services are not totally unbiased. They offer other GIS services and products that will later be needed to implement the GIS. These include GIS software and hardware products, and a variety of related services, such as surveying, photogrammetry, and GIS database construction.

Most often, the revenue from GIS planning services constitutes only a small portion of their total business in GIS-related products and services. These firms naturally have a bias toward the purchase of their other offerings. This interest can influence their recommendations during the initial planning stages. A truly impartial GIS consultant has no financial interest in the products and services clients purchase to implement a GIS.

No Substitution of Personnel

Large consulting firms will present the qualifications of their most experienced personnel, usually their top managers, but actually perform much or all of the work using more junior staff. This is not necessarily bad, and can actually be more cost effective for the client, provided the more senior manager is properly supervising and contributing to the work. However, beware of the "bait and switch," where the proposed senior manager never gets involved in the project. Insist that "who you see is who you get."

Credibility

Look for a consultant with established credibility in the industry. Such a consultant may have spoken at conferences on GIS and computer mapping, written featured articles on GIS issues in national publications, written a book on GIS, or served in leadership positions for trade associations or GIS user groups.

Two Cautions About GIS Consultants

Having worked as a GIS consultant, I must admit that there are some disadvantages and potential pitfalls to hiring a consultant. Obviously, a consultant and the study process will cost you both time and money, as will the consultant selection process. However, this money is well spent if your program moves ahead smartly and you receive beneficial advice.

However, there are two primary things to watch out for with regard to a consultant. First, avoid paying for lengthy reports that state the obvious and are destined to gather dust on a shelf because they are too large to wade through or too theoretical to mentally digest. You can be sure they contain long sections of "boilerplate" text the consultant uses time and again. Second, be wary of "analysis paralysis." Consultants make their money doing studies, so there is a temptation to study the problem to death and never get around to making concrete recommendations so that you can get on with the program.

Summary

Many organizations have had the staff time and the expertise available in-house to develop a credible GIS plan on their own. Others have begun by hiring a GIS specialist to lead the planning process and eventually manage GIS operations. Still others have found help through regional planning agencies, schools, and federal or state government programs. But for many organizations, the best solution has been to seek the assistance of a qualified GIS consultant to help them through the critical initial planning process.

All of these approaches can work well. The most important point is that a thoughtful plan for the GIS be developed before major commitments to GIS purchases are made.

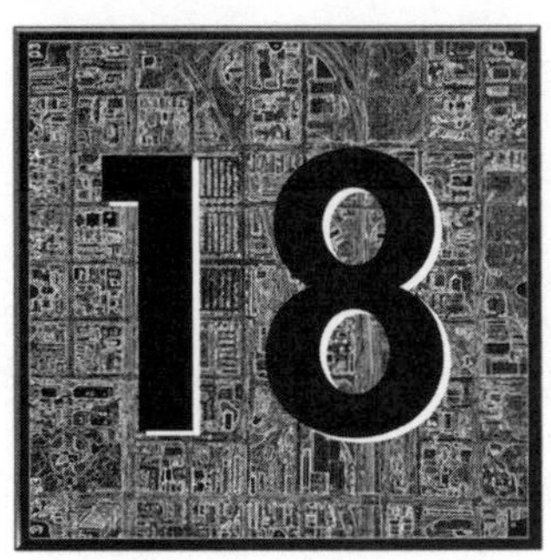

Managing and Staffing a GIS

The Functions and People Needed to Make a GIS Work

In This Chapter. . .

There are two primary aspects to maintaining a GIS after it has been implemented. One is the operation of the system itself; the other is the maintenance of the GIS database. This chapter first discusses the basic functions of GIS system management, and then presents job descriptions of the staff typically needed to perform them. The following chapter explores GIS database maintenance.

System Management

System operations can be broken down into two basic tasks: hardware and software maintenance and user training. These two tasks are discussed in the sections that follow.

Hardware and Software Maintenance

Hardware and software maintenance tasks are usually shared between the GIS support staff and GIS vendors. The support staff is normally the "first line of defense." GIS users count on the support staff to perform the following functions.

- Answer general questions
- Purchase and install new software and hardware
- Troubleshoot unusual problems
- Operate the data communications network and file servers

- Communicate with vendors regarding problems that cannot be solved in-house
- Purchase and replace printer and plotter supplies

Although the GIS support staff can handle most routine problems, GIS hardware and software companies also offer maintenance programs for their products. The cost of these programs varies but generally runs about 10% of a product's purchase price. Terms and conditions also vary, but the following are normally included in agreements.

- A warranty period
- An "800" telephone help desk
- "Bug" fixes and upgrades for software
- A formal program for evaluating and responding to customer-requested software enhancements
- Off-site hardware repair (customer ships the product to the vendor for repair)

They usually cost more, but optional support programs usually include on-site hardware repair. Note also that although software vendors may provide telephone support to virtually any user with a question, they generally provide product upgrades and "bug" fixes only to customers under a maintenance contract, making it well worth the investment.

User Training

Because of personnel turnover, training is an ongoing task. Like system maintenance, training is usually shared between the GIS support staff and GIS product vendors. Most organizations have a large number of "casual" GIS users that only need, for example, to learn a basic GIS data viewing tool. They can usually be taught by the support staff in one relatively short session.

These organizations also typically work with a few GIS "analysts," who need to use a more sophisticated GIS product. Once again, the support staff is usually capable of providing this training. On the other hand, the support staff itself will most often take advantage of vendor training classes because they need to master all of the GIS products in use, many of which are very complex.

GIS Staff Positions

Many of the GIS functions described in the following material will be new to an organization. Existing data processing departments may already have staff positions that perform similar functions in support of management information systems (MIS). Sometimes more than one of these functions can be assigned to the same position. For instance, it may be possible (or necessary) to combine the GIS manager and GIS database manager functions into one position.

On the other hand, sometimes more than one person is needed to perform a function. For example, when a great deal of database maintenance work is required, it may be necessary to establish more than one staff cartographer position to keep pace with the database updates. The basic requirements of these GIS staff positions are described in the sections that follow.

GIS Manager

The position of GIS manager requires a broad, general knowledge of GIS applications and the needs of the organization. The GIS manager can design GIS databases, as well as the analytical procedures required to support GIS applications. He knows how to plan production steps for GIS applications. He will be the chief liaison between the GIS and its users, and must command their respect to be effective. He should be a strong communicator who can listen to user requests and translate them into methods and procedures for generating GIS products.

The GIS manager is also ultimately responsible for getting GIS products to users, and he seeks useful feedback from them. He is their primary educator about GIS. The individual must also be experienced in personnel management and be able to keep the GIS staff motivated and working efficiently. He must know how to estimate resource needs and costs, as well as how to obtain funding and commitment in support of his facilities.

GIS Database Manager

The GIS database manager has a knowledge of GIS database design, and has experience with the organization's GIS applications. She also knows and understands land-survey mapping principles. She must organize data sets into layers and descriptive fields, and choose the data sources, levels of resolution, coordinate systems, and mapping procedures to be used.

The GIS database manager also designs attribute databases and associates this descriptive data with map features. She develops and manages automated map libraries and data dictionaries, and ensures the quality of the data. She directs all data automation and maintenance, and knows how to use the GIS software, hardware, and data to generate products, as directed by the GIS manager.

Cartographer

The position of cartographer is known by other names, including GIS operator, CAD operator, and GIS technician. This person has experience in mapping and cartography. He also has experience in general GIS operation and data entry. He has responsibility for GIS data-entry operations, including digitizing, map editing, and key entry of descriptive data.

The cartographer also compiles map data from many sources, including existing maps, aerial photography, satellite imagery, and existing digital map files. The cartographer is responsible for map production. He designs maps to be displayed or plotted by the system. He is detail-oriented and likes to work on well-defined tasks; moreover, he can concentrate on the task at hand. He works under the direction of the database manager.

System Manager

The position of system manager requires knowledge and experience in managing a computer system, including all hardware, software, peripherals, and supplies. This individual is responsible for the operations of the computer system. This includes setting access privileges for users, backing up files and maintaining the backup library, installing new software, helping users, troubleshooting problems, and coordination with hardware and software vendors. She usually reports to the GIS manager.

Programmer

The programmer has experience with GIS systems and is proficient at programming in C and similar languages, but most especially the application programmer's interface (API) of the GIS. He develops application software that meets very specific needs, such as special data-conversion programs. He also develops and maintains user macro commands and custom command menus. He may be called upon to troubleshoot problems with computer files and programs. His supervisor is usually the system manager.

Summary

All of the functions and personnel positions discussed in this chapter are important to the success of a GIS. Users cannot be fully productive, even with the best GIS software and hardware available, if the system does not have an adequate support staff. Failure to provide this type of support will seriously handicap a GIS program. There are the dangers that users will be dissatisfied, that the GIS database will become hopelessly out of date, and that the system will not provide the full range of benefits management sought when it approved the GIS program.

When the organization cannot hire full-time employees to fill these positions, this type of support can be obtained from the GIS vendor, GIS consultants, or other contractors. Without exception, the most successful GIS operations have adequate system management, database management, user support, and system support in place, either in the person of full-time employees or contracted personnel.

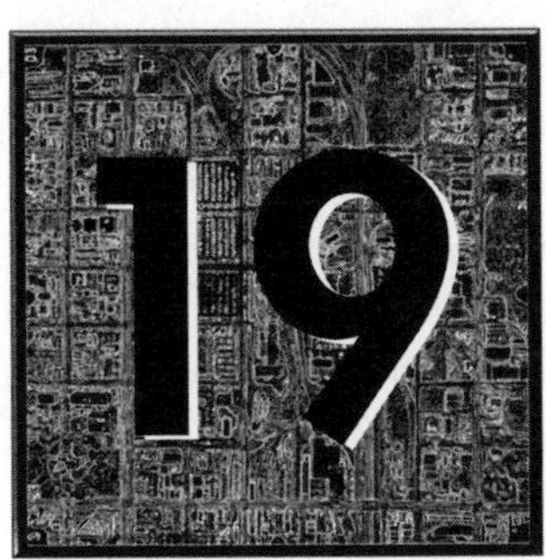

Managing and Maintaining a GIS Database

Using GIS Throughout the Enterprise

In This Chapter...

The greatest return on an investment in GIS occurs when GIS data is available to as many projects and departments in an organization as possible. This chapter explores basic strategies for "enterprise-wide" GIS database management and maintenance, and discusses the political and technical pros and cons of both areas.

Enterprise GIS

It is common for the use of GIS to begin with a single project or department and gradually expand within that department. This is typically a technical department of some sort, such as engineering, planning, or environmental. Managing and maintaining the GIS database in a single department is relatively straightforward because there is one "owner" of the data. There is an inherent control in this situation over both access to and maintenance of the database.

However, most organizations eventually move toward an "enterprise GIS," making GIS data available to users in non-technical and other departments, or even to everyone "throughout the enterprise." It is also common that several technical departments in the organization are using GIS. Usually, each of these departments is responsible for a particular type (i.e., map theme or coverage) of GIS data. Moreover, they usually have independent GIS operations. To implement an enterprise GIS program, therefore, the organization must manage and maintain GIS data "owned" by several departments. This usually proves to be a complex problem.

Keep in mind that just fifteen years ago computer systems typically consisted of proprietary data and graphics terminals linked to mainframe and mini computers over dedicated lines. The advent of local area network (LAN) data communications, "open systems" standards for data storage and communication, "client-server" computer architecture, and distributed computing strategies has made possible the various approaches to "enterprise" GIS data management discussed in the sections that follow.

Centralized GIS Databases

The obvious approach to enterprise GIS is to store all data on a single server and make it available to the entire organization over its LAN. Although several departments are responsible for maintaining their respective GIS themes, they can still access and edit (share) this data over the LAN. The illustration that follows shows an example of a centralized GIS database strategy. The chief advantages of this approach are the following.

- Only one database needs to be managed (backed up, archived, and so on).
- Only one server and server software license must be purchased.
- All data is under the physical control of only one department. This is often, but not necessarily, the organization's information technology (IT) department.

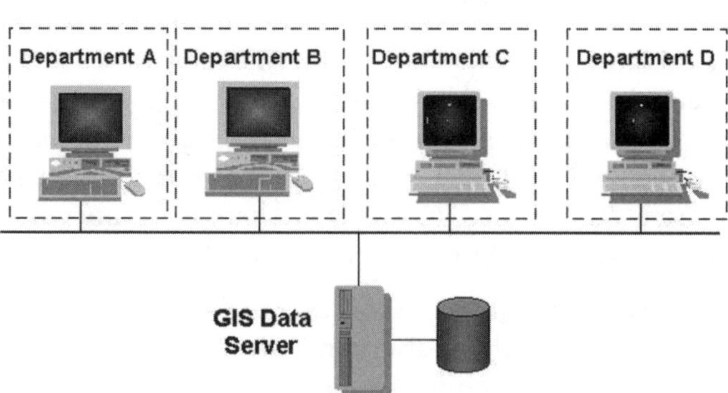

Centralized GIS database strategy.

The chief disadvantages of the centralized GIS database strategy are the following.

- Departments that "own" certain GIS data themes do not have physical control of them.

- Data "traffic" on the organization's LAN backbone will increase significantly. The heaviest GIS users are departments that "own" GIS data. Because all editors must access their data over the LAN backbone, its data transmission volumes will increase accordingly. Whereas vector GIS data files are typically only somewhat larger than normal "office" data files (e-mail, word processing, and so on), raster GIS data files are typically very much larger. Therefore, the transmission of GIS data, especially raster data, may severely tax the available bandwidth of the LAN backbone.
- As discussed in material to follow, sophisticated file utilities must be in place to manage the data editing process, controlling the editing of numerous files by numerous GIS data editors.

Distributed GIS Databases

Another approach is to physically distribute GIS data on multiple servers, one for each department that "owns" (i.e., maintains) the respective data themes. In this configuration, each department has access to its data over a departmental LAN, and other departments also have access to it over the LAN backbone. The illustration that follows depicts the concept of a distributed GIS database. The chief advantages of this approach are the following.

- Departments that "own" certain GIS data themes have physical control of them.
- The heaviest GIS data users access their department's data over its own LAN, and therefore GIS data transmission volumes on the organization's LAN backbone are significantly lower.
- The data editing process is less complex, because editors work with data under the control of their own department.

A distributed GIS database scheme.

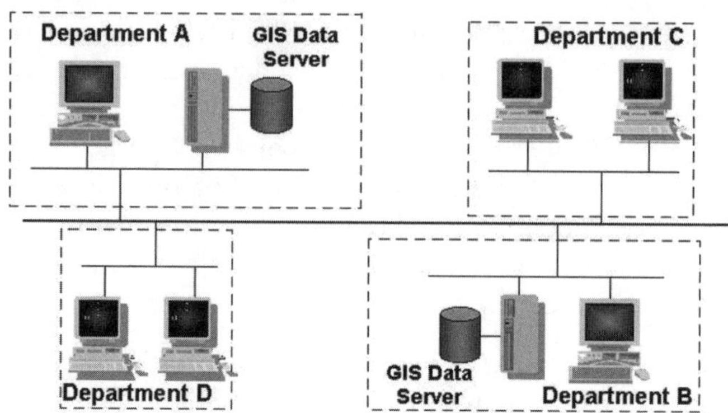

The chief disadvantages of the distributed GIS database scheme are the following.

- Each department must manage its own database, which requires duplicate personnel skills and training.
- Multiple servers and server software licenses must be purchased.
- GIS data is under the physical control of several departments, and therefore effectiveness may vary.

Master and Working GIS Databases

There is yet another approach to GIS database configuration that combines the best features of the two previously discussed. However, this setup is usually the most costly of the three. Under this strategy, departments work with their respective GIS data files located on their own "working" data server, and copies are stored on a "master" data server that is accessed by the entire organization. "Snapshot" copies of the working data are uploaded to the master data server on a periodic basis. The illustration that follows shows an example of the use of master and working GIS databases. The chief advantages of this approach are the following.

- Departments that "own" certain GIS data themes have physical control of them, whereas the master database, upon which the entire organization relies, is under the physical control of only one department. Once again, this is often the organization's IT department.
- The heaviest GIS data users access their department's data over its own LAN, and therefore GIS data transmission volumes on the organization's LAN backbone are lower.

- The data editing process is less complex, because editors work with data under the control of their own department.

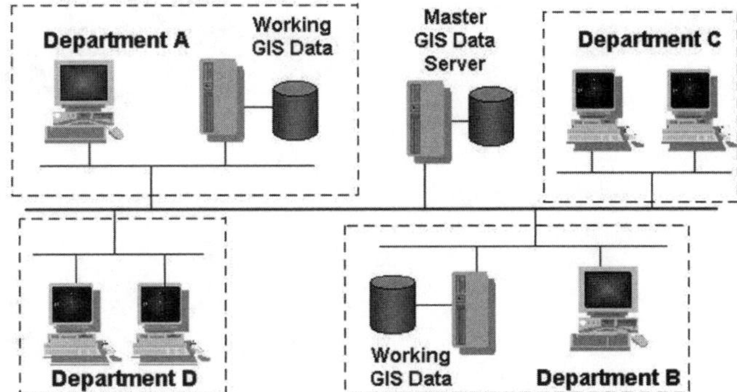

Employment of master and working GIS databases.

The chief disadvantages of the master/working database strategy are the following.

- This approach requires the greatest investment in training and personnel.
- This approach requires the greatest investment in data servers and server software.
- When changes are made to the working databases, the master database will not be current until it is refreshed.

Data Maintenance Issues

There is yet another important issue in enterprise GIS, regardless of which of the foregoing strategies is chosen. This is management of the data editing process. If the organization has only one GIS data editor, an example of which is shown in the first of the following illustrations, this is not a problem. This editor logs in to a file, makes changes, and then logs out. However, if several editors must access the same GIS data server, an example of which is shown in the second of the following illustrations, significant problems can result.

Single GIS data editor.

Multiple data editors accessing one server.

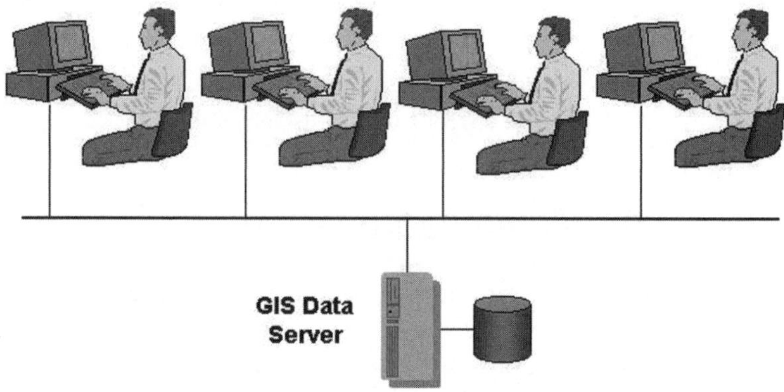

A problem occurs when two users must work on the same file. Suppose both editors are permitted to simply "check out" the file to their workstation, make changes, and then save the file again. The changes made by the first editor who saves a file will be lost when the other editor saves his or her copy of the file. The general rule is, "The last one who saves wins."

This problem is more significant, for example, for the large utility company that has several dozen GIS data editors in one department constantly making changes, as opposed to the small municipality with one or two editors in a department.

Although the fact that there are numerous GIS files reduces the frequency of such conflicts, the problem must nonetheless be planned for and dealt with.

There are two approaches to solving this problem. The following illustration depicts the strategy of "file-level" locking. Under this strategy, the GIS permits only one user to edit, or to copy a file for editing, at a time. When another user tries to access the same file for editing, the system replies that the file is not available for editing or copying at that time and that he should try again later.

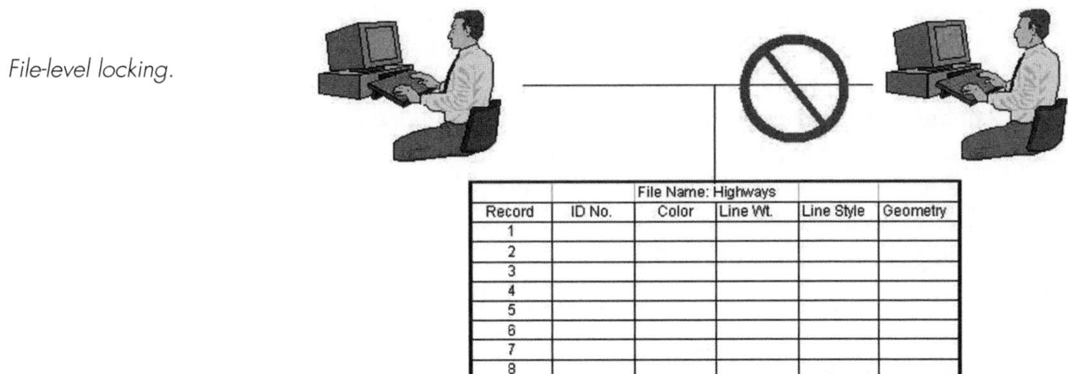

File-level locking.

This approach may work very well, for example, for the small municipality, but will probably not be satisfactory for the large utility. The large entity will instead prefer a second strategy, known as "feature-level" locking, depicted in the following illustration.

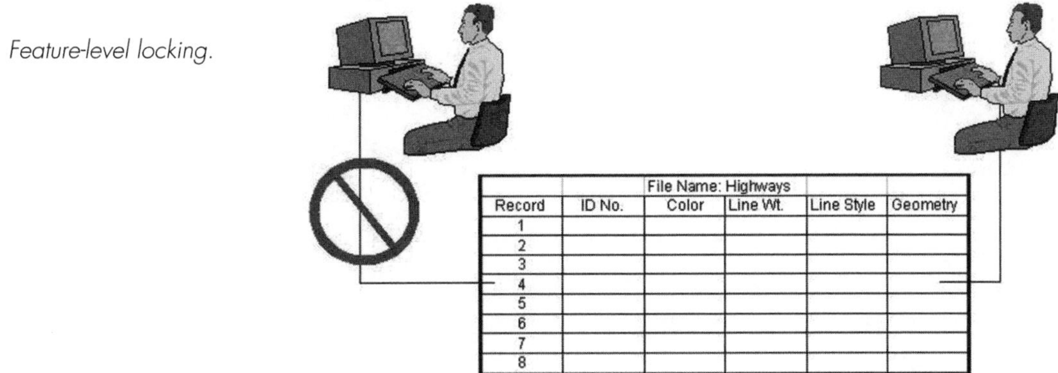

Feature-level locking.

Under feature-level locking, the GIS permits two or more editors to access the same file, but prevents them from simultaneously editing the same feature in that file. This effectively solves the large organization's problem and enables a number of GIS data editors to work at will with very little chance of conflict.

Summary

An organization's GIS may need file-level or feature-level locking, regardless of which database distribution strategy it chooses among centralized, distributed, or working/master. The need is determined by the number and location of GIS data editors working simultaneously. This is usually a function of the volume and frequency of changes to the organization's GIS data, which is dictated by the nature of the organization's mission. Once again, the large utility that is constantly maintaining its "outside plant" and physical assets will generally need a large number of editors to maintain its GIS database, as opposed to the small utility, municipality, college campus, or military base.

Each organization that employs enterprise-wide GIS must ultimately select a data management strategy that balances the pros and cons discussed in this chapter. It will have to consider not only these technical options, but its own particular needs, resources, and constraints. These other considerations include the types of GIS users and their applications, as well as GIS user support and infrastructure. The ideal strategy is one that provides the greatest return on the organization's GIS investment given the available resources to support it. For example, an organization may have ample GIS support personnel, but limited hardware or data communications, or vice versa.

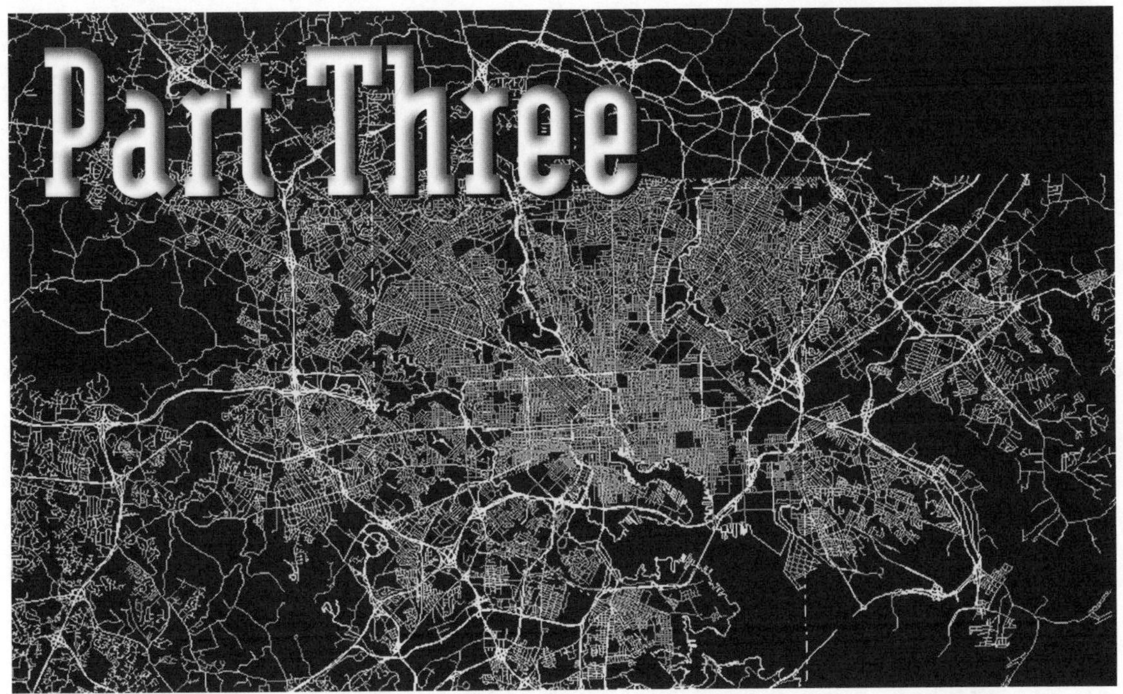

Considerations for Making GIS Decisions

Part Three deals with other important considerations for the implementation and management of a GIS. Included are discussions of the financial analysis of GIS, data quality, data integration, legal issues, and government and trade association involvement in GIS. Part Three also presents three case studies in GIS implementation and management. The chapters are:

- *Chapter 20, Financial Justification of GIS,* presents a detailed model for the financial analysis of GIS, including implementation and maintenance costs, as well as cost savings. It relates published, empirical data on productivity improvements resulting from GIS. It also shows how to determine the hourly cost of running a GIS.
- *Chapter 21, The Economics of GIS Base Map Accuracy,* describes factors in GIS base map accuracy in regard to the needs of various departments and the costs associated with greater accuracy.

- *Chapter 22, GIS Data Quality,* relates the importance of data quality in avoiding the sources of data errors the chapter examines in detail.
- *Chapter 23, Getting CADD Data into a GIS,* looks at the issues and common problems associated with incorporating CADD data. It also discusses the principal strategies for translating CADD data into a GIS format.
- *Chapter 24, Legal Aspects of GIS,* is an overview of legal issues surrounding the use of GIS, including access, pricing, privacy, liability, and copyright. The chapter also examines legal issues regarding the practice of land surveying.
- *Chapter 25, Government and Industry Involvement in GIS,* examines the role government at all levels and associations are playing in the realm of GIS. The chapter provides profiles of major associations, including contact information.
- *Chapter 26, A Case Study in GIS Implementation: Clinton Township, MI,* tells the story of how one organization implemented a GIS.
- *Chapter 27, A Case Study in GIS Implementation: Prince William County, VA,* also presents a case study in GIS implementation, and describes principal benefits derived and key lessons learned, as recounted by the man who managed the process.
- *Chapter 28, A Case Study in Maintaining GIS Data: Aberdeen Proving Ground, MD,* tells another story of GIS implementation, emphasizing the process and the importance of maintaining an accurate GIS database.

Financial Justification of GIS

Looking at the Economics of a GIS Investment

In This Chapter...

It is important to take a look at the "bottom line." the costs that can be expected when implementing a GIS, and how to justify them. This chapter looks at several types of costs and cost savings for a GIS, as well as some of the many other benefits not easily quantified. It presents a model for the financial justification of GIS, as well as sample calculations for determining the hourly cost of operating a GIS.

GIS Costs

Costs can be broken down into two types, which coincide with the two principle phases of the GIS program: implementation and maintenance.

Implementation Costs

There are three general types of costs associated with the implementation of a GIS:

- Services
- Hardware and software purchases
- Database creation

Services needed to implement a GIS may include:

- Consulting (e.g., requirements analysis or implementation planning)
- Training
- Software development or customization

- Systems integration

Hardware costs may include the purchase of:

- Workstations
- File servers
- Data storage devices
- Digitizing tables and/or scanner
- Plotters
- Printers
- Data communication networks

Software costs may include the purchase of:

- Operating system
- GIS application programs
- Database management system

These costs also include ancillary charges such as taxes, insurance, shipping, and installation. It is important to carefully plan purchases to be sure that all items have been included. It is not uncommon to find that miscellaneous items such as extra cabling, extra memory, miscellaneous data communication devices, workstation tables, chairs, tape racks, supplies, and so forth have not been considered in a system purchase. Table 20-1, which follows, presents typical cost ranges for GIS hardware and software purchases.

Table 20-1: GIS Hardware and Software Costs

	First Seat[a]	**Added Seats**[b]
Personal Computer	$8–20,000	$3–10,000
Personal Workstation	$13–25,000	$5–15,000
Engineering Workstation	$18–30,000	$8–20,000

a. Includes software license, plotter, and large disk drive.

b. Includes GIS software licenses.

Database creation is usually the largest cost component of a GIS program. This category comprises all costs required to create the GIS database. It may include:

- Surveying ground control
- Aerial photography
- Digital topographic mapping
- Conversion of existing paper maps and drawings
- Attribute data collection and entry
- Digital orthophotography
- Satellite imagery
- Checking and correcting data collected in-house
- Checking data provided by a contractor or other source

Other themes—including land use, zoning, district boundaries, utilities, natural resources, and environmental overlays—are usually scanned into a raster digital format from existing hard-copy maps, then manually traced into vector format. Attribute data associated with these maps are usually keyed in from existing records, such as index cards, log books, and other types of manual filing systems. A tax parcel overlay is typically constructed using precision placement of parcel geometry from deed descriptions and recorded plats.

Two costs of database creation are often overlooked. The first is the cost of collecting and "scrubbing" all data needed to populate the attribute database. Utility records are especially notorious for being incomplete. The second is the expense of checking and correcting the data after it has been converted to a digital format. Table 20-2, which follows, presents typical cost ranges for GIS database creation.

Table 20-2: Database Creation Costs

Database Type	Cost Range
Topographic base map	
1 inch = 200 ft., 5-ft. contours	$5–10 per acre
1 inch = 100 ft., 2-ft. contours	$10–20 per acre
Tax map overlay[a]	$2–5 per parcel
Tax map overlay[b]	$10–20 per parcel
Other themes	$1–5 per parcel

a. Digitized from existing tax maps.

b. Researched from land records and digitized by precision placement.

Between 1991 and 1995, Edwards Air Force Base, California, built a comprehensive GIS database. This included new 1 inch = 200 feet topographic base mapping for the entire base, new 1 inch = 50 feet topographic base mapping in the cantonment ("built-up") area, new digital aerial orthophotography, eleven utility systems, cultural resources, natural resources, and environmental data. Nine departments currently benefit from the system, which includes 49 workstations and 3 gigabytes of data. The following illustration shows the distribution of GIS implementation costs. Note that approximately 70 percent of the cost is in base mapping and data collection; the remainder is in equipment, software, and services.

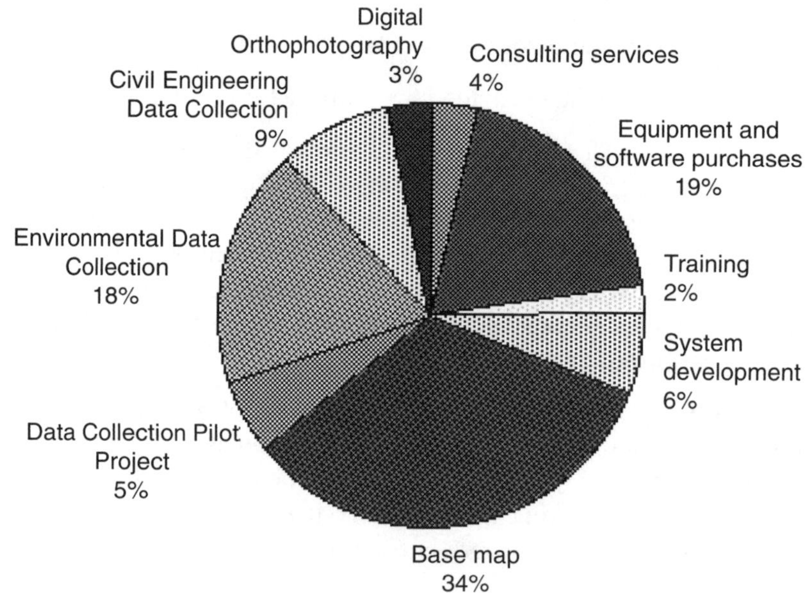

Edwards Air Force Base GIS implementation costs.

Ongoing Costs

Ongoing costs include the cost of operations and the maintenance of the digital database following system implementation. These include:

- Hardware and software maintenance
- Data management
- Data maintenance

- User training
- System support
- User support
- System development
- Supplies (inks, printing and plotting media, magnetic storage media)

Many of these costs will escalate over time, especially the salaries for GIS staff, including system managers, digital cartographers, and computer programmers. Maintenance costs may also include rent and utilities, depending on the organization's accounting practices. Similarly, salary costs may or may not include employee benefits, such as insurance, social security, vacation and sick leave, and retirement benefits, again depending on the cost accounting practices of the organization. *The cost of maintaining the GIS should allow for the periodic replacement or upgrade of hardware, generally every four to six years.*

Cost Savings

GIS cost savings are primarily found in four areas: improved management decision making, reductions in facility construction costs, employee productivity improvement, and employee cost avoidance. Each of these is discussed in the material that follows.

More Cost-effective Management Decisions

A GIS can be a tremendous support to executive decision making, especially when these involve capital-investment facilities. For example, consider a public works director allocating funds and planning for right-of-way maintenance operations. These operations include resealing and repaving roadways, repainting roadway markings, repairing sidewalks and curbs, cleaning sewers, replacing outdated water meters, and so forth.

In the past, the director might typically fly by the seat of his pants to allocate funds and resources for this maintenance program, largely because he lacked data needed to make more informed decisions. Instead, he mainly relied on verbal reports, memory, and judgment. His worst nightmare was to tear up a street recently repaved in order to replace the water line.

A GIS is a powerful tool for data storage and analysis that can make the public works director better informed. It can store utility line locations, maintenance records, condition reports, and planned improvements within the public right-of-way. Using the GIS, the public works director will be able to make better decisions. This should lead to more efficiency and real cost savings in carrying out the annual maintenance program. Similarly, using a GIS to support the siting of a new school or water treatment plant could produce significant one-time savings. Many organizations use GIS to locate underground utilities before excavating for new construction projects. One such organization is the subject of a case study presented later in this book: Aberdeen Proving Ground, Maryland. There, the annual cost of repairs to underground utilities accidentally damaged during construction projects has gone from an average of $100,000 per year to zero.

Unfortunately, the potential savings in this area are very difficult to quantify. Several managers of large operations that use GIS for their estimates report that their organizations saved from 1 to 5 percent of the annual facility operations and management cost.

Reduced Costs for Facility Construction

GIS can significantly reduce construction costs of capital facilities by reducing construction change orders. Design engineers must reference existing facility *as-built* records when planning a new project. Change orders on new construction projects frequently occur when the actual field conditions turn out to be different from those shown on existing paper records. Because the information shown on as-built drawings is generally known to be questionable, designers often perform a new field survey of existing conditions during the design process. However, a field survey is still not able to document all existing conditions, especially regarding hidden, buried, or abandoned utility systems.

One military installation built a water supply project that included nine miles of new water transmission main, two new wells, and the renovation of an existing pump station. The construction contract award was just under $4.5 million, but the project required 18 construction change orders to complete. These were all due to the fact that designers "did not know what was out there," according to the project manager. He cited several problems, among which were that the assumed locations of several sewer line crossings and a parallel water line were all incorrect (the existing water line location was off by more than 60 feet), requiring several field adjustments to the new water line location. Also, the contractor ran into an old building foundation, subsurface granite, and a quantity of asbestos rubble, none of which were known to the designers.

These change orders resulted in a cost overrun totaling nearly $600 thousand, or 12 percent of the initial award. Conceding that this was an extreme example because of the "linear nature" of the construction project (i.e., more problems would be encountered digging nine miles of trench than erecting a building of equal cost), the project manager estimated that, on average, 6 percent of the cost of new construction projects at the installation was spent on cost overruns resulting from inaccurate as-built data.

A GIS can contain more and better information regarding existing facilities infrastructure than was previously available on paper records. This information should be provided to design engineers for new construction projects, reducing the incidence and cost of construction change orders.

Improvements in Employee Productivity

The third area of cost savings is employee productivity improvement. A GIS makes it easier and cheaper to keep maps up to date, to create custom maps and reports, and to analyze land data for studies and reports. Professional literature is full of papers describing GIS projects. Unfortunately, relatively few of these include detailed information on costs and benefits. A bibliography of several examples is presented at the end of this chapter. Perhaps the best single reference that attempts to include detailed benefit and cost data is that of the *Joint Nordic Project —Community Benefit of Digital Spatial Information.*

The Nordic Project's Report No. 3, "Digital Map Data Bases—Economics and User Experiences in North America," (see References at the end of this chapter; Joint Nordic Project, 1987) describes 16 projects in North America and two in Italy. The team included representatives from both government and industry in Norway, Sweden, and Finland, as well as a representative from a major U.S. accounting firm. The team visited the projects during the fall of 1985 and collected as much information as could be obtained on costs, benefits, applications, and so on.

These projects range from those focused simply on automating mapping and revision, engineering, and design purposes, to those that integrate all information systems of the organization. The report provides brief descriptions of each of the projects and then does a fairly detailed benefit/cost analysis of the following projects:

- City of Edmonton
- City of Bellevue
- Transalta Utilities
- Edmonton Power (part of the City of Edmonton)

- Peoples Natural Gas Company

The Nordic Project report presents the following findings for the benefit/cost ratio (B/C):

- A digital system used only for computer-aided mapping and updating returns an entire initial investment. (B/C 1:1)
- If the system is also used for planning and engineering purposes, an initial investment will be doubled. (B/C 2:1)
- Research reports published in Norway and Sweden show that the benefit-cost ratio for automating conventional maps is greater than 3:1.
- If you manage to create a common system in which information can be shared among organizations, your return on an initial investment will be quadrupled. (B/C 4:1)
- For organizations with a poor system for manual map production, the automated system has given benefit-cost ratios up to 7:1. As an average, the study found a reduction in time of 50 percent in map production.
- For special effects, automation has derived an efficiency rate of 20:1 compared to manual production.

In its detailed evaluations, the Nordic Project looked at several cost factors, including equipment and software purchases and accompanying maintenance, leasing of equipment not purchased, operating cost (especially training personnel), and the cost of converting existing graphic records and maps to digital form.

In looking at the benefits, they included benefits in the initial mapping, benefits in the revision of maps and other data, reductions in the duplicate mapping of the same area by different organizations, savings in salaries, and improved quality of data and decisions resulting from the use of data.

The Nordic Project report concluded that the City of Edmonton mapping program produced a benefit/cost ratio of 3.2:1 for drafting functions, and in the range of 5–20:1 for various updating functions using GIS. The City of Bellevue, Washington, GIS program produced a benefit/cost ratio of 3.7:1. Antenucci et al. (1991) cite several instances of cost savings from GIS programs. They report that the City of Long Beach, California, uses an automated mapping system to draft maps at twice the speed of manual mapping. Updates to these maps are now accomplished four times more quickly than with traditional drafting techniques.

The Denver Water Department implemented engineering and planning applications on a GIS and reduced the labor and calendar time needed to create cross-sectional drawings of underground pipes. Before automation, drafters spent two months to produce drawings for 100 cross-sectional maps. With automation, a similar effort was accomplished in less than two days.

Other papers address the issue of costs and benefits, but usually in terms of specific amounts saved. One paper (Miller, 1985) describes the benefits of a GIS system to Detroit Edison. This project was begun in 1980 with the installation of an Intergraph system, and it is now an operational system. All of the data conversion has been completed. The paper describes some of the applications and savings in their operational use of the system.

Maintenance of the mapping database is the most significant single source of savings for Detroit Edison. Each year approximately 21,000 changes are made to the mapping database. The system saves 13,000 hours annually, as the automated mapping has proven to be three times as efficient as manual map updates. These range from simple line extensions to new subdivisions. Other major savings are in the area of facility maintenance and engineering design. Applications described in Miller's Detroit paper include street light audits, engineering statistics for distribution loadflow analysis, and pole tests to determine maintenance requirements and schedule the work. The GIS saved 1,500 man-hours in compiling the information for the pole test. These three applications combined require the preparation of nine to ten thousand special maps that would have been very time-consuming if done manually.

Smith and Tomlinson (1992) examined the benefits of GIS for the City of Ottawa. They estimated that the time required to research and respond to requests for information would be reduced by 50 percent through ready access to digital data using a GIS. They regarded this as a conservative figure because they considered only the benefits of having the data in digital form and did not assess the benefits of improved data quality. Note that "nearly all" of the Ottawa City staff indicated that these savings would be reallocated to meet the demands for services. That is, city officials expected that many of the savings would be used to allow them to do more tasks and would not necessarily result in reductions in expenditures.

On the other hand, another paper (Chambers, 1985) describes the experience of Houston Lighting & Power Company. One specific benefit is cited in the paper. In 1978, there were 27 draftsmen performing manual mapping, and this was projected to grow to 33 by 1984. In that same year, contract expenditures for mapping were \$225,000 and expected to grow to \$675,000 by 1984. In 1984, after the installation of the automated mapping system, \$200,000 was spent for contract mapping and the staff was reduced to 22 drafting and digitizing personnel.

The productivity improvement for a given GIS operation can range from 10:1 (or more) for spatial analysis, to 3:1 for producing customized maps and 2:1 for simple data queries. The actual percentage of improvement will depend on the organization and how it intends to employ the GIS. However, GIS savings do not start immediately. Instead, they begin sometime after the installation of the GIS. This is because users need a period of training and initiation to the system. A user typically requires from three to six months to become proficient in the operation of a GIS. An entire department may require up to a year to show appreciable cost savings.

Employee Cost Avoidance

The fourth area of cost savings is cost avoidance. It is common for government agencies to be understaffed in those areas that maintain and use land data. The budget does not allow enough staff positions to accomplish all of the work. There are usually numerous unapproved requests for new staff. These positions will eventually be filled if the operating funds become available for them.

However, a GIS can make the existing staff so much more productive that many of these new positions are no longer needed. The existing staff is able to furnish more and better quality map products, reports, analyses, and other services. This often postpones or entirely eliminates the need for new hires in these areas.

Increased Revenues

It is also possible to generate new and added revenues by selling GIS products. The value of data available from the GIS is significantly greater than that of traditional hard-copy data. Although the content may be largely the same, the format is far easier to manipulate and research. Moreover, the data is often more complete, accurate, and current than data shown on hard-copy maps and related files. Potential customers include utilities and government agencies, land developers, real estate firms, consulting engineers, and land surveyors.

Additional Benefits

Chapter 5 discussed several general reasons for using GIS. Many of these benefits will definitely result in cost savings, but these are difficult to quantify. They deserve mentioning again, in a brief summary.

- The data is better organized.
- The data is stored in a central place, eliminating duplicate map sets and duplicate map changes.
- GIS users have constant access to the most current data available. They can

also retrieve it faster and more easily.

- Maps and reports are more consistent and have a higher graphic quality.
- Graphic (spatial) data and nongraphic (attribute) data are explicitly linked and can be analyzed simultaneously.
- Users have flexibility in selecting the data they want to view, analyze, or present.

A Model for the Financial Justification of GIS

Working as a consultant to municipalities and utilities several years ago, I developed a model for the financial analysis of GIS. While specific assumptions may vary from organization to organization, this method employs sound and widely accepted principles of financial analysis. The technique is straightforward and has been accepted by numerous decision makers as a reasonable approach.

The analysis involves three primary steps: first, estimate the cost of the GIS program; second, estimate the cost savings that will result; and third, employ one or more standard techniques of financial analysis to determine the attractiveness of the investment, given the expected returns.

GIS Cost Savings

Estimating GIS cost savings is a two-step process. First, determine the current costs related to the map and attribute data that will be placed in the GIS database. Second, estimate the portion of this cost that will be saved through use of the GIS.

Current Costs

A fast and easy way to estimate the current cost of using map data is to survey the supervisors of all departments in the organization that will benefit from the GIS. Make sure that the interviewee understands which of his or her existing data sources will be included in the GIS database. Then ask, "What percent of their time do employees in your department currently spend using this data?" Explain that by "using" you mean collecting, maintaining, retrieving, analyzing, producing, or distributing these data. Obtain an estimate for each employee in the department.

Given the portion of an employee's time spent using map data, the cost is derived by applying this percentage to the employee's total compensation. To arrive at the total compensation, start with the employee's salary. It may be easiest and best to use the average, or mid-range, salary for the employee's pay grade. This data is usually available from the human resources department. Increase the employee's direct salary by a burden rate for indirect benefits (e.g., Social Security, Medicare, Medicaid, hospitalization insurance, life insurance, educational assistance, retirement or profit sharing, and paid leave, including vacation, holiday, sick, and other leave).

This indirect fringe benefit rate within the U.S. federal government is about 46 percent for both civilian and military employees. Total the resulting number for each employee to arrive at the department's total. Total the departmental figures to arrive at an estimate of the organization's current cost of using the data that will be automated in a GIS. (By the way, top management usually finds this figure surprisingly large.) Table 20-3, which follows, presents an example of the current cost of using map data in one department. Of course, more sophisticated methods of work and cost assessment can be used. These may produce more accurate results, but will involve considerably greater effort.

Table 20-3: Example of the Cost of Using Map Data in a Department

Employees	Number	Title	Pay Grade	Avg. Salary for Grade	Indirect Salary	Total Pay or Contracted Cost	% Time Using Facility Data	Cost of Using Facility Data
T Baybrook	1	Comm. Planner	GS-13	$59,480	$27,480	$86,960	20%	$17,392
R Michael	1	Comm. Planner	GS-12	$50,027	$23,112	$73,139	75%	$54,855
C Waagen	1	Proj. Programmer	GS-12	$50,027	$23,112	$73,139	30%	$21,942
S Daniels	1	Proj. Programmer	GS-11	$41,739	$19,283	$61,022	20%	$12,204
D Csucsai	1	Proj. Programmer	O-2	$33,169	$15,324	$48,493	20%	$9,699
L Martin	1	Proj. Programmer	O-2	$33,169	$15,324	$48,493	20%	$9,699
Total current cost of using facility data								$125,790

It may be necessary to estimate current costs by task or function, rather than by employee. For instance, a municipal real estate tax department may know the cost of doing an annual property reassessment study. This type of data can also be used, but you must be sure not to double count employee-oriented cost estimates in task-oriented estimates. You should also document the costs of tasks contracted out—costs that could be reduced by the GIS. For example, a city may contract its civil engineering design to outside consultants. New, accurate digital topographic mapping developed for a GIS could be provided to the consultants by the city, eliminating the cost of site-specific topographic surveys.

Cost Savings

The second step in calculating GIS cost savings is to estimate how much the current costs of using map data will be reduced by the use of the system. Given the findings of the reports cited earlier in this chapter, it seems reasonable to assume that implementation of the GIS will result in an overall productivity improvement of 2:1 over current activities involving map and map-related data. That is, employees using the GIS would be able to accomplish these tasks in half the time. Obviously, this 50-percent reduction in the time required for these tasks can be expressed as a 50-percent cost savings, relative to the current cost of using map data.

It is also reasonable to expect that this cost savings is achieved in phases. Users need a period of training and orientation to the system. A full-time GIS user typically requires three to six months to become proficient. Once proficiency is gained, cost savings begin to accrue. Therefore, one might assume a 10-percent cost savings in a given department during the first year after implementation of the GIS, 25 percent in the second year, and 50 percent during the third and subsequent years.

Although the GIS should generate a significant productivity improvement for employees who use map data, this time savings will not necessarily lead to a corresponding reduction in payroll costs. Instead, employees who enjoy these productivity gains will have more time available for other tasks. Essentially, the time and cost saved in the use of map data will be reallocated to other activities.

Current contracted costs, including the cost of change orders for facility construction, may also be reduced through the use of GIS. I have not found any published reports on this topic, so estimating the amount to be saved in this area is a case-by-case judgment. Savings due to better decision making in facility operations and management are very difficult to quantify, much less support. Table 20-4, which follows, presents a hypothetical set of GIS cost and savings data.

Table 20-4: GIS Cost and Savings

GIS COSTS		FY 91	FY 92	FY 93	FY 94	FY 95
Implementation consulting services		$100,000				
Equipment and Software Purchases		$250,000	$250,000	$250,000	$250,000	$100,000
Equipment and Software Maintenance			$25,000	$50,000	$75,000	$100,000
Data Maintenance			$100,000	$200,000	$300,000	$300,000
Training		$40,000	$80,000	$80,000	$40,000	$40,000
System and User Support		$100,000	$100,000	$200,000	$200,000	$200,000
System Development		$200,000	$100,000	$100,000	$100,000	$100,000
Supplies		$2,500	$5,000	$7,500	$10,000	$11,000
Base Map Photography		$50,000				
Ground Control		$200,000				
Analytical Triangulation		$100,000				
Photogrammetry		$1,000,000				
Field Data Collection			$100,000	$100,000	$100,000	
Environmental Map Conversion			$200,000	$200,000	$200,000	
Engineering and Utility Map Conversion			$200,000	$200,000	$200,000	
Digital Orthophotography			$200,000			
Total annual cost		$2,042,500	$1,360,000	$1,387,500	$1,475,000	$851,000
GIS SAVINGS	Current Cost of Using Map Data	FY 91	FY 92	FY 93	FY 94	FY 95
Environmental Department	$350,000			$35,000	$87,500	$175,000
Cultural Resources Department	$250,000			$25,000	$62,500	$125,000
Natural Resources Department	$200,000			$20,000	$50,000	$100,000
Air Quality Department	$100,000			$10,000	$25,000	$50,000
Planning Department	$125,000			$12,500	$31,250	$62,500
Maintenance Department	$100,000			$10,000	$25,000	$50,000
Operations	$150,000			$15,000	$37,500	$75,000
Engineering Department	$400,000			$40,000	$100,000	$200,000
Contracted costs	$2,500,000			$250,000	$625,000	$1,250,000
Total Costs of Using Map Data	$4,175,000	$4,175,000	$4,175,000	$3,757,500	$3,131,250	$2,087,500

GIS COSTS		FY 91	FY 92	FY 93	FY 94	FY 95
Total Annual Savings		$0	$0	$417,500	$1,043,750	$2,087,500
Net Annual Savings (cost)		($2,042,500)	($1,360,000)	($970,000)	($431,250)	$1,236,500
Cumulative Savings (cost)		($2,042,500)	($3,402,500)	($4,372,500)	($4,803,750)	($3,567,250)
Present Value Factor (Discount Rate = 3%)		1.0300	1.0609	1.0927	1.1255	1.1593
Present Value of Net Annual Savings		($1,983,010)	($1,281,930)	($887,687)	($383,160)	$1,066,616
Net Present Value	$1,608,771					

GIS COSTS		FY 96	FY 97	FY 98	FY 99	FY 00	FY 01
Implementation consulting services							
Equipment and Software Purchases		$100,000	$50,000	$50,000	$50,000	$500,000	$50,000
Equipment and Software Maintenance		$110,000	$120,000	$120,000	$120,000	$120,000	$120,000
Data Maintenance		$300,000	$300,000	$300,000	$300,000	$300,000	$300,000
Training		$40,000	$40,000	$40,000	$40,000	$40,000	$40,000
System and User Support		$300,000	$300,000	$300,000	$300,000	$300,000	$300,000
System Development		$100,000	$100,000	$100,000	$100,000	$100,000	$100,000
Supplies		$12,000	$12,000	$12,000	$12,000	$12,000	$12,000
Base Map Photography							
Ground Control							
Analytical Triangulation							
Photogrammetry							
Field Data Collection							
Environmental Map Conversion							
Engineering and Utility Map Conversion							
Digital Orthophotography							
Total annual cost		$962,000	$922,000	$922,000	$922,000	$1,372,000	$922,000
GIS SAVINGS	Current Cost of Using Map Data	FY 96	FY 97	FY 98	FY 99	FY 00	FY 01
Environmental Department	$350,000	$175,000	$175,000	$175,000	$175,000	$175,000	$175,000

GIS COSTS		FY 96	FY 97	FY 98	FY 99	FY 00	FY 01
Cultural Resources Department	$250,000	$125,000	$125,000	$125,000	$125,000	$125,000	$125,000
Natural Resources Department	$200,000	$100,000	$100,000	$100,000	$100,000	$100,000	$100,000
Air Quality Department	$100,000	$50,000	$50,000	$50,000	$50,000	$50,000	$50,000
Planning Department	$125,000	$62,500	$62,500	$62,500	$62,500	$62,500	$62,500
Maintenance Department	$100,000	$50,000	$50,000	$50,000	$50,000	$50,000	$50,000
Operations	$150,000	$75,000	$75,000	$75,000	$75,000	$75,000	$75,000
Engineering Department	$400,000	$200,000	$200,000	$200,000	$200,000	$200,000	$200,000
Contracted costs	$2,500,000	$1,250,000	$1,250,000	$1,250,000	$1,250,000	$1,250,000	$1,250,000
Total Costs of Using Map Data	$4,175,000	$2,087,500	$2,087,500	$2,087,500	$2,087,500	$2,087,500	$2,087,500
Total Annual Savings		$2,087,500	$2,087,500	$2,087,500	$2,087,500	$2,087,500	$2,087,500
Net Annual Savings (cost)		$1,125,500	$1,165,500	$1,165,500	$1,165,500	$715,500	$1,165,500
Cumulative Savings (cost)		($2,441,750)	($1,276,250)	($110,750)	$1,054,750	$1,770,250	$2,935,750
Present Value Factor (Discount Rate = 3%)		1.1941	1.2299	1.2668	1.3048	1.3439	1.3842
Present Value of Net Annual Savings		$942,589	$947,658	$920,056	$893,259	$532,399	$841,982
Net Present Value	$1,608,771						

Options for Financial Analysis

Two general principles form the basis of a GIS financial justification analysis: "payback" and "net present value." Either or both of these "methods" can be employed to assess the economic impact of a GIS.

Payback

The GIS program involves an initial investment in items such as new equipment, software, database construction, and training. Some time will elapse before the new technology produces cost savings. However, these savings will steadily increase and will eventually be larger than the ongoing costs of operating the system. As the savings continue, the initial investment is gradually recovered. The length of time required for the initial investment to be fully recovered is known as the payback period. The shorter the payback period, the more desirable the investment in terms of cost recovery. The following illustration is a chart of the cost and savings data from the previous table, and indicates that the payback period for this hypothetical GIS program is about seven years following the start of its implementation.

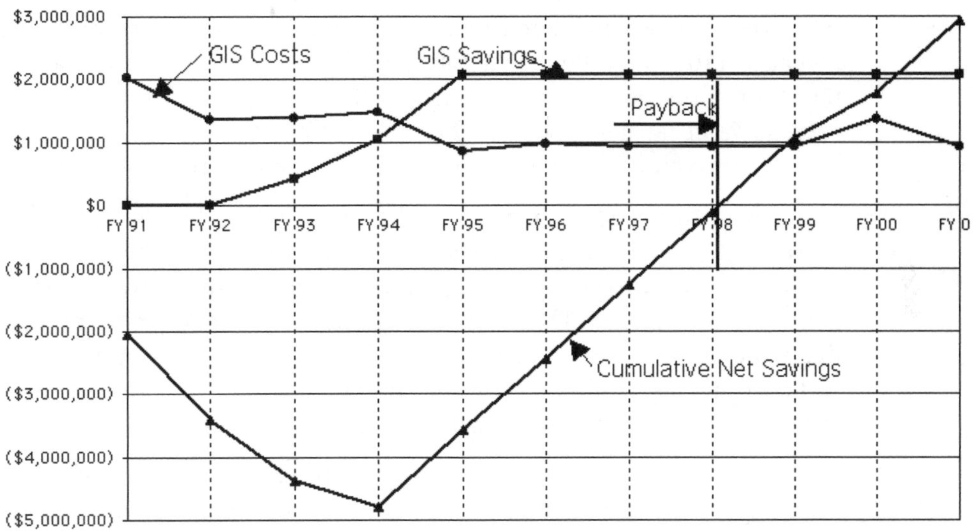

GIS costs and savings.

Net Present Value

A net present value analysis can also be used to examine the financial aspects of a GIS. Investment decisions frequently involve spending and receiving various amounts of money at different times in the future. Net present value analysis essentially transforms future cash flows into current-year dollars, allowing a comparison of the economic benefits of alternative investments that may be different in scope and dollar investment. The net present value of each of these cash flows is used to determine their relative attractiveness. Net present value analysis is somewhat more complicated than a payback computation, but it is more precise because it accounts for the "time value" of money.

Money received today is worth more than an equivalent amount received in the future. The difference in value increases over time. The present value of future receipts can be determined by discounting them back to the present at an appropriate interest rate. However, future cash flows for this type of investment will vary from year to year, making it impossible to compute a simple rate of interest returned. Present value analysis converts all future cash flows to equivalent amounts at a common point in time. The interest rate used for the calculations is referred to as the discount rate. If the net present value computed at the desired discount rate is positive, the investment is justified. A negative net present value indicates that the investment will not yield the desired rate of return.

The risk associated with the investment primarily determines the discount rate. If the investment is considered relatively risky, a greater return on the investment is demanded to justify the higher risk. In other words, the greater the risk, the higher the desired discount rate. This means that a higher discount rate is used in the present value analysis. The higher the discount rate, the greater the difference between the value of money received now and money received in the future.

The U.S. Office of Management and Budget (OMB) Circular A-94 (1992 revision), presents guidelines for the analysis of capital investments by federal agencies. This publication recommends using a real discount rate of 3 percent for capital investments of this nature. This rate, as opposed to a market rate, reflects the cost of financing without regard to inflation. Using this criterion, all figures for GIS costs and cost savings would be expressed in constant (current) dollars, not adjusted for inflation. This permits comparisons across years without regard to inflation.

This technique assumes any increases in prices are due to inflation and not to changes in price. The hypothetical GIS cost and savings data in the previous table includes a computation of the present value of annual costs and savings using a 3-percent discount rate. As you can see, it is positive, so the investment is justified. The actual rate of return is that discount rate that yields a net present value of zero. For the data presented in the previous table, this is 8.5 percent.

A financial analysis of this sort would also normally account for the "residual value" of the investment at the end of the period under consideration. This value could be subject to some debate, however. Although this may be a significant amount, one might choose to ignore the investment's residual value, as does the previous table, in order to be conservative and to avoid the question of how it should be determined.

Other Business Factors

As the GIS is implemented, the actual benefits can be evaluated against the projected benefits. This evaluation will ensure that effective management and cost control measures have been employed in the implementation and use of GIS. Such an evaluation will require a comparison of current business conditions with the conditions existing at the time of the evaluation in order to be valid.

Changes in these basic business conditions may have an impact on the use of GIS and must be considered in the evaluation. For instance, if a utility is contracting less design work to outside consultants, the total amount of time required by utility employees for planning and design functions may increase relative to today, even though automation has reduced the amount of time required for individual tasks.

Similarly, the utility is likely to be serving more customers. This will likewise increase the total amount of work to be done, even though the GIS may have made the utility employees more productive.

Obtaining the Greatest Return on a GIS Investment

The initial investment in GIS hardware, software, and database conversion is substantial, but this initial cost is recovered through savings that accrue each time the data is used. A GIS can provide users throughout the organization access to a common file of spatially referenced data. This strategy integrates and coordinates the operations of all GIS users, and offers the greatest opportunity for repeated use of the data. Therefore, making a common GIS database available to the largest possible number of users produces the greatest return on investment. Likewise, the sooner GIS data is made available to users, the sooner cost recovery begins.

The Hourly Cost of Operating a GIS

The hourly operating cost for a GIS is meaningful to professional service firms because they normally express billing rates and salaries on an hourly basis. In that GISs are thought of as productivity improvement tools, it is useful to consider the hourly cost of operating a GIS system relative to the hourly labor savings it produces. Moreover, many professional service firms are providing map digitizing services to municipalities and utilities that have initiated GIS programs like the one previously described.

The hourly operating cost of the GIS is important in computing the fee to these clients. Begin with these assumptions:

- A GIS system has ten workstations.
- One workstation is used for training and programming; nine are used for production work.
- The system is operated one shift per day and each production workstation is used an average of six hours per day for billable work.
- The workstations are a mixture of engineering workstations, personal workstations, and personal computers purchased at various times over the last five years.
- The total cost of the system, including software, plotters, and other peripheral devices, equates to $20,000 per workstation.

- The purchase costs are being amortized over five years using straight-line depreciation.
- The annual maintenance cost is 10 percent of the purchase cost, including hardware maintenance and software upgrades.
- Two full-time personnel are required to provide system management and system operation support.
- Four new employees are trained on the GIS every year at a cost of $2,000 each.
- Employee benefit costs are 50 percent of salaries.
- Supplies cost $200 per workstation each year.
- Space rental costs $2,000 per workstation each year.
- Utilities cost $500 per workstation each year.
- The organization allows ten days of vacation, five days of sick leave, and eight holidays each year, reducing the number of days available for operations from 260 (52 weeks x 5 days per week) to 237 per year.

The hourly cost of operating this ten-workstation system can be derived as follows:

- Total purchase cost:
  10 workstations x $20,000 per workstation = $200,000
- Annual depreciation charges:
  $200,000/5 = $40,000
- Annual maintenance charges:
  $200,000 x 0.1 = $20,000
- Annual salary charges for system management and support:
  2 employees x $35,000 average salary = $70,000
- Annual employee benefits for system management and support:
  $70,000 salaries x 0.5 = $35,000
- Annual training charges:
  4 employees x $2,000 each = $8,000
- Annual supply charges:
  $200 per workstation x 10 workstations = $2,000

- Annual space rental charges:
  $2,000 per workstation x 10 workstations = $20,000
- Annual utility charges:
  $500 per workstation x 10 workstations = $5,000
- Total annual charges = $200,000
- Annual billable time:
  6 hours per day x 9 production workstations x 237 days per year = 12,798 hours
- Operating cost:
  $200,000/12,798 hours of production = $15.62 per hr.
- Hourly operating cost = about $16.00 per hr.

Assume the average compensation for GIS cartographers is $25,000 per year, or about $12 per hour. Allowing 50 percent for the employee's indirect benefits, the average cartographer costs about $18 per hour to employ. Thus, an employee using the GIS system costs the organization about $34 per hour ($18 for the employee and $16 for the system). This needs to be considered carefully when computing the hourly rate to charge internal projects and clients for work on the GIS.

An important final consideration is the risk of underutilizing the system. We assumed the production workstations would be billed six hours every working day. The organization must evaluate the probability of slow periods. If they are likely, it should adjust the annual billable time accordingly. Of course, this adjustment will have a direct impact on the hourly operating cost.

Summary

A GIS program requires a significant start-up expense. This is needed for the purchase of the hardware and software, and for creation of the database. Initial cost is followed by ongoing system operation and maintenance costs. These annual operating expenses are significantly lower than the initial start-up costs, but they escalate over time because of increases in staff salaries.

GIS savings, although low in the beginning, increase steadily over time until they exceed the operating costs by a substantial margin. Eventually, the initial costs for system purchase and database conversion costs are fully recovered. Thus, the system's payback occurs sometime later, often several years later. The payback period for a typical municipal government will be in the order of 5 to 15 years, depending on the extent of the GIS program, the type of base mapping done, the condition of the existing records, and the efficiency of existing operations.

The GIS program at Edwards Air Force Base, California, was mentioned earlier in this chapter. It was analyzed using the model for financial analysis presented in this chapter. GIS managers and user departments were interviewed to determine the costs of implementing and managing the GIS, as well as the potential cost savings and other benefits that would result from the use of a GIS. The analysis found that Edwards Air Force Base should fully recover the cost of implementing GIS, eight years after project initiation and four years following full implementation of the GIS database.

Based on conservative assumptions and ignoring inflation, the investment yields a 7.5-percent real rate of return over 10 years. The largest source of expected savings is a reduction in construction costs due to construction change orders. This will result from providing better data on existing facilities to new project design engineers and architects. The next most significant cost savings will be increased employee productivity when working with automated facilities data in the GIS.

Chapter References

Antenucci, J., Brown, K., Croswell, P., and Kevany, M., 1991. *Geographic Information Systems —A Guide to the Technology.* Van Nostrand Reinhold, New York, New York.

Chambers, John, 1985, "Automated Mapping System, Houston Power & Lighting Company." In *Proceedings of AM/FM Keystone Conference VIII,* Snowmass, Colorado, August 1985.

Joint Nordic Project, 1987. "Report No. 3, Digital Map Data Bases—Economics and User Experiences in North America." The Joint Nordic Project, Community Benefit of Digital Spatial Information, Viak A/S, Norway.

Miller, Eugene E., 1986. "Tracking AM/FM Benefits in an Electric Distribution Environment—Adding PEP to Your System." In *Proceedings of AM/FM Conference IX*, Snowmass, Colorado, August 1986.

Smith, D., and Tomlinson, R., 1992. "Assessing Costs and Benefits of Geographical Information Systems: Methodological and Implementation Issues." *International Journal of Geographical Information Systems*, Vol. 6, No. 3. Taylor & Francis, Ltd.

The Economics of GIS Base Map Accuracy

Trade-offs in Scale Versus Cost for Base Map Production

In This Chapter...

One very important factor in the success of a GIS implementation is base map accuracy. How does the scale of the base map affect the cost of creating it? This chapter looks at several fundamental issues regarding base maps.

Features in a Topographic Base Map

Creating a GIS database normally accounts for as much as three-quarters of the total cost of a GIS project. A base map is usually the most expensive of all map topics included in the database. Moreover, the land data for other map themes are placed by reference to the base map. Thus, the positional accuracy of all data in the GIS depends on the accuracy of the base map, which serves as the foundation of the GIS database. (See the following illustration.)

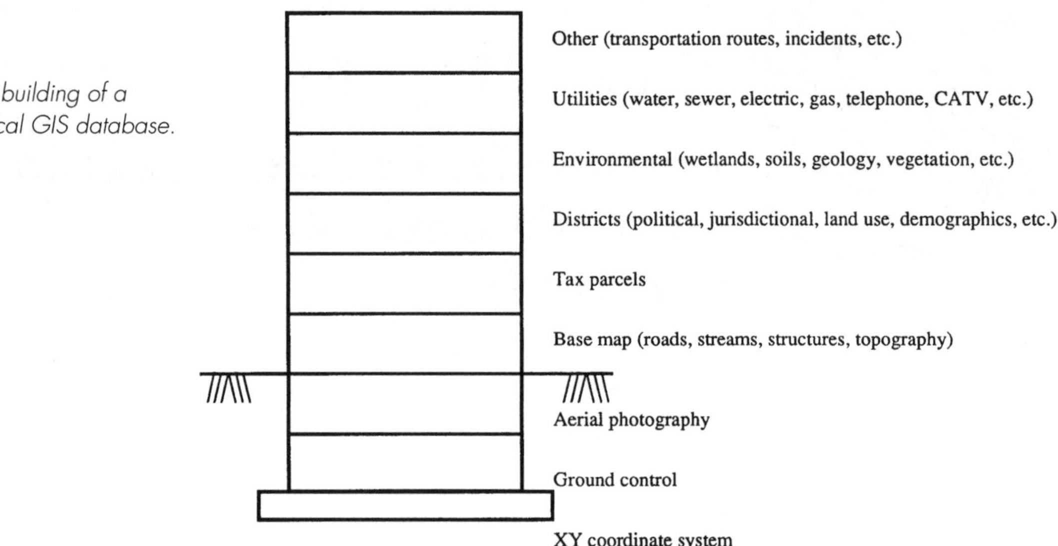

The building of a typical GIS database.

A topographic base map generally includes the following features.

- Roads and railroads
- Rivers, streams, and shorelines
- Contour lines and spot elevations
- Structures
- Vegetation
- Survey control and grid coordinates

Other map themes registered to the base map usually include the following.

- Tax parcels
- Zoning
- Election, census, and school districts
- Political boundaries
- Environmental data
- Utilities

Base Map Accuracy Standards

The base map is ordinarily created from aerial photography by a photogrammetric surveyor. The horizontal accuracy of the base map depends on its mapping scale.

National Map Accuracy Standards (NMAS) are the most widely accepted principles for map accuracy used in this country. These standards require that at least 90 percent of well-defined points fall on the map itself within 1/30 of an inch of their true position on the ground.

Consequently, on a map drawn to a scale of 1 inch = 100 feet, 90 percent of all points tested must lie within ± 3.3 feet (1/30th of an inch at map scale) of their true position. A map drawn to a scale of 1 inch = 200 feet is said to have an accuracy of ± 6.7 feet. Likewise, a map drawn to a scale of 1 inch = 400 feet has an accuracy of ± 13.3 feet. This standard applies to *large-scale* maps. Large-scale maps are commonly used for municipal and utility mapping, and generally range in scale from 1 inch = 50 feet to 1 inch = 1,000 feet. The following illustration shows a detail from a large-scale base map.

Detail of a large-scale map.

Most civil engineers and land planners are familiar with the 7.5-minute quadrangle map series published by the U.S. Geological Survey. These maps have been drawn at a scale of 1 inch = 2,000 feet and are generally referred to as *small-scale* maps. They also conform to NMAS. However, the horizontal accuracy standard for mapping at this scale is 1/40 of an inch at map scale. Therefore, these popular maps have an accuracy of only ± 50 feet. The following figure shows a detail from a small-scale base map.

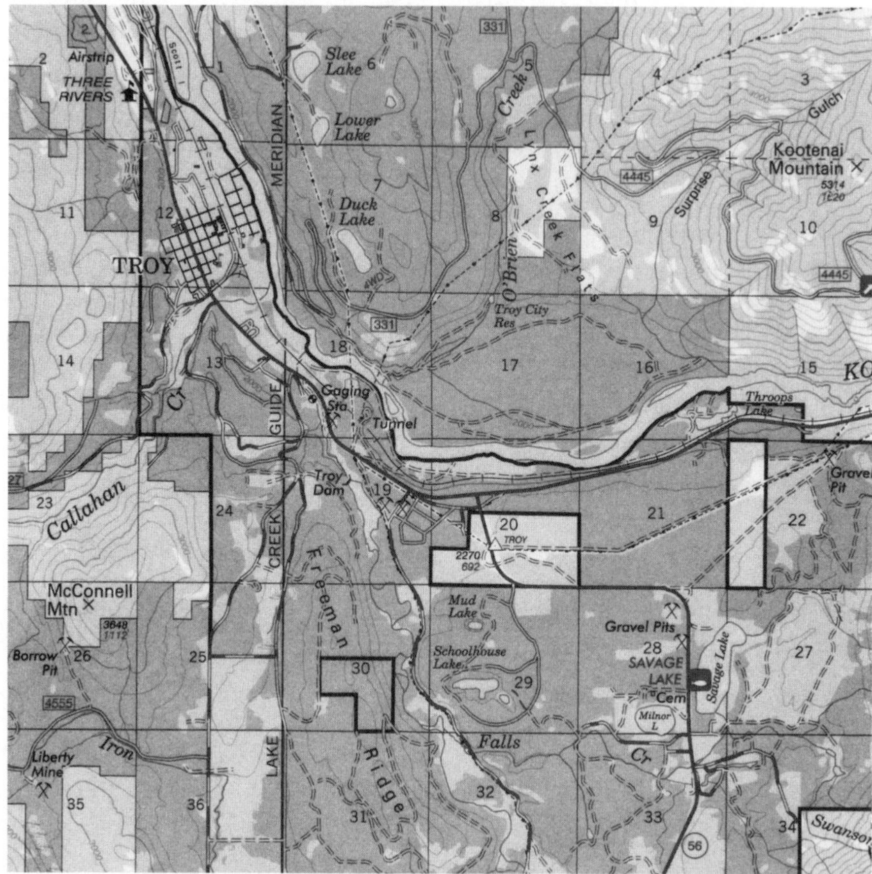

Detail of a small-scale map.

With regard to vertical accuracy, NMAS dictate that at least 90 percent of the elevation points tested shall fall within one-half of the contour interval of the map. Accordingly, a map with a 2-foot contour interval must have a vertical accuracy of ± 1 foot. USGS quadrangle maps with a 10-foot contour interval have a vertical accuracy of ± 5 feet. Large-scale topographic maps with a contour interval of 5 feet have a vertical accuracy of ± 2.5 feet.

Map accuracy is not the only thing influenced by map scale. The cost of a mapping project is also significantly affected. A photogrammetric surveyor will normally increase his or her fee by about 120 percent to map an area at 1 inch = 100 feet, as compared to 1 inch = 200 feet. The previous chapter presented typical ranges for base mapping costs.

There are several reasons the cost of base mapping increases according to the scale and accuracy of the map. First, the surveyor has to plot more information from aerial photographs. This is not only because more details can be seen and thus plotted, but because there is more information to trace. The photogrammetrist traces linear features (such as contour lines, vegetation boundaries, and drainage features) along their entire length. The larger map scale requires more time to plot this information.

Another reason is that the larger map scale requires more aerial photographs and more stereo models to cover the project area. Each stereo model requires an initial setup operation before the photogrammetrist begins map compilation.

Finally, it is common practice to decrease the contour interval for larger map scales. For instance, 5-foot contour intervals might be specified for a 1 inch = 400 feet map, whereas 2-foot contours would be required for a 1 inch = 100 feet map. The smaller contour interval means that more contour lines will have to be plotted.

Other Base Map Cost Factors

Two other factors unrelated to map scale have an impact on the cost of the base map. These are the type of terrain being mapped and the degree of land development in the project area. Hilly areas require more contour lines to depict their relief than do flat areas. Urbanized areas have more man-made features to be plotted than rural areas. Photogrammetry firms will adjust their fees accordingly.

The needs of GIS users are perhaps the most important aspects in this issue of base map cost versus base map scale. A municipal planning department might be well satisfied using maps drawn to a scale of 1 inch = 800 feet and having an accuracy of ± 26.7 feet. However, the engineering and public works departments would normally desire mapping at a scale of 1 inch = 100 feet (or larger), with its greater accuracy and increased detail. This level of accuracy and detail is necessary for site planning and construction activities. Surveyors would want the greatest possible accuracy to support the establishment of property boundaries and for construction stakeout.

One other problem to keep in mind is the false suggestion of greater accuracy a GIS often creates. A map-scale ruler is normally used to measure distances on a map drawn to scale. The precision of these measurements is limited by the scale of the map. For instance, the distances scaled from a 1 inch = 100 feet map can normally be read to the nearest whole foot. Distances scaled from a 1 inch = 1000 feet map can be read to the nearest ten feet.

On the other hand, a GIS can provide distance measurements in decimals of a foot, regardless of the map scale. This precision may imply a map accuracy far greater than the standard under which the map was created. There is a real possibility that someone will believe that GIS data is more accurate than it really is. Therefore, municipalities and utilities that furnish GIS data to the public need to be aware of this possibility and provide the public with adequate education to prevent it. The GIS database can include a qualifier for each record. Such a qualifier would indicate the source of the data, its reliability, and its accuracy.

Summary

Selecting a scale for a GIS base map can be difficult because not every user needs the same level of accuracy. Planners and natural resource managers generally need less accurate mapping than do engineers and utility managers. In the previous chapter you saw that the cost of topographic base mapping increases 100 percent or more for each incremental increase in typical mapping scale. For example, whereas mapping an area at 1 inch = 200 feet may cost in the range of $5 to $10 per acre, mapping the same area at 1 inch = 200 feet will cost in the range of $10 to $20 per acre.

The difficulty often boils down to the following issue. Mapping at a less accurate (smaller) scale lowers the cost of creating the base map and satisfies the community's planners, resource managers, and, in many cases, the tax assessment operation. On the other hand, the engineering and public utilities departments need more accurate (and more expensive) mapping, yet their operations usually represent a much larger share of the municipal budget. Giving them the greatest possible use of the GIS presents the largest potential for cost saving.

A key consideration to keep in mind is that once the topographic base map is compiled from its source documents, its accuracy cannot be increased. More accurate mapping requires the compilation of a new map. Moreover, other GIS mapping themes are registered to the features of the base map. The location of map features on a new, more accurate base map will very likely not match the features of the older base map. Therefore, a new base map will likely require the adjustment of many, if not most, of these other types of data to fit the new base map's features. The cost saved by choosing to initially map at a smaller, less accurate scale can easily be exceeded in the process.

GIS Data Quality

How Errors Creep In, How Mistakes Are Made

In This Chapter. . .

The phrase "garbage in, garbage out" has often been used to describe the simple fact that the quality of data provided by a computer system is largely affected by the quality of the data put into it. In this chapter you will see that although this cliché certainly applies to GIS data, there are other issues to be considered. The first section defines important terms that apply to the topic of GIS data quality. Subsequent sections describe sources and types of error in GIS data.

Definitions

Many of the following terms and concepts will be familiar to readers who have a background in science, engineering, or land surveying. Nonetheless, before you delve into data quality issues, you need an understanding of the terminology involved.

Data quality: The degree to which GIS data accurately represent the real world, the suitability of these data for a certain purpose, and the degree to which the GIS data meet a specific standard for accuracy.

Data errors: Differences between GIS data and a feature in the real world represented by the data. GIS data errors may pertain to a single feature or be spread throughout the database. If widespread, they may be inconsistent or consistent in their degree of error (see also the definition for "Data bias").

Data accuracy: The extent to which a measured or estimated value approaches its true value. "Accurate" GIS data truly represent the real world within specified tolerances, especially with respect to spatial (graphical) GIS data.

Data precision: The level of detail in GIS data. This is usually the number of decimal places used to store the coordinate values of points. Precision also applies to the measurement of lengths, areas, and coordinates, usually in terms of the number of decimal places in the readout. The following illustration points out the difference between GIS data accuracy and precision.

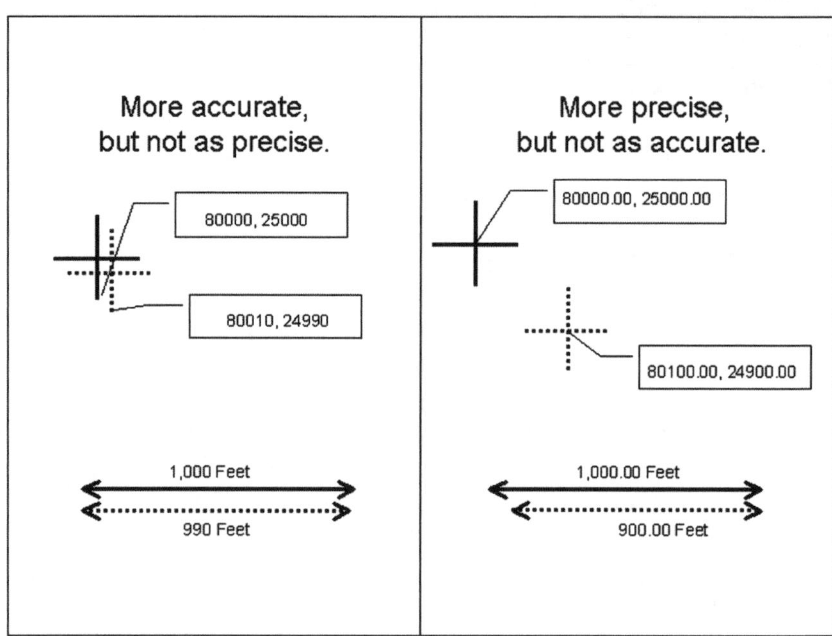

Difference between GIS data accuracy and precision.

Data bias: A consistent error spread throughout the GIS database. GIS data that exhibits a systematic variation from real-world values.

Data resolution: Usually applied to raster GIS databases, this is the size of the raster grid cells. It is, therefore, the measurement of the smallest map feature that can be stored or displayed by the raster GIS.

Complete GIS data: A GIS database that satisfactorily describes the area of interest at the specified time or during the specified period. Because the real world is constantly changing and GIS data represents only "snapshots" in time, it is important to specify a point in time or period of interest.

Compatible GIS data: GIS databases that can be compared or merged in a way that produces sensible results. If two GIS databases having very different degrees of accuracy (e.g., data digitized from 1 inch = 2,000 feet USGS quad sheets and data digitized from a county's 1 inch = 100 feet topo maps) are merged, the inconsistencies in map features (e.g., road edges from the USGS maps running though structures on the county topo maps) will be so glaring as to make the resulting data set useless.

Consistent GIS data: A GIS database that exhibits a uniform level of accuracy throughout. Typically such a GIS database is created from consistent (or very compatible) data sources, and is collected, edited, and processed in a consistent manner (these procedures are described in more detail later in the chapter).

Applicable GIS data: GIS data that is well suited for a specific application. GIS data derived from USGS quad sheets may be applicable for environmental or land-use analyses, but it may be totally unsuited for utility mapping in a densely developed urban area.

Sources of GIS Data Errors

Given a set of terms for discussing GIS data quality issues, you can now examine when and how errors creep into GIS databases. The sections that follow explore the following three major areas of potential data error.

- Source data
- Data entry
- Data analysis

GIS Source Data

Only recently has it become commonplace to collect GIS data directly in the field. This can be done using field survey instruments that download data directly into a GIS, or using GPS receivers that directly interface with GIS software on a portable PC. Using these techniques can eliminate the need for GIS source data.

However, over the past twenty years or so, GIS data has most often been digitized from several sources, including hard-copy maps, rectified aerial photography, and satellite imagery. Hard-copy maps (typically paper, Vellum, and plastic film) may contain unintended production errors. These forms can also contain unavoidable or even intended "errors" in their presentation. The sections that follow discuss these types of "errors."

Map Generalization

Cartographers often deliberately misrepresent map features, usually due to the limitations they encounter working at a given map scale. Complex area features, such as industrial buildings, may have to be represented as simple shapes. Linear features, such as roads, may have to be represented by parallel lines that are wider when measured on the map than the actual road. Curvilinear features, such as streams, may have to be represented without their smaller twists and bends. The following illustration is an example of map generalization.

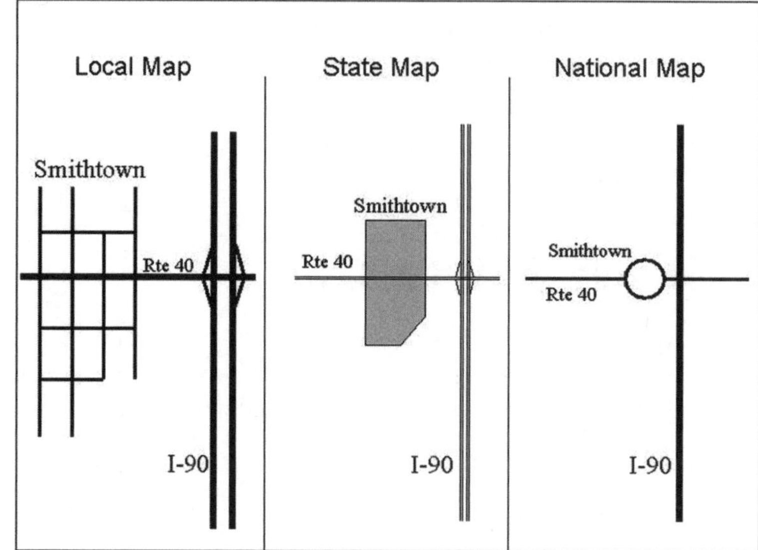

Map generalization.

Indistinct Boundaries

Indistinct boundaries typically include the boundaries of vegetated areas, soil types, wetlands, and land-use areas. Whereas in the real world such features are characterized by gradual change, the cartographer represents these boundaries with a distinct line. Some compromise is inevitable.

Map Scale

Cartographers and photogrammetrists work to accepted levels of accuracy for a given map scale. The following illustration is of a table of map scale and accuracy taken from the National Map Accuracy Standards (NMAS). This means that the actual location of a map feature may well disagree with its actual location on the ground, even though this error is likely to fall within the specified tolerance. Of course, this problem is compounded by limitations in making linear map measurements, typically about 1/100th of an inch on a map scale.

NMAS map scale and accuracy table.

Scale	Contour Interval	Horizontal Accuracy	Vertical Accuracy
"Large-scale mapping"			
1"=100'	2'	±3.3'	±1.0'
1"=200'	5'	±6.7'	±2.5'
1"=400'	5'	±13.3'	±2.5'
"Small-scale mapping"			
1:24,000	10'	±50'	±5.0'
1:50,000	25'	±100'	±12.5'

Map Symbology

It is impossible to perfectly depict the real world using lines, colors, symbols, and patterns; therefore, cartographers must work with certain accepted conventions. As a result, the facts and features represented on a map must often be interpreted or interpolated, a process that inevitably produces error. For example, terrain elevations are typically depicted using topographic contour lines and spot elevations. One must interpolate the elevation of the ground between these lines and spots. Likewise, an area symbolized as "forest" may well include open areas among trees that have not been depicted. All of the errors in the source data, unintended or not, are likely to be reproduced in the GIS database.

GIS Data Entry

As previously mentioned, GIS data is typically created from source data such as hard-copy maps, aerial photographs, and satellite imagery. The process is often called "digitizing," referring to the fact that the source data is converted to a computerized (digital) format. Human digitizers can compound errors in source data, as well as introduce new errors when creating the GIS database. There are three primary methods of digitizing hard-copy source data. These are discussed in the sections that follow.

Manual Digitizing

Typically, a map is affixed to a digitizing table and registered to the GIS coordinate system, amd then "traced" into the GIS. The digitizing table has a fine grid of wires embedded in its surface that sense the position of the crosshair on a hand-held cursor. When the cursor button is depressed, the system records a point at that location in the GIS database. The operator also identifies the type of feature being digitized, or its attributes.

Photogrammetric mapping is, in a sense, also a manual digitizing process. Through an exacting and rigorous technical process of aerotriangulation, overlapping pairs of aerial photographs are registered to one another and viewed as a 3D image in a stereoplotter. In a process called stereocompilation, the photogrammetrist traces the map features in the image, which are encoded directly into the database.

Scanning and Keyed Data Entry

In scanning, source data is mechanically read by a device that resembles a large-format copy machine. Sensors encode the image as a large array of dots, much like a facsimile machine scans a letter. High-resolution scanners can capture data at resolutions as fine as 2,000 dots per inch (dpi), but maps and drawings are typically scanned at 100 to 400 dpi. The resulting raster image is then processed and displayed on the computer screen. Further manual digitizing on screen is usually needed to complete the data entry process. This is often called "heads-up digitizing," for obvious reasons. If the source data contains coordinate values for points, or the bearings and distances of lines (e.g., parcel lines), these map features can be keyed into the GIS with great precision.

Data Entry in General

Accurate manual digitizing is not an easy task, and requires certain basic physical and visual skills, as well as training, patience, and concentration. There are many opportunities for error in manual digitizing because the process is subject to visual and mental mistakes, fatigue, distraction, and involuntary muscle movements. Obviously the "setup" (registration with the GIS coordinate system) of the map on a digitizing table or the scanned raster image can produce errors. In addition, the cell size of the scanned raster image can affect the accuracy of heads-up digitizing.

With either method, the person digitizing must accurately discern the center of the line or point, as well as accurately trace it with the cursor. This task is especially prone to error if the map scale is small and the lines or symbols are relatively thick or large. The method of digitizing curvilinear lines will also affect the accuracy of the result. "Point mode" places sample points at selected locations along the line in order to best represent it in the GIS.

The process is totally subject to the judgment of the person digitizing, who selects both the number and placement of the data points. "Stream mode" digitizing collects data points at a preset frequency, usually specified as the distance or time between data points. Every time the operator strays from the intended line, a point digitized at that moment will be inaccurate. This method also collects many more data points than may be needed to faithfully represent the map feature. Therefore, post-processing is often used to "weed out" unneeded data points, using a specific tolerance.

Heads-up digitizing is often preferred over table digitizing because usually it yields better results and is more efficient. Keyed data entry of land parcel data is the most precise method. Moreover, most errors are fairly obvious because the source data has usually been carefully computed and thoroughly checked. Most keyed data entry errors show as obvious mismatches in the parcel "fabric."

The following illustration points out the types of GIS data errors that can occur during the digitizing process. GIS software usually includes functions that detect many types of errors in a GIS database. These error-checking routines can find mistakes in the topology of the data, including gaps, overshoots, dangling lines, and unclosed polygons.

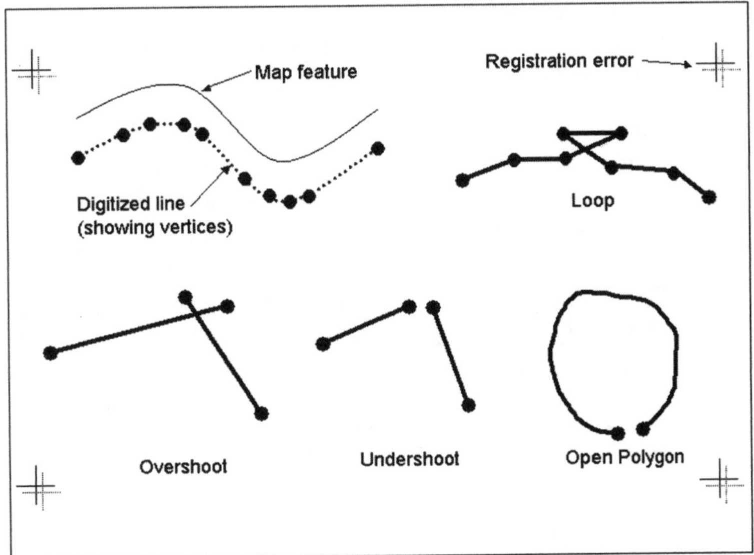

Data errors during digitization.

The operator sets tolerances that the routine uses to inspect for these errors. Of course, the effectiveness of the system in finding these errors depends on setting the correct tolerances. For instance, a tolerance that is too small may pass over an unintentional gap. A tolerance that is too large may improperly remove a short dangling line or small polygon that was intentionally digitized.

GIS Data Analysis

Even if it is "accurate," the manipulation and analysis of GIS data can create errors. These errors can be introduced within the data itself or produced at the time data is displayed on the computer screen or plotted in hard-copy format.

The concepts of complete GIS data, compatible GIS data, consistent GIS data, and applicable GIS data previously defined apply at this stage. The user must consider whether a selected GIS data set is complete, consistent, and applicable for the intended use, and whether it is compatible with other data sets used in the analysis. The problem that results from merging GIS data derived from 1 inch = 2,000 feet USGS quad sheets with that derived from 1 inch = 200 feet county topographic maps was previously cited as an example of such considerations.

The phrasing of spatial and attribute queries may also lead to errors. Likewise, the use of Boolean operators can be complicated, and the results can be decidedly different, depending on how a data query is structured or a set of queries are executed. For example, the query "Find all structures within the 100-year flood zone" yields a different result than "Find all structures touching the 100-year flood zone." The former question will find only those structures entirely within the flood zone, whereas the latter will also include structures that are partially within the zone.

The overlay of data sets is a powerful and commonly used GIS tool, but it too can yield inaccurate results. For instance, to determine areas suitable for a specific type of land development project, one might overlay several layers of GIS data, including natural resources, wetlands, flood zones, existing structures and land uses, land ownership, and zoning. The result will usually narrow the possible choices down to just a few parcels. These would be investigated more carefully to make the final choice. The final result of this analysis will reflect any errors in the original GIS data. Its accuracy will only be a good as the least accurate GIS data set used in the analysis.

It is also common to overlay and merge GIS data layers to form new layers. In certain circumstances this process introduces a new type of error, the polygon "sliver." Slivers often appear when two GIS data sets that have common boundary lines are merged. If these common elements have been digitized separately, the usual result will be sliver polygons, an example of which is shown in the following illustration. Most GIS software products offer routines that can find and fix such errors, but once again the user must be careful in setting the search and correction tolerances.

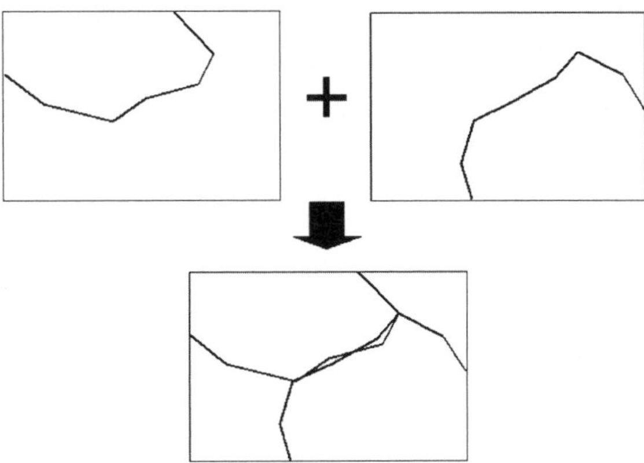

Sliver polygons.

Summary

Errors in GIS data are almost inevitable, but they can be controlled and their negative impacts kept to a minimum. Knowing the types and causes of GIS data errors is half the battle. Many errors can be avoided through the proper selection and "scrubbing" (i.e., preparation and editing) of source data before it is digitized. Moreover, a structured, well-managed digitizing process (backed up by a thorough, disciplined quality control process) can catch and eliminate most data entry errors. Finally, proper training for all GIS analysts, and even casual GIS users, is needed to avoid the misuse of GIS data and the misapplication of analytical GIS software.

Getting CADD Data into a GIS

What to Consider When Translating CADD Data into a GIS Format

In This Chapter...

CADD systems are often used to capture GIS data. This chapter presents requirements and options for translating CADD data into a GIS format.

The Need for CADD Data Translation

CADD has brought about a revolution in the way engineers, architects, surveyors, and planners do their work, especially their drafting work. This revolution began in earnest in the early 1980s, fueled by the advent of the personal computer.

GIS technology attracted relatively little attention during the five years following the first surge of interest in CADD. Consequently, there are many organizations using CADD systems instead of a GIS for mapping, although this situation is changing rapidly in favor of GIS. Moreover, many other organizations that make maps have a greater need for automated design and drafting functions than they do for spatial analysis. They have little need for a full GIS, although they may need to produce files for someone else's GIS.

For instance, many organizations receive data from engineers and surveyors that will be added to their GIS. Examples are subdivision plats, site plans, and as-built road and utility plans. These organizations include municipalities, utilities, and military installations. Likewise, at the outset of the design or survey project, they may give the surveyor or engineer a copy of the data in their GIS database to provide survey control or as-built data for the design.

For these reasons, it is appropriate to address the issue of transferring CADD data to and from a GIS. There are two primary considerations in this regard: data structure and data format.

Data Structure

Part I of this book discussed the differences in the structure of CADD and GIS data. It pointed out that data elements in a CADD system are not related to one another except by reference to a common coordinate system, and perhaps by layer. There are no restrictions on placing elements in a CADD file, as long as they lie within the same drawing plane (3D space).

In a GIS, however, the conventions of data topology must be applied. This places restrictions on how elements may be placed in the GIS file. Failure to follow these restrictions results in errors in file processing. Therefore, to successfully translate CADD data to a GIS, the CADD data must first be structured in accordance with the conventions of data topology. Otherwise, the CADD data will produce errors in GIS processing that must be corrected after the data is translated to the GIS format.

These considerations do not apply to the transfer of GIS data to a CADD system. The elements of data topology will be eliminated in the process of translating GIS data to the CADD format. The following guidelines should be followed when preparing CADD data files that will be translated to a GIS format. These procedures will greatly simplify the translation process, mainly by reducing the need for "cleanup" of the resulting GIS data.

Edge Matching

All digitized map sheets must be edge matched, both visually and by coordinates, with all adjacent sheets. No edge match tolerance may be allowed. The layer assignments for adjoining features must also be identical. The following illustration shows the edge matching of two map files.

Edge matching GIS data.

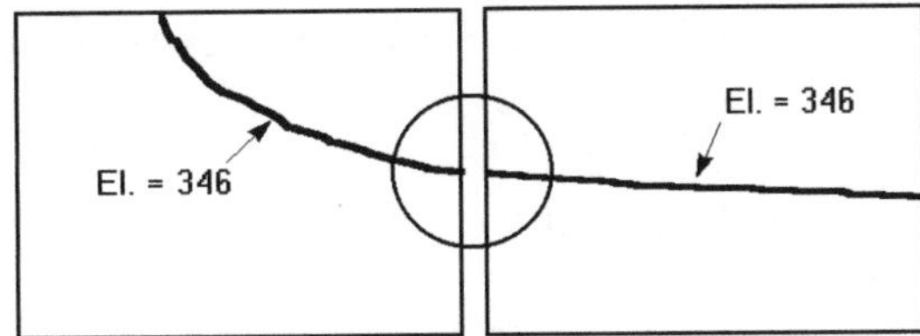

Common Features

All graphic features that share a common boundary, regardless of digital map layer, must have exactly the same digital representation of that boundary feature. Moreover, all features within a map theme that have the same boundary must be represented by a single graphic element. The attributes associated with this feature will distinguish the boundary types. The following illustration shows two files with common map features.

GIS data with a common boundary.

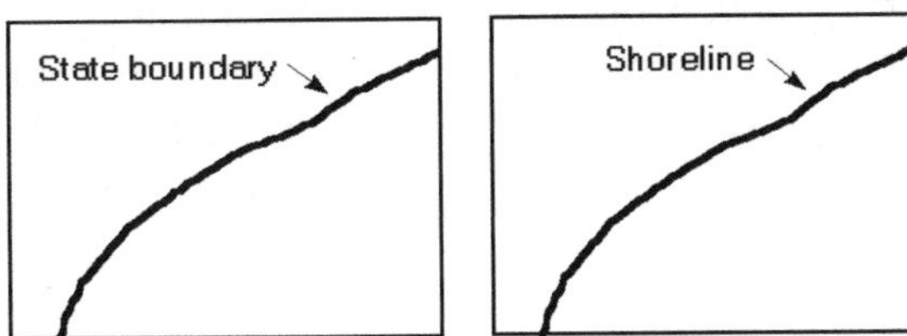

Networks

The digital representation of linear elements must reflect the actual structure of any network they represent. Lines representing the same type of data must not cross except at an intersection. Furthermore, an element should not be broken or segmented unless that segmentation reflects a visual or attribute code characteristic, or unless the break is forced by database limitations.

Polygons

The last coordinate pair of a digitized polygon must exactly match the first coordinate pair. No line or line string may cross a polygon of the same data type except to join it at an intersection. Each polygon must have a single, unique centroid to which attributes can be associated. Polygons of a single data layer must cover the area of interest completely and be exclusive within that area. There must be no holes in the polygon net and no overlaps in a layer.

Lines and Line Strings

All digitized lines must be topologically structured. Lines and line strings may not cross except at an intersection. There must be no zero-length segments, and a line may not cross back upon itself. Straight lines should be defined only by the two points that represent the beginning and ending nodes. Curvilinear features should be represented by the minimum number of points needed to provide a smooth appearance when plotted. Where appropriate, programs should be used to eliminate unnecessary data points.

Connectivity

All graphic elements that connect visually must exactly connect digitally. That is, the intersections of all line and polygon data must be mathematically exact. No *overshoots, undershoots,* or *offsets* may be permitted. Lines that connect to polygons must intersect those polygons precisely. That is, the end point of the line must be connected to an intersection point of the polygon. The following illustration shows some common errors in the topological structure of networks, polygons, and line strings.

Errors in topological structure.

Overshoots

Undershoots

Gaps

Loops

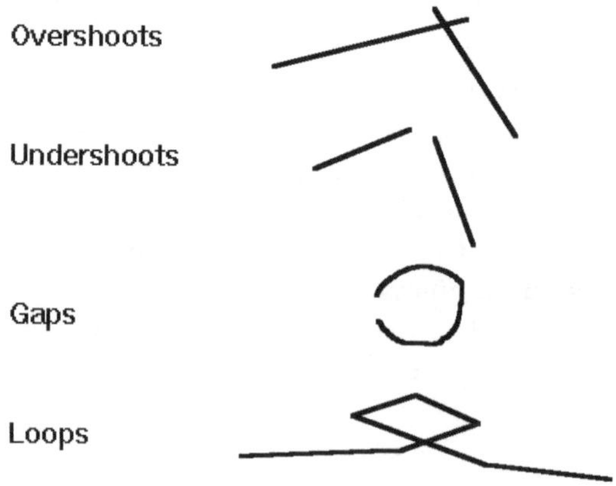

Data Format

The second primary consideration in translating CADD data to or from a GIS is data format. Whereas some GIS programs make direct use of CADD files (e.g., Intergraph's MGE, and ESRI's ArcCAD), many employ a unique file format. CADD files must first be translated to this format before the GIS software can process them. This translation process can be accomplished in two ways: *direct translation or indirect translation.*

Direct translation is the fastest method. This process uses a translator program that reads the input file, makes the appropriate translations, and then directly creates the output file. Because the program starts with a binary file and produces a binary file, it is the most efficient method. Unfortunately, there are relatively few direct translator programs available commercially, and these have been written for only the most popular CADD and GIS vendor formats.

On the other hand, many organizations have developed proprietary direct translator programs when they must routinely move a lot of data between systems. Here, the cost of custom programming is offset by the savings in processing time over the long run. *Indirect translation* uses a neutral ASCII file format to move the data between systems. This neutral file format has been made public so that all vendors can write translator programs for this format. Two programs are needed.

The first translator reads the CADD or GIS file and outputs a file in the neutral format. The second translator reads the neutral format file and translates it to the GIS or CADD format. Because two steps are required, this process is slower than direct translation.

On the other hand, there are many more translator programs available for this technique. In fact, nearly all vendors serious about the commercial marketplace offer the capability to move their data to and from popular neutral file formats. The following are two commonly used neutral file formats.

- *Digital Line Graph (DLG),* U.S. Geological Survey, Reston, Virginia
- *Drawing Exchange Format (DXF),* Autodesk, Inc., Sausalito, California

Summary

Successful translation of CADD data to a GIS format requires planning and coordination in advance. Several trial runs may be required to iron out all the bugs. Nonetheless, once the proper data structure has been achieved, the translation process should produce a satisfactory result.

Legal Aspects of GIS

Unique Issues of Data Access, Pricing, Privacy, Liability, Copyright, and Land Survey

Federal, state, and local government agencies are among the very largest groups of GIS users. This is understandable, given one observer's estimate that 80 percent of the data collected by a governmental entity involves a geographic component. However, the nature of GIS data presents some unique legal issues for government agencies. This chapter first looks at the value of GIS information and the role of government in GIS data collection, management, and dissemination. It then discusses six legal issues regarding GIS data: access, pricing, privacy, liability, copyright, and the practice of professional land survey.

The Law and Public Information

A fundamental precept of democracy has been that citizens have a right of access to public information. The underpinning of this precept is a belief that access to public information is essential to self-government, an important check in the system of checks and balances. The adage "knowledge is power" implies that "access to information is power" as well. In our information-oriented society, access to information enhances both economic and political power, as well as the power to affect society itself.

Our society has used laws and regulations to prevent the abuse of these powers, as well as to promote their beneficial application. Information ownership, control, management, and dissemination are a few of the issues these laws and regulations address. They seek to protect the rights of data owners, to protect the rights of the public that pays for the data, to protect the privacy of individuals about whom information has been collected, to protect national and organizational security, and to rightly assign liability for damages arising from inaccurate or improperly used information.

The Value of GIS Data

Building a GIS database may cost hundreds of thousands to millions of dollars. However, the value of a GIS database is not limited to the expense of building it. Although much of the information contained in the database may have already existed on paper or in computer databases, the total is much greater than the sum of its parts. The GIS database is better organized and is easier to access and use than paper records. In addition, data from a variety of sources is "virtually" centralized.

Although the data may be physically disbursed around a computer network, it is logically integrated and universally accessible to users in such a way that all of the data now appears to be "at your fingertips." In addition, the data is much easier to analyze and present. In fact, not only are the analyses that were done before easier and less expensive to conduct, entirely new types of analysis are performed that were simply not feasible before.

Moreover, building a new GIS database usually involves the collection of much information that was not recorded previously. Thus, the GIS database not only represents a significant investment, it offers more and better information and a powerful tool for using and working with it.

The Role of Government

Federal, state, and local government agencies must collect and analyze data to carry out their various missions, much like any other organization. Many of these missions involve land data, including the management and regulation of land, land use, real estate tax assessment, land title records, the environment, emergency response, public safety, public works, public utilities, and transportation systems. Some of these missions involve demographic data that can be related to geography, including age, income, family, racial, and educational statistics.

Although there are many existing laws and regulations regarding data collection, management, and access, they typically do not adequately address the role of government as a disseminator of information. Moreover, although there has been much talk of a public "information utility," there is little legal guidance for such an expanded role of government.

Typically government agencies use the data to support their various missions but view data dissemination as a "passive" service. If a citizen wants the data, she can come and get it. On the other hand, there are many private providers of value-added information services. These companies obtain data from public agencies, and then reformat and repackage the data. They often provide tools and services to make better use of the data. These private data resellers aggressively market and sell their product. Data distribution is not an ancillary activity; it is their business.

Understandably, many of these companies express concern about the possibility of increased competition from public agencies. They and their trade associations argue that the government should provide the raw data for a reasonable fee, but the government should not provide the tools for manipulating or analyzing the data, nor data analysis services. (As discussed further in material to follow, the method of arriving at a "reasonable fee" is subject to some debate.)

Access to GIS Data

The federal government has traditionally collected and made available to the public tremendous amounts of information in a wide variety of data types. Consequently, access to data under the control of federal agencies is well defined. The Printing Act of 1895, the Depository Library Act of 1962, the Freedom of Information Act (FOIA) of 1966, and the Paperwork Reduction Act of 1980 all provide for the distribution of federal information.

The Federal Freedom of Information Act

The FOIA in particular was enacted to ensure that the public had access to information concerning the functioning of the government. Under the FOIA, federal agencies are required to publish substantive rules, statements of policies, and information on agency organization and procedures in the Federal Register. Agencies must also make final policy statements and rulings not published in the Federal Register, as well as staff manuals and instructions, available for inspection and copying. Agencies must also make available upon request other records that do not fall into these two categories. The FOIA exempts from disclosure data relating to the following areas.

- National security
- Agency personnel matters
- Matters exempt from disclosure by another statute
- Commercial secrets
- Agency deliberations
- Private personal matters
- Law enforcement investigations

- Financial institution examinations
- Geological surveys

Under the FOIA, the agency is obligated to furnish access to an agency record, provided the request is reasonably specific and provided the request complies with the rules and procedures of the agency. The agency may establish a uniform fee schedule for responding to these requests, but the fee may only recover the "direct costs" of searching and copying the information.

Although the FOIA was enacted before the electronic age, subsequent federal guidelines hold that information stored in a computer-readable form is considered an agency record under the FOIA, just as paper documents are agency records. If computer-readable information is available and requested, the agency must release its records in an electronic format if other formats (e.g., paper) would be significantly more difficult to use.

On the other hand, if the available non-computer information is virtually identical to the computer information, the agency need not release the information in electronic form. Moreover, agencies view computer programs as either tools or records, thus drawing a distinction between programs required to interpret or retrieve a record and programs needed to analyze or manipulate the data.

State Legislation

State and local governments generally disseminate only limited types and relatively small quantities of data. The State Open Record and State Freedom of Information acts have generally been based on the federal FOIA, but they vary considerably from jurisdiction to jurisdiction. A review of information laws indicates that the following actions have been taken by most states to effect information access and individual privacy.

- All have open record laws.
- Most define public records to include electronic data.
- Nearly all make no distinction for access based on format.
- Most states require that copies be provided.
- Nearly all have some fee to be charged for copying public records.
- Most have enacted specific GIS legislation.
- Many have created specific privacy laws.
- Some now prohibit the use of public information for commercial purposes.

- Some have established offices of information practices.

GIS Data Sales and Pricing

The commercial value of many records, particularly GIS records, far exceed their regulatory value. In these days of financially strapped government, charging more for public records to help fund new and costly information systems is extremely tempting. Although the majority of all costs associated with developing a GIS are in the building of the database, the cost of distributing the data is marginal.

The fee charged for the production of information in response to an open-records request has been typically limited to a minimal cost for copying the information. However, recent state legislation has allowed the charging of fees for information based on the cost of providing the electronic service and a portion of the cost of building and maintaining the database.

Although some states allow only "appropriate fees," others provide that the custodian of the record may take into consideration the commercial value of the information in establishing the fee. In many instances, fees vary based on a range of factors, including staff time; program creation; whether the information is on disk, tape, or printout; computer time; and similar requirements.

One State's Approach

Florida's Joint Committee on Information Technology (JCIT) conducted a detailed study of policy issues raised by the use of new technologies in government record keeping. With respect to fee schedules for information requests, the committee staff found three alternatives to Florida's current "actual cost of duplication" approach: commercial value, statutory exemption, and statutory fee. The findings of this study and the judgment of the JCIT illustrate the types of arguments that can be made for and against these different approaches.

Commercial Value

Those in favor of charging commercial value for public records offer two possible approaches: fees based on the value of the record requested, or fees based on the motivation of the individual requesters. Both approaches result in a higher fee, whether based on commercial value or commercial motivation.

This is something akin to a user fee, imposed under the rationale that those who stand to benefit or gain financially from their use of public records should pay to help recover the cost of developing and maintaining public records and the corresponding information systems. Although not disputing the possible economic benefit of the commercial value alternative, the JCIT rejected this approach, concluding that such an alternative was not feasible due to serious constitutional problems inherent in distinguishing between requesters of public records.

Statutory Exemption

Another possible alternative was to create a statutory exemption from the inspection and copying requirements of Florida's Public Records Law for specific types of technology (all records maintained on a GIS, for example), or an exemption for the commercial use of such records. There are two possible rationales for this alternative. First, an exemption for commercial use of GIS records would effectively pull those records from the requirements of the Public Records law, thus allowing agencies to charge commercial users more for GIS records than for other records. The second rationale is to protect the subjects of certain public records (crime victim and motor vehicle accident records, for example) from commercial exploitation.

Although a number of states have recently amended their statutes to adopt this approach, the JCIT rejected it as being too problematic, the major concern being the difficulty in predicting the path of the rapid advances in information technology. Furthermore, there was concern that if an exemption were created for a specific type of technology, it would be relatively easy to circumvent the access requirements of the Public Records law simply by utilizing that technology.

Statutory Fee

The third approach considered was the establishment of a flat statutory fee for certain records in an attempt to recoup some of the overhead costs associated with providing access to those records. Under Florida's Public Records law, the custodians of a public record may not charge more than the actual cost of duplicating a record, plus an extensive use charge if applicable.

The problem with the flat fee approach is in determining the fee. Many questions arise, such as what factors are to be considered, and how much is too much. Additionally, allowing a flat fee per record may become unreasonable if a requester is seeking an entire database in an electronic format.

The JCIT ultimately concluded that the more reasonable approach would be to continue to follow the example provided by a provision in Florida's Public Records law that allows a reasonable charge for the labor and overhead associated with duplication of county maps, including GIS maps and aerial photography.

The Need for Legislation

In addition to the public, governmental, and private interests in GIS data previously described, there are the needs of educators, researchers, and journalists for access to information. These conflicting interests have hindered the formulation of adequate public policy regarding GIS data dissemination. Moreover, the legislative process inevitably lags behind the rapid advances in GIS technology. As a result, state and local agencies and legislators especially need to continue to refine the laws and regulations dealing with governmental policy and the dissemination of GIS data.

Data Privacy

As previously stated, federal and state governments have enacted laws to ensure that the public has access to information concerning the functioning of the government. They have also enacted laws to ensure that such access does not invade their privacy. But how does this issue pertain to GIS?

It used to be impractical for anyone to maintain information on the location of all individuals in a community along with detailed personal information. Today, local governments and utilities routinely use GIS databases to organize, manage, and access information about privately owned properties, their owners, and/or their residents.

Similarly, the Census Bureau employs a spatially referenced information system to support the demographic information it collects on U.S. households. Such data files of household or personal information can easily be linked to or cross-referenced against GIS data. GIS thus serves as an effective integrating technology, allowing the user to tie databases from disparate sources in a common, spatially referenced framework. GIS also offers powerful tools for spatial, social, business, political, and economic analyses that simply did not exist until recently. Therefore, data privacy is also an important legal aspect of GIS.

The Federal Privacy Act

The U.S. Privacy Act was written in 1974 to safeguard individuals against invasions of personal privacy by federal agencies. It has served as a model for similar acts in state houses of legislation. It can be summarized as a law that does the following.

- Establishes principles of information practices
- Defines specific individual rights
- Defines administrative procedure for compliance

- Provides remedies for statutory violations
- Addresses systems that contain personal information
- Establishes limitations of disclosure of Social Security numbers

The Act provides that no agency can disclose any records on an individual except pursuant to a written request by, or with the prior written consent of, the individual. There are twelve exceptions to this non-disclosure rule. Disclosure of information on an individual is permissible under the following conditions.

- To employees of the agency that maintains the record and have a need for the record
- As required under the Freedom of Information Act
- For a "routine" use of a record
- To plan or carry out the U.S. Census
- For statistical research or reporting, provided the record as transferred is not individually identifiable
- To the National Archives as a record with sufficient historical or other value
- To another agency for civil or criminal law enforcement
- To a person showing compelling circumstances affecting the health or safety of the individual, provided the individual is notified of such disclosure at his last known address
- To the Congress
- To the Comptroller General in performing his duties as head of the general accounting office
- Pursuant to a court order
- To a consumer reporting agency

Under this Act a citizen may protect certain data subjects in the following ways.

- Refuse to provide personal information
- Learn of possible dissemination of personal information
- Inspect information
- Obtain copies of documents containing personal information at reasonable cost for search and duplication

- Challenge, correct, or receive an explanation about the information

The Privacy Act provides instructions to agencies that collect information. When an agency disseminates that information into the hands of someone else for whatever reason, two things are required.

- Requirements and limitations placed on the information must accompany the information
- Security and usage must be specified at the time of transfer

Administrative procedures provided for in the Act also call for maintaining accurate records on individuals requesting and gaining access to public information, including the purpose for making the request. The Act also requires that agency employees be informed of the requirements of the Act and understand what it means. Furthermore, the agency must establish safeguards to see that the requirements for employee awareness and compliance are carried out.

State Privacy Legislation

Some states have provided additional safeguards in their privacy legislation. For instance, Oregon recently introduced legislation that provides for "opting-out." This allows citizens, whenever they engage in a transaction where information about them is transferred, to indicate that such information will not be used for commercial advertising purposes.

It is interesting to note that although the dissemination of private information by federal and state agencies is regulated, the commercial sector has fairly free reign on what data it collects about individuals and on how it can use, exchange, and integrate that information. Unlike the United States, in much of Europe the law applies the principles of privacy protection to the commercial sector. Thus, the greatest threat to personal privacy in the United States is from the private sector.

GIS and Data Privacy

GIS enables a commercial market researcher to link household and personal data to a parcel-level digital map of the community. This may include addresses and phone numbers from telephone directories, drivers' license databases (often accessible through the state open records laws discussed previously), zip codes, Social Security numbers, credit card information, and retail store product purchases (recorded by the bar code scanning device at the checkout counter and linked to the customer through the store's credit card or check cashing account number). This is tremendously useful information for the analysis of product lines and pricing, delivery routes, store locations, direct mail campaigns, and so forth.

The question then arises, where is the dividing line between the appropriate and permissible exchange of facts to pursue a commercial goal and an impermissible intrusion on personal privacy? As you can see, federal and state legislators have sought to ensure that public data is accessible, while at the same time protecting the individual citizen's right to privacy. However, as technology continues to enable the creation of highly value-added information products, and as demand for the products increases, concerns become far more complex than merely a quest for balance between access and privacy.

Liability

In general terms, legal liability can arise from a number of situations and relationships, some controllable and others not. Three principal types of liability are pertinent to GIS: contractual liability, statutory liability, and tort liability.

Contractual liability can arise from a number of situations, including a breach of contract. A breach of contract may result from the failure to pay for a product or service, or from the delivery of goods or services that do not conform to the agreement. Private companies that provide GIS products and services are therefore liable for full compliance with the requirements of their contract with the customer.

Statutory liability arises when specific statutes enacted by the state and federal legislatures are violated. With regard to GIS, statutory liability can arise under the open records laws and freedom of information acts described previously. It can also arise under the privacy acts previously discussed, where there has been a release of information required to be kept confidential.

A tort is committed when one person owes a duty to another person but the duty is breached or violated, resulting in injury to the one that has a right to that duty. This breach of duty may involve harming or injuring a person or property, withholding property, or harming or injuring a legally protected relation such as employer and employee, or business person and customer. Such harm may be inflicted intentionally, through neglect, or through the improper use of one's property, such as where a person creates a nuisance.

Intentional torts include assault and battery, false imprisonment, trespass, and the intentional inflicting of mental distress. However, the majority of torts in the courts today involve negligence, such as in the case of automobile accidents. Torts arising from negligence can also include the dissemination of inaccurate or misleading information, invasion of privacy (including the public disclosure of private facts), or interference with a business expectation or relationship.

Of course, a prominent news topic today is the expansion of torts related to product liability. When a person commences a tort action, he or she is seeking compensation for damages suffered, so the purpose is to compensate the victim, not punish the offender, as in the case of criminal law.

Liability for GIS Data

Private companies that provide GIS products or services are responsible for some level of competence in their work. If others are injured or damaged by mistakes, the courts have reasoned that the product or service provider should bear at least some of the responsibility. However, the courts also hold that the provider is only responsible for those damages that result from proven negligence or other error based on legally established statutes defining the duty of the provider. Establishing this duty is typically accomplished under the previously described theories of negligence law or contract law. Where appropriate contract language is absent, the court resorts to tort principles, negligence being the most likely tort principle to arise.

Where a private company has supplied incorrect or insufficient data, this does not in itself prove negligence. In general, the data supplier must do that which another "reasonably prudent" company would have done under the same or similar circumstances. However, this standard causes problems because it can be broadly interpreted and it is applied after the fact.

GIS data and service providers can avoid this uncertainty through the use of properly written contracts. These contracts should spell out the requirements for the accuracy and completeness of the data to be delivered, or the requirements for the service to be provided. What the "reasonably prudent" company would have done becomes irrelevant when the contract constitutes the entire agreement between the parties.

Government agencies also supply GIS products and services. Mapmaking has traditionally been viewed as a function under which governments generally have a right to claim sovereign immunity against errors. However, the key distinction in determining whether or not a governmental authority is liable for its undertakings is whether or not the action is "discretionary." Sovereign immunity will not be applicable where discretionary activities or functions are undertaken. In these cases, the government will be held liable for its negligence, and for the negligence of its employees while doing their jobs.

Today, marketing GIS products and services is not considered a pure governmental function but a proprietary function of the government. Not only does the government have the discretion to choose whether or not to market and sell GIS products and services, it has the right to determine to what extent it will do so.

An aggressively marketed GIS database could certainly be considered to go beyond the traditional bounds of government authority. Therefore, if legal action were taken against a government agency for its GIS products or services, a court would most likely consider claims of negligence or product liability. As in the case of the private corporation, if a government agency sells GIS products and services it must also exercise a proper level of due care.

Steps to Limit Risk

The nature of GIS creates an important dilemma regarding the apparent or implied accuracy of the data. For example, even though property, easement, and ROW boundaries might have been digitized from relatively inaccurate hard-copy tax assessor maps, the GIS can calculate and display highly precise values for bearings, distances, and geographic coordinates.

When parcel boundaries are shown on a hard-copy tax map, the scale of the map implies a certain level of inaccuracy. One could scale the length of a parcel boundary line shown on a 1 inch = 200 feet tax parcel map to, say, + 4 feet. However, after the tax map has been traced into the GIS, the length of that same boundary line can be measured to decimal places. The same can be said for the bearings of lines or the geographic coordinates of points. The data is no more accurate than when it was shown on the tax assessor map, but the precise readout from the GIS implies a level of accuracy that simply does not exist.

Moreover, government agencies are making GIS data very easy to access and use. In the past, hard-copy tax assessment maps were usually obtained at a public information counter from a government employee who could answer questions and explain their limitations. Now the public has nearly unlimited access to ROW, easement, and parcel boundary data stored in the GIS, and although the public may have direct access to the GIS sitting on the public information counter, local governments can now distribute GIS data over the Internet to their homes.

For this reason, it is imperative that cadastral GIS data include some indication of its source and accuracy. This can be done in several ways. One is to place GIS land base data on different layers (or in different coverages) of the GIS database to distinguish their source and accuracy. Another is to distinguish the GIS data obtained from various sources using graphic symbology (e.g., differing colors and line styles) and a map legend. A third is to include metadata ("data about the data") in the GIS database. This metadata would describe the source of the data, its accuracy, how it was collected, and so on. In this way, the public can be afforded a minimum level of protection against the misuse or misapplication of GIS data.

Likewise, any sale of GIS products or services to the public should include express disclaimers of liability, whether delivered on CD-ROM, through a workstation sitting on the public information counter, or over the Internet. Such a disclaimer should describe the limits of the accuracy and reliability of the data and warn the user that the information may not be perfect and that it should be viewed accordingly.

However, even disclaimers may not be enough if damage or injury occurs. Therefore, the best protection for both private and government GIS data and service providers is a well-written contract coupled with competent and complete performance of the required work. The fewer mistakes made, the less likelihood those mistakes will cause damages, and the less liability exposure will exist. Quality control procedures that catch mistakes before they leave the office are critical.

Copyright Protection

Copyright protection is another important legal issue that relates to GIS. Although the federal Copyright Act has always protected maps, a GIS may be viewed as simply an electronic collection of map and map attribute data. Under a recent Supreme Court ruling, such a collection would be classified as a "compilation" rather than a "map," and would therefore be exempted from copyright protection. A subsequent lower court ruling has upheld a hard-copy mapmaker's copyright but failed to expressly address the GIS dilemma.

In general, a copyright is a federal grant of an exclusive monopoly for a period of time, so that the distribution of ideas and information will not be restricted indefinitely. It is intended to give the public the greatest possible access to a creative work. It gives the copyright holder the right to market the creative work in hopes of receiving income. It includes the right of the copyright holder to reproduce the copyrighted work; to prepare derivative works; to sell, lease, or rent the work; and to perform or display the work publicly. Thus, it allows the creator to reap the benefits from his or her creation.

Maps and the Copyright Act

The Copyright Act has always protected maps. Nonetheless, the protection of map copyrights has presented some problems. These problems arise from the fact that although maps themselves can be protected by copyright, copyrights do not protect facts, systems, or ideas.

Traditional maps are pictorial representations of geographic and demographic facts, organized to allow the user to readily understand and easily extract them. However, the factual information itself is not protected by a map's copyright. For instance, one cannot copyright the location of a political boundary line. Moreover, the method of presenting the information is also usually an unprotected system or idea. For instance, one cannot copyright the symbology used to depict political boundary lines on a map. As a result, many maps will apparently contain only unprotected elements. However, the map "as a whole" can be protected, even though all elements that constitute the map remain unprotected.

Pictorial maps are classified as "pictorial, graphic, and sculptural works" under the 1976 Copyright Act. Therefore, a map is generally protected when it is the product of its maker's intellectual effort; that is, when it is not copied from another map. This includes original survey maps, as well as original compilations of preexisting maps.

Thus, although the factual information portrayed on a copyrighted map is generally not protected, the copyright protects against certain methods of appropriating the information, such as photocopying. However, another map compiler may independently collect the information and produce another map. The original copyright is not infringed, even if the latter mapmaker was borrowing an idea from the copyrighted map.

GIS and Map Copyrights

The advent of digital maps creates new dilemmas. A GIS database is a collection of spatial and nonspatial attribute data. The data can be used to display or plot a map, but the database itself is merely a list of geometric elements (i.e., nodes, lines, and areas) and their attributes, referenced to some spatial coordinate data system. Under traditional copyright law, these data are considered "compilations," rather than "maps."

On the other hand, if a copyrighted pictorial map is digitized and stored in a GIS database, it would not lose its copyright protection. A "copy" is any representation of the work from which the work may be viewed or reproduced, either directly or with mechanical assistance. Consequently, a digital representation of the copyrighted map is simply a copy of the original pictorial map.

However, many GIS databases include data that was not derived from pictorial maps. It is difficult to find a copyright classification for such a collection of data other than "compilation," and as such they do not enjoy copyright protection. Moreover, data stored in an electronic database have no discernible "arrangement" that can serve as the basis for copyright originality. Therefore, when GIS data are collected according to some standard classification scheme, copyright protection may not apply. In that case, anyone having access to the data may legally download all of it without liability for copyright infringement.

The Courts and Map Copyright

The 1866 English directory case of Kelly v. Morris provided the foundation for most twentieth-century map cases. The court reasoned that there is one objectively verifiable reality that the mapmaker must present in his work: "there are certain common objects of information." Therefore, each cartographer who does his job properly will reach the same result: "certain common objects of information... must, if described correctly, be described in the same words." Because there is no originality in the object or presentation of the words, protection goes only to the originality of the effort: "(i)n the case of a roadbook, he must count the milestones for himself."

Kelly v. Morris was later quoted in the 1922 case, Jeweler's Circular Publishing Co. v. Keystone Publishing Co., which is the principal authority for the leading "modern" cases on maps. For instance, in 1951, Amsterdam v. Triangle Publications, Inc., clearly set forth the "direct observation" rule, which held that maps would be protected only for the labor involved in their creation. The statement in Kelly v. Morris that "(the cartographer) must count the milestones for himself" was echoed in the court's 1951 finding that maps are protected "only when the publisher of the map in question obtains originally some of the information by the sweat of his own brow."

One of the few cases to give any significant attention to the pictorial, graphic form of maps occurred more than two decades later. United States v. Hamilton was a criminal prosecution for copyright infringement, decided under the 1909 Copyright Act, and thus prior to the 1976 Act that classified maps as "pictorial, graphic, and sculptural works." In Hamilton, the court rejected the "sweat of the brow" approach of the Amsterdam case and held that the synthesis of terrain features from prior maps was an "element of originality" entitling the map to copyright protection, independent of any "direct observation" by its creator.

Hamilton recognized the similarity of cartography to other creative activity and drew an analogy between mapmaking and photography, noting that a photographer's "selection of subject, posture, background, lighting, and perhaps even perspective alone" are granted copyright protection. Hamilton thus found that a "(s)imilar attention" should be given the cartographer's art: "the elements of authorship embodied in a map consist not only of the depiction of a previously undiscovered landmark or the correction or improvement or scale or placement, but also in selection, design and synthesis... Expression in cartography is not so different from other artistic forms seeking to touch upon external realities that unique rules are needed to judge whether the authorship is original."

However, a 1991 Supreme Court ruling in the case of Feist Publications, Inc. v. Rural Telephone Service Co. emphasized that the facts contained in a compilation of data can never be original. The high court rejected the "sweat of the brow" or "industrious collection" doctrines as a basis for protecting compilations under copyright laws. The fact that Feist rejected the "sweat of the brow" doctrine as the basis for the protection of fact-based works raised special questions concerning the future copyright protection of maps.

Although the Feist ruling killed copyright protection for compilations based solely on the efforts of the compiler, in 1992 the Fifth Circuit Court nevertheless held in Mason v. Montgomery Data, Inc., that maps are pictorial, graphic works that should be protected for their expressive transformation of facts. The court found that the plaintiff's maps also had sufficient creativity to warrant protection "as pictorial and graphic works of authorship."

Although acknowledging that courts have historically treated maps "solely as compilations of facts," the court took note that the 1976 Copyright Act categorizes maps not as factual compilations but as "pictorial, graphic, and sculptural works." Citing Hamilton, the court held that maps, unlike factual compilations, "have an inherent pictorial or photogenic nature that merits copyright protection."

Conclusion: "Confusion"

The Mason ruling upheld the mapmaker's copyright, apparently answering the question of the viability of map copyrights following the Feist decision. Nonetheless, there is a definite distinction between the hard-copy maps involved in the Mason case and the digital map data stored in a GIS. This leaves an important question unanswered, "Is a GIS database a map deserving of copyright protection or merely a compilation of factual data with no such protection?"

Dennis J. Karjala, Professor of Law at Arizona State University, writes, "'Confusion' is the best description of the state of post-Feist copyright protection of maps, especially digitized geographic information systems. The lower courts are struggling to avoid the strictures of Feist, because they see that costly and economically valuable products are vulnerable to misappropriation. These courts cannot directly contradict Feist, and they do not wish to contradict other long-standing principles of copyright... Courts cannot simultaneously follow Feist, remain faithful to these long-standing copyright principles, and protect all the works they wish to protect."

Clearly, new legislation is needed that specifically addresses the copyright protection of GIS data. This legislation must go beyond the traditional thinking that a map is a visual representation of selected geographic and other data.

GIS and the Practice of Land Survey

State legislatures have chosen to enact laws regarding the practice of engineering and land survey because both professions affect public health, safety, and welfare. This is true of many other trades, including law, medicine, and real estate sales. The law typically restricts the practice of those trades to licensed professionals who have met specified minimum requirements for education and experience, and have passed one or more required formal examinations.

The National Council of Examiners for Engineering and Surveying (NCEES) promotes legislation and policies regarding the practice of professional engineering and land survey. The following illustration shows the definition of the practice of land survey presented in the NCEES Model Law on the Licensure of Land Surveyors. The wording of this model law has generally been adopted by the states.

National Council of Examiners for Engineering and Surveying (NCEES)
Model Law

CHAPTER II. LICENSURE OF LAND SURVEYORS
SECTION 2. DEFINITIONS

"**(4) Practice of Land Surveying** ... The practice of land surveying includes, but is not limited to, any one or more of the following:

(a) Locates, relocates, establishes, reestablishes, lays out, or retraces any property line or boundary of any tract of land or any road, right of way, easement, alignment, or elevation of any of the fixed works embraced within the practice of land surveying.

(b) Makes any survey for the subdivision of any tract of land.

(c) Determines, by the use of principles of land surveying, the position for any survey monument or reference point; or sets, resets, or replaces any such monument or reference point.

(d) Determines the configuration or contour of the earth's surface or the position of fixed objects thereon by measuring lines and angles and applying the principles of mathematics or photogrammetry.

(e) Geodetic surveying which includes surveying for determination of the size and shape of the earth utilizing angular and linear measurements through spatially oriented spherical geometry.

(f) Creates, prepares, or modifies electronic or computerized data, including land formation systems, and geographic information systems, relative to the performance of the activities in the above described items (a) through (e)."

Definition of the practice of land surveying.

In essence, this law defines land survey as pertaining to the determination of property boundaries, rights of way (ROW), easements, property subdivisions, survey monuments, topography, the location of fixed objects on the ground, and geodetic data, as well as the creation, preparation, or modification of computerized data relative to the performance of these activities. Now look at the types of data typically stored in a GIS. The example shown in the following illustration is a listing of GIS themes maintained by Prince William County, VA.

BOUNDARIES
Census Areas
Chesapeake Bay Intensely Developed Areas
Council of Governments Areas
County Parks
Inspection Districts
Political (Precinct, Town, City, County, State, Federal)

BUILDINGS
Building Footprint
Public Facilities

CADASTRAL
Comprehensive Plan Future Land Use
Easements
Parcels
Right-of-way
Subdivision Boundaries
Zoning
Assessments Splits and Consolidations
Tax Levy Areas

CULTURAL
Cultural Sites
Historic Districts
Land Use
Landmarks
Miscellaneous Structures
Recreational

GEODETIC
Control Monuments
Virginia State Plane Grid (1927)
Virginia State Plane Grid (1983)

NATURAL
Chesapeake Bay Resources
FEMA Flood Hazard Areas
Hydrography
Topography
Vegetation
Watersheds
Wetlands

NETWORKS
Streams
Street Centerline

TRANSPORTATION
Air Traffic Corridors
Airfields
Bridges
Airfields
Parking Lots
Railroads
Roads (Edge of Pavement)
Trails and Bikeways
Traffic Analysis Zones

UTILITIES
Sanitary Sewer
Storm water Management Inventory
Water Supply
Water Supply Service Levels

GIS themes maintained by Prince William County, VA.

Typical GIS Layers

A number of these GIS layers represent data that clearly fall within the definition of land survey, including property, ROW, and political boundary data (collectively these are commonly called the "cadastral layer"), as well as building locations, topography, and survey control points. However, there are also a number of data layers that clearly fall outside this definition, including census areas; a variety of regulatory districts; land use and tax assessment areas; environmental, natural, and recreational resources; and vegetation, watersheds, wetlands, and streams.

Moreover, the previous illustration pertains to a county GIS. Consider the GIS used by a corporation to manage sales territories or locate retail outlets, a GIS created by a scientist to support her research, or the GIS used by a politician to help plan campaign strategy. Clearly, the law could not be interpreted to relate to all data in any GIS.

However, note that the definition of the practice of land surveying in the NCEES Model Law includes this activity: "(f) Creates, prepares, or modifies (GIS data), relative to the performance of (land surveying)." There seems to be no question that electronic data (COGO, GIS, or otherwise) that is created, prepared, or modified relative to the performance of land survey must be done so by or under the supervision of the licensed professional.

Entering Land Survey Data into a GIS

On the other hand, suppose a recorded subdivision plat has been referenced to the state plane grid. Suppose further that a GIS technician uses that system's commands to enter the parcel geometry shown on the plat into the GIS, placing lot corners at grid coordinates shown on the plat. Is that process of data entry or the display of that data in a GIS "land survey"? For that matter, suppose the land surveyor delivers a COGO file of the recorded subdivision to the government agency, which then reads the data directly into the GIS.

The GIS automatically builds the parcel geometry and places it in the state plane coordinate system. Is that "land survey"? Personally, I would say no. The professional land surveyor has already performed all of the measurements and calculations and made his determinations, and the GIS is simply displaying the results of that work.

On the other hand, suppose the recorded subdivision plat has not been referenced to the state plane grid. Instead, the land surveyor used an assumed local coordinate system. Now, suppose a technician uses GIS commands to enter the parcel geometry from that subdivision plat into the GIS. To place that subdivision within the state plane grid, she then uses GIS commands to translate and rotate the parcel geometry of the entire subdivision to fit available reference information.

This reference data might be features on a planimetric map or digital orthophoto already in the GIS. In effect, the GIS technician has made a determination of the location (i.e, the coordinate values) of the parcel corners within the state plane grid. This is something the surveyor had not done. Does that process of data adjustment in the GIS fall within the legal definition "land survey"? I would say absolutely yes.

Moreover, I expect a professional land surveyor would have executed the process differently. He would prefer to use accurate coordinates, probably surveyed in the field using GPS, to ascertain the coordinates of two or more property corners of the subdivision. He could then confidently locate all of the property corners in the state plane grid.

Managing Land Survey Data Stored in a GIS

There is, however, some debate over the question of whether or not the entry, storage, and display of such data in a GIS should be done so by or under the supervision of a professional land surveyor. Some hold that all of the data created, prepared, or modified in connection with land survey and stored in a GIS should be the result of work performed by a licensed, professional land surveyor. As you might expect, many professional land surveyors support this view.

Others argue that simply including "land base" data in a GIS does not fall within the legal definitions of the practice of land survey. Instead, they hold that the GIS data simply makes reference to the official land records, and that this action does not constitute land survey. They draw a distinction between land data that has been developed during the performance of land survey and the same land data contained in a GIS database.

Therefore, they argue that public agencies should not be required to place GIS data under the supervision of professional land surveyors. Again, as you might expect, many people who manage GIS databases but are not professional land surveyors support this argument. URISA has created the web site *www.urisa.org/gispolicy.htm* to support a dialog between the two sides in this debate, hoping to resolve it to the satisfaction of both.

Summary

In regard to GIS data, information, and technology, this chapter has taken a look at the right to public information access, data privacy law, liability, and copyright protection. Although maps have always been protected under copyright, the advent of digital maps has created confusion and questions regarding specific points of the law. Similarly, questions, confusion, and even some controversy surround the issue of regarding the practice of land surveying and the use of GIS. These questions continue to be sorted out, possibly calling for new legislation.

Chapter References

GIS Law, Volume 1, Number 4, GIS Law and Policy Institute, Harrisonburg, Virginia, (540) 434-3307.

GIS Law, Volume 2, Number 1, GIS Law and Policy Institute, Harrisonburg, Virginia, (540) 434-3307.

GIS Law, Volume 2, Number 2, GIS Law and Policy Institute, Harrisonburg, Virginia, (540) 434-3307.

"GIS and Privacy," Harlan J. Onsrud, November 1993, GIS/LIS Proceedings, ACSM/ASPRS, Bethesda, Maryland, (301) 493-0290.

Jurimetrics: Journal of Law, Science and Technology, Volume 35, Number 4, Summer 1995, Section of Science and Technology, American Bar Association and the Center for the Study of Law, Science and Technology, Arizona State University College of Law, Tempe, Arizona.

"Legal and Liability Issues in Publicly Accessible Land Information Systems," Harlan J. Onsrud, 1989, GIS/LIS Proceedings, ACSM/ASPRS, Rockville, Maryland.

Legal Aspects of Geographic Information Systems, Steven G. Wright, Esq., September 1993, Isaacson, Rosenbaum, Woods & Levy, P.C., Denver, Colorado.

Government and Industry Involvement in GIS

How States and Associations Are Promoting GIS

In This Chapter. . .

State governments and their agencies are not just using geographic information system (GIS) technology. They are actively setting GIS policies and influencing GIS usage, not only within their own state but at the federal and local levels. Likewise, a number of professional and industry trade associations have shown great interest in GIS. These groups recognize that GIS is becoming more important (and in some cases vital) to their membership. Their members, be they individuals or organizations, are asking them to provide information, education, initiatives, standards, and other types of leadership relative to GIS.

State Government GIS Activities

This section first takes a look at an organization of states that is focused on GIS, the National States Geographic Information Council (NSGIC). This is followed by a discussion of typical areas and examples of state-level GIS activity.

National States Geographic Information Council

According to the NSGIC, its purpose is "to encourage effective and efficient government through the coordinated development of GIS and related technologies to ensure that information may be integrated at all levels of government." Members of NSGIC include delegations of senior state GIS managers from across the United States. Other members come from federal agencies, local government, the private sector, academia, and other professional organizations. NSGIC includes nationally and internationally recognized experts in GIS and information technology (IT) policy, and is governed by a nine-member board.

GIS has many applications in state government, including economic development, delivery of health and human services, environmental protection, facilities management, taxation, education, emergency government, and transportation. NSGIC recognizes that the potential benefits of GIS technology and data can be fully realized only through intergovernmental and private sector cooperation, coordination, and partnerships. NSGIC's efforts and focus include the areas discussed in the following sections.

Policy

The NSGIC provides a unified state voice on GIS issues, advocates state interests, and supports the membership in their individual initiatives. The NSGIC reviews legislative and agency actions, promotes positive legislative actions, and provides advice to public and private decision makers. The NSGIC has also influenced the development of GIS policy on a national level. NSGIC members have served on a variety of federal task forces and working groups relative to the National Spatial Data Infrastructure (NSDI). The NSGIC also influences policy by providing speakers and education to help states form GIS policies.

Liaison and Networking

The NSGIC promotes interaction and cooperation among its members; among federal, local, and regional governments; and among professional associations and public and private sector groups. NSGIC publishes a quarterly newsletter to keep members abreast of its activities and developments. The newsletter provides a forum for state activities, technical issues, and general interest. The Council also maintains a bulletin board, accessible via the Internet and modem. The bulletin board provides internal communication and posting of articles of interest.

Research

The NSGIC studies and provides a forum for examining GIS issues. The NSGIC also provides resources and personnel that facilitate the research and testing of GIS concepts, applications, policies, and coordination mechanisms. The NSGIC has conducted several surveys, and has written issue papers and proposals in a number of technical and policy areas.

Education and Public Relations

The NSGIC develops, and helps others develop, a variety of educational programs and materials to promote discussion of GIS management and integration. The NSGIC annual conference is an educational program, as well as a working session, where the Council develops policy, works on technical issues, and provides in-depth analysis of issues and opportunities.

One example of the NSGIC's activities is a cooperative project with the Federal Geographic Data Committee (FGDC). The NSGIC is developing information about the status of data that can contribute to the NSDI. The first phase of the project will provide an initial "snapshot" of data available from state, regional, and local governments throughout the United States that could contribute to the NSDI.

The second phase of the project will develop and test guidelines for developing and maintaining ongoing inventories of data, and for reporting this information through the National Geospatial Data Clearinghouse. Ultimately, the NSDI will provide several basic themes of data commonly needed to support GIS applications, including geodetic control, elevation and bathymetry, digital orthoimagery, transportation, hydrography, boundaries of governmental units, and cadastre.

The NSGIC has likewise compiled a State Agency GIS Managers Directory that provides the name, address, phone, fax, and electronic mail information for state officials leading the GIS activities in their respective state agencies. NSGIC membership is open to individuals, public sector and nonprofit organizations, and commercial companies. Contact information follows.

National States Geographic Information Council
167 W. Main St., Suite 600
Lexington, KY 40507
Phone: 859-514-9208
Fax: 859-514-9188
E-mail: *nsgic@amrinc.net*
Internet site: *http://www.nsgic.org/*

State GIS Activities

The NSGIC has surveyed its members on their GIS activities. The resulting report, "NSGIC State Summaries," describes the principal GIS activities of each state and lists its GIS coordinating agency and points of contact. Although the full report is available only to NSGIC members, the names and contact information for state GIS coordinators are available on the NSGIC web site. State GIS coordinating bodies typically consist of representatives from state agencies that rely on GIS technology and data, as well as representatives from local government, academia, related professions, and private industry.

In general, statewide GIS activities include those discussed in the sections that follow. Traditional GIS applications within a specific state agency (such as the preparation of emergency response plans, the creation of a state highway map, bus routing analysis, and so forth) are not included.

Planning

Many states are preparing strategic plans for the implementation of statewide GIS programs and databases. These plans usually describe statewide mapping and GIS needs, set forth goals, and present an action plan for achieving these goals.

Coordination

Most states are taking an active role in the sharing of statewide GIS databases, programs, and resources. They are also focused on eliminating duplication of efforts in collecting and maintaining GIS data.

Legislation

Many state agencies and statewide GIS coordinating agencies are making policy recommendations to their state lawmakers that will produce or influence state legislation on issues related to GIS. GIS legislative issues typically include the privacy, public access, copyright, and sale of GIS data. Many state GIS coordinating bodies were created and funded by public law. State legislatures have also funded grant programs and created grant program boards to promote GIS technology.

Standards

Many states are developing and implementing standards for state agencies to follow in developing and sharing GIS databases. These standards typically address the definition, format, map projection, content, and conversion of GIS data sets. They also address issues of GIS data accuracy.

GIS Data Creation

Where certain GIS data themes are not already available, some states are funding programs aimed at creating new statewide GIS data sets, usually derived from satellite imagery or high-altitude aerial photography. Data sets typically include satellite imagery, planimetric and orthophotographic base maps, land use data, land cover data, wetlands data, and so forth.

GIS Data Clearinghouse

Many states are funding the creation and maintenance of data clearinghouses (also known as GIS libraries or data warehouses). The clearinghouse may be a central site for researching and acquiring GIS data, or simply references to distributed sources of GIS data. This process usually begins with a survey of government agencies to locate appropriate GIS data sets, followed by an effort to collect and maintain appropriate data. These clearinghouses usually include metadata (i.e., "data about the data" that describes the source, date of creation, accuracy, and so on) for each data set.

Education and Information Exchange

To increase the general awareness of and support for GIS, most states are sponsoring statewide GIS conferences, forums, and seminars. Likewise, they are publishing GIS newsletters, reports, web sites, and job openings, as well as surveys of GIS usage. They are likewise publicizing GIS training centers and college courses, as well as setting curriculum standards. Some states are sponsoring GIS laboratories in public schools and colleges.

Partnerships and Demonstration Projects

Many states have entered into various types of partnerships to achieve the foregoing activities. Their GIS partners include other states, federal agencies, local governments, utilities, and the private sector. Some states have funded pilot demonstration projects within both state agencies and local governments, whereas others have won federal grants to support their GIS activities.

Associations Promoting GIS

The sections that follow describe the GIS-related activities of the following associations.

- American Public Works Association (APWA)
- National Association of Counties (NACo)

- National Association of County Surveyors (NACS)
- Urban and Regional Information Systems Association (URISA)
- Women Executives in State Government (WESG)

American Public Works Association (APWA)

The APWA is a voluntary association of public and private sector professionals. Most of its 25,000 members work in city, county, and state governments or private companies that provide public works services. APWA has 67 chapters in the United States and Canada.

One of the primary organizational objectives of the APWA is to provide information, materials, and education programs that serve the needs of individuals, firms, and agencies delivering public works services. This objective is accomplished through publications, satellite videoconferences, and technical sessions at annual meetings. The APWA has used all of these mediums to provide information on GIS.

For example, the publication *Implementing Successful Geographic Information Systems* was developed through the cooperative efforts of 16 member agencies of the APWA, the Quebec Chapter of APWA, the American Gas Association, and the American Water Works Association. This publication serves as a basic guide for agencies interested in implementing GIS, and includes chapters such as "Organizing for Success," "Resolving Implementation Issues," "Making GIS Happen," "Maintaining the System," "Sample GIS Applications," "Case Studies," and "Examples of GIS Products." In addition to this publication, the APWA publishes two to three articles per year on GIS in its monthly magazine *The APWA Reporter.*

Another example occurred in 1997 when the APWA sponsored a satellite videoconference titled "GIS Infrastructure Management for Public Works and Utilities." It was viewed live by an audience estimated at more than 6,000 public works professionals. Tapes of this program were produced for those who were not able to view it live, which are still available. In addition, since 1995 seven technical sessions relating to GIS implementation and applications have been presented at the APWA's annual meetings. Each of these technical sessions has been exceptionally well attended, with "standing room only" crowds. Contact information follows.

John McMullen
American Public Works Association

2435 Grand Boulevard
Suite 500
Kansas City, MO 64108
(816) 472-6100
www.pubworks.org

National Association of Counties (NACo)

NACo was created in 1935 to represent the nation's counties in Washington, D.C. It is a full-service organization that provides legislative, research, technical, and public affairs assistance to its members. The association acts as a liaison with other levels of government, works to improve public understanding of counties, and serves as a national advocate for counties, providing them with resources to help them find innovative solutions to the challenges they face.

County officials and their staff also participate through NACo's 29 affiliate organizations and caucuses. These organizations consist of officials who share similar responsibilities, interests, or knowledge areas. NACo's membership totals nearly 1,800 counties, representing over 85 percent of the nation's population.

To keep county officials informed, NACo publishes a biweekly newspaper, with a circulation of 26,000, that focuses on issues and actions in Washington, D.C., and throughout the country. NACo has also developed an Information Technology Center, which keeps county officials abreast of changes in advanced technologies.

NACo is involved in a number of special projects that deal with such issues as the environment, sustainable communities, volunteerism, and intergenerational studies. NACo's research department ensures that county leaders have access to related data and statistics. NACo's committees, task forces, and focus groups evaluate issues and policies.

This policy development process leads to the publication of the *American County Platform*, which NACo uses as a guide to deliver county governments' message to the nation. NACo has two additional committees, the Member Programs and Services Committee and the Information Technology Committee, which make recommendations on programs and projects for NACo to pursue that will help counties.

NACo is currently focusing on GIS as a decision-making support tool for the next decade. It provides a "GIS Starter Kit" to counties as a stimulus to accelerate their use of GIS. NACo has also established a GIS committee, and NACo members sit on the Federal Geographic Data Committee (FGDC) committees, and participated in FGDC hearings on the National Spatial Data Infrastructure Strategy (NSDI) last year.

The NACo GIS Committee meets at its annual and legislative conferences. These conferences have featured numerous GIS workshops and training sessions. NACo is also working with the NSGIC to conduct a nationwide survey of GIS use in county government. Contact information follows.

Director for Information Technology Services
National Association of Counties
440 First Street, N.W.
Washington, D.C. 20001
(202) 393-NACO
www.naco.org

National Association of County Surveyors (NACS)

The NACS is an affiliate of NACo and was established in 1989 to represent county surveyors across the nation. The county surveyor is a vital link in the development and maintenance of GIS. The NACS has focused its attention on promoting GIS at the local, state, and national levels.

Working with NACo through their committees and on their board of directors, the president of the NACS acts as a delegate. One area of involvement is providing the voice and position of the county surveyor, on behalf of all counties, to federal subcommittees connected with the building of the NSDI. Members of the NACS are currently involved on the Cadastral, Bathymetric, and Geodetic Control subcommittees of the FGDC. These memberships provide vital county surveyor points of view from across the nation.

The NASC has also joined with the National Society of Professional Surveyors (NSPS) in a collaborative effort to enhance the County Surveyors Forum, held during the annual American Congress of Surveying and Mapping (ACSM) conference. The NASC has moved its annual meeting to coincide with ACSM annual conference. The Association conducts its necessary business and provides, along with NSPS, an entire day of seminars pointed directly at the business and professional needs of all county surveyors. The NACS is a young association and is continually looking for new ways of promoting one of the oldest and most critical governmental positions, the county surveyor. Contact information follows.

Vaughn Butler, Past President
(801) 468-2028
R. Charles Pearson, President
(503) 640-3405

Urban and Regional Information Systems Association (URISA)

URISA was one the nation's first industry associations to promote GIS, and is a leader in this area. Self-described as "The association for spatial information management professionals," URISA's mission is to facilitate the use and integration of spatial information technologies to improve the quality of life in our urban and regional environments. Its members are from many disciplines and are dedicated to advancing the effective application of information technology in decision making.

To carry out this mission, URISA provides high-quality professional education, fosters communications among spatial information resource management professionals, and encourages a multidisciplinary approach to the design and use of urban and regional information systems.

URISA members come from the public and private sectors, as well as research, academic, and educational institutions. It also relies on other organizations that routinely interact with public organizations to provide better products and services to the community, and other professional organizations whose missions complement that of URISA.

URISA relies on the active participation and volunteer activities of its members and is governed by a board of directors that provides leadership, vision, and policy direction. It is advised by an Executive Committee, a Strategic Planning and Policy Committee, and an Presidents' Advisory Council. URISA is structured into operating divisions, each directed by a URISA board member for effective management and operation. The Education Division focuses on educational product development, delivery, and quality control.

The Publications Division provides high-quality journals, reports, books, newsletters, and web-based information. The Operations Division oversees important operational functions, including membership, awards, elections, and external relations. The Outreach Division oversees programs and communication with URISA chapters, Special Interest Groups (SIGs), industry, the public, and government. URISA is supported by a professional executive director and staff.

URISA's chapters support organizational activities within smaller geographic areas. URISA chapter responsibilities include the development of programs and activities that advance URISA's mission and principles at the local, state/provincial, and regional levels; participation in the development of URISA educational products and services; identification of marketing opportunities and potential partnerships within their regions for URISA products and services; and coordination of chapter activities with the URISA Board of Directors through the Chapter Relations Committee.

URISA SIGs encompass a particular field of interest within the spatial information community. SIG responsibilities include participation in the development of URISA educational products and services, identification of marketing opportunities and potential partnerships within their field of interest for URISA products and services, and coordination of SIG activities with the URISA Board of Directors through the SIG Relations Committee.

URISA's goals include establishing a sustained strategic planning and policy process. URISA also hopes to expand the number and types of its program products and services, build stronger relationships with the vendor community, and expand partnerships with other information technology organizations. It will also launch a local/regional level workshop program, hold two to four regional conferences, and launch a partnered seminar program. It also plans to deliver two to four new publications and leverage current materials for other educational uses. Contact information follows.

David Martin, Executive Director
Urban and Regional Information Systems Association
1460 Renaissance Drive
Suite 305
Park Ridge, IL 60068
(847) 824-6300
www.urisa.org

Women Executives in State Government (WESG)

The National Leadership Conference of Women Executives in State Government (WESG) is a national organization for women serving in the highest-level appointed and elected positions in the executive branch of state government. WESG, formed in 1983, provides appointed and elected leaders of state government agencies with a network for discussing state governance issues, sharing perspectives as leaders and managers, seeking advice, and exchanging innovative ideas on management and leadership. The WESG promotes excellence in government through innovation and public-private partnerships, and by helping develop future leaders.

As part of its mission, the WESG focuses on technology as a key aspect of innovation in state government. For example, the WESG's upcoming annual conference in September will focus on technology and government. Conference participants will have the opportunity to learn more about GIS directly from several GIS vendors who are WESG corporate sponsors.

In addition, many of the WESG's members are using new technologies to improve service delivery and to make their state agencies run more effectively, and they are sharing this valuable information with other WESG members. For instance, WESG member Sharon Priest, the Secretary of State for Arkansas, is currently working to implement GIS technologies within her office. The Secretary of State's office maintains Arkansas' legislative boundaries, a task which has in the past been done manually. A digital methodology for creating and updating this information is being developed to make this process less time consuming and more efficient.

Likewise, the Executive Office of Environmental Affairs (EOEA) of Massachusetts, headed by WESG member Trudy Coxe (EOEA Cabinet Secretary), has made GIS technology an integral part of how the agency operates. MassGIS, developed by EOEA, offers digital data distribution and map production services.

These services are available to both the public and private sectors. MassGIS develops and distributes environmental data that is shared with other state and local agencies such as the EPA, regional planning agencies, and the highway department, as well as many nonprofit organizations. MassGIS is currently developing digital orthophotographs that can be used to extract information on wetlands and streams. Contact information follows.

Jane Moya, Director of Member and Public Relations
Women Executives in State Government
1225 New York Avenue NW, Suite 350
Washington, DC 20005
Phone: (202) 628-9374
www.wesg.org

Summary

This chapter has reviewed governmental and other organizations that are leaders in the promotion of the range of GIS technologies. These organizations work individually and in coordination to disseminate information and to help users manage their involvement in this field. Readers are encouraged to become aware of their state's GIS activities and to join one or more of these associations, as appropriate to their work, and to get involved in their GIS-related activities.

A Case Study in GIS Implementation: Clinton Township, MI

How a GIS Established for One Application Wound Up Benefiting an Entire Organization

In This Chapter...

This chapter presents a case study in the implementation and use of GIS at the Township of Clinton, Michigan. The information was provided by Spalding, DeDecker & Associates, Inc. (SD&A), of Madison Heights, Michigan. Jennifer Scrutton of SD&A and Bruce White of CMD & Associates, Inc., provided the information that follows. CMD & Associates is the Generation 5 Technology dealer for Michigan and has worked closely with SD&A and the township throughout the GIS project.

This medium-size Michigan Township used a consultant to help with the implementation of its GIS. The initial application was to support the municipal water and sewer system operations. The system produced many unforeseen benefits in other departments as well.

The Consultant

SD&A is a consulting engineering and surveying firm that has provided services to communities throughout Michigan since 1954. The firm offers services in site planning, civil engineering design, and surveying, as well as construction engineering and administration. SD&A was founded by Vernon Spalding and Frank DeDecker and remains a privately held corporation. The firm employs 60 persons at its Madison Heights headquarters.

SD&A provides many of its services to municipal customers. These include site plan review and the design of streets and water, sewer, and storm water management systems, as well as computerized mapping of utilities, streets, and tax parcels. SD&A also provides surveying, planning, design, and construction administration for highways, parking facilities, bridges, and railroad spur lines. In addition, the firm provides land development services, including feasibility studies, site planning, engineering design, and land surveying. SD&A employs GPS surveying on many of its projects.

SD&A formed its municipal mapping department ten years ago, when it first set out to automate survey data collection and plotting. Since then the department has provided GIS services to the Townships of Clinton and White Lake and the City of Berkley.

The Client

The Charter Township of Clinton is located in south central Macomb County, Michigan, northwest of Detroit. It is the tenth largest township in the state and is the state's most heavily populated township. It covers 28 square miles of gently rolling land and includes approximately 32,000 tax parcels.

During the past three decades, Clinton Township has grown from a semi-rural community of 25,000 to a progressive, active Detroit suburb with more than 90,000 citizens. About 80 percent of the township is now developed. Nearly half of this development is residential, but commercial, medical, and professional office development has surged in recent years.

The Initial Applications

SD&A has been the township engineer for Clinton Township for the past 40 years. The firm first approached the township in 1981 with the idea of developing a GIS. The project began in 1989 with funds made available in the budget of the Water & Sewer Department. In each year since, the township has made funds available to expand the GIS. The work was done by SD&A under its contract as the township engineer. Several initial meetings were held with key water and sewer officials to establish plans and expectations for the GIS project. The plan was to create a comprehensive, accurate map of Clinton Township's tax parcels, water system, and sewer system.

SD&A began the GIS project by digitizing a variety of road right-of-way maps to create a digital base map of the township. Property lines were then digitized from a variety of sources, including tax maps, certificates of survey, recorded plats, right-of-way maps, and field survey notes. The property lines were registered to the right-of-way data on the base map. Street and waterway names, subdivision names, and lot frontage distances were added as text.

All of this map digitizing work was done in AutoCAD (Autodesk, Inc., Sausalito, CA). After each map was digitized, it was loaded into a seamless spatial database using Geo SQL (Generation 5 Technology, Inc., Denver, CO). Therefore, users could then access data and create new maps in a single step without the need for combining the information from hard-copy maps. Geo SQL requires AutoCAD for its graphics "engine."

Parcel identification numbers were entered as attribute data attached to tax parcels. Parcel attribute data was also extracted from the township's tax assessment database. This data was loaded into an Oracle database and linked to parcel graphics in Geo SQL using the parcel identification number. This tax parcel database includes the following attributes for each tax parcel.

- Identification number
- Sidwell number
- Owner's name
- Property address
- Subdivision name
- Lot number

> **NOTE:** *Sidwell numbers represent a standard numbering system Michigan townships and counties use to uniquely identify a parcel. The number is a combination of the township number, section number, block number, and parcel number.*

SD&A added water mains, water valves, water meter pits, sanitary sewer lines, and sewer utility holes to the database by digitizing their position from as-built drawings. These drawings are maintained by SD&A as the township's engineer. SD&A also maintains an Oracle database of attributes for sewer lines and structures. The Township Water & Sewer Department maintains an Oracle database on sewer leads (building service connections). SD&A linked the data in both of these files to the appropriate sewer lines and structures in the graphics database using unique identification numbers. The water and sewer attribute data includes the following.

- Pipe diameter
- Utility hole rim and invert elevation
- Sanitary sewer lead diameter, as-built file number, material type, and year of installation

The following are some of the pertinent statistics for the effort.

- 312 miles of water lines
- Over 7,000 hydrants and water valves
- 345 miles of sanitary sewer lines
- Over 3,000 sanitary sewer manholes
- About 19,500 sanitary sewer leads

The township also uses Geo SQL for utility data management. SD&A customized the user interface to enhance its ease of use by township employees. Four months were needed to add the water and sewer information on top of the base map.

All data and the GIS software were installed on two PCs equipped with digitizing tables and printers. These workstations are situated in the Water & Sewer Department at the Civic Center and in the Water & Sewer Maintenance Building.

SD&A also entered water system taps (building services) into an Oracle database linked to the GIS graphics. This phase began in the Fall of 1991 and was completed in the Fall of 1993. There are 22,500 water services in Clinton Township. The data includes the following.

- Tap number
- Tap size
- Tap length
- Year installed
- Stop box location

Water & Sewer Department utility crews now use the GIS to call up a map of an area that requires repair work. They print out the map to the laser printer in just a few minutes before leaving the office in the morning. They use the map to locate the site and the affected utility lines and structures, note any changes on the map, and then revise the GIS database when they return to the office at the end of the day.

One of the other many benefits of the GIS is that the Water & Sewer Department must collect and add data on the water and sewer systems more carefully. In the past, newly obtained as-built information could be "shoehorned" into open spaces left on existing paper maps or added as free-standing lines. The GIS requires that connections be shown exactly as they occur in the real world. Thus, the data in the system is constantly being checked, corrected, and updated.

The Other Applications

In 1991, other departments became interested in the GIS. SD&A added zoning classification data for the Planning Department. Geo SQL also provides the planning department with spatial analysis capabilities. The GIS software and data were initially installed on PCs in the planning department. A plotter was added to one of these workstations.

When the township is considering zoning changes, the planning department uses the GIS to create and plot custom maps of the affected areas at any desired map scale. The Planning Department can also quickly determine what utility services are available to the area in question by referring to the water and sewer data stored in the GIS.

The Planning Department also gained the ability to automatically generate notification lists for planning and zoning cases. The master notification letter is first prepared in WordPerfect. The GIS user then enters the appropriate parcel number and the system finds all properties lying within 300 feet of this parcel. The GIS then reads the owners' names and addresses from the assessor's database and automatically merges them with the form letter. The entire process of creating notification letters used to require about two days. Now it now takes just a few minutes.

The fire department also took notice of the GIS. SD&A added building footprints to the digital base map between the Fall of 1992 and the Spring of 1993. An attribute indicating whether the building is a government, commercial, or residential structure was tagged to each footprint. SD&A also digitized fire hydrant locations. When the department purchased a new computer-aided dispatch system, it used the locations of buildings and hydrants, as well as street names and addresses, extracted from the GIS database.

Similarly, when the Assessing Department wanted to begin using the GIS, several GIS workstations were placed there. SD&A created a link to the township's assessment database. The Assessment Department was then able to access water and sewer information. In turn, it added debt service data to its assessment records, which the Water & Sewer Department needs.

System Management

GIS users have access to all GIS data on the network and can receive updates to other databases over the network. SD&A and CMD & Associates worked closely with the township and its GIS users to implement the GIS, and provides them with ongoing support. For this reason, the township has not required any new staff or organizational changes to implement and use the GIS. The GIS is managed by the department heads of the respective user departments.

At the outset of the GIS installation, SD&A provided one week of training for water, sewer, and planning personnel. The training covered the use of AutoCAD and Geo SQL and was conducted at the offices of SD&A. CMD & Associates provides introductory training to new Geo SQL users. This training requires approximately two to three hours.

Summary

A significant benefit of the GIS program has been the identification of discrepancies between the township's various databases and files. For instance, the tax parcel records maintained by the township assessor were cross checked against water taps digitized from the water maps. Discrepancies were reconciled and corrected as needed. Similarly, street addresses and hydrant locations checked for the fire department were matched to water and sewer data files. Thus, the GIS has facilitated the correlation of data from a variety of sources. As a result, the township has significantly improved the quality and consistency of its data.

The initial goal of the Clinton Township GIS was to create a comprehensive, accurate map of the township's tax parcels, water system, and sewer system. This has been accomplished and the GIS is currently moving to more advanced applications, such as debt service calculations and utility network analysis. The system has proven to be a powerful tool for improving efficiency, accuracy, and communications throughout the Clinton Township government.

A Case Study in GIS Implementation: Prince William County, VA

How One Municipal Government Went About the Process of Implementing GIS, What It Learned, and How It Has Benefited

In This Chapter...

This chapter describes the implementation and use of a GIS in Prince William County, Virginia. It begins with the history of the county's GIS program. It then describes eight lessons learned during the process, as well as six principal benefits derived from using the GIS. This historically rural county, located on the outskirts of the Washington, D.C., suburbs, began the process of automating its land records in 1984. The action was primarily in response to pressures for rapid suburban land development.

Prince William County, Virginia

Prince William County is located approximately 20 miles south of Washington, D.C. The county covers 348 square miles and contains over 86,000 tax parcels. Today, about 50 percent of the county is rural farmland and open space. Most of the remainder of the county has been developed in just the past 20 years, including a variety of commercial uses, garden apartments, townhouses, and single-family homes. The single-family lots range in size from one-quarter to five acres.

Approximately 3,000 new parcels are added through subdivisions each year. Spill-over from the suburban sprawl of the Washington metropolitan area increased the county's population from under 145,000 in 1980 to nearly 260,000 in 1996, an increase of 80 percent in just 16 years.

History of GIS in Prince William County

The history of the county's GIS can be traced to 1984, when a committee was formed to study land information systems (LIS). One of two key recommendations of the committee was that the county establish an office of mapping, which would report directly to the county executive. This new department would consolidate all of the existing mapping resources throughout the county. Mapping resources had been dispersed across the department of public works, the planning department, and the county's water and sewer authority. This department was established in July of 1985. Chuck McNoldy was hired as the first director of mapping in October of that year.

A few years later, the Office of Mapping was renamed the Office of Mapping and Information Resources. This name change reflected the expanded role of the department that resulted from the use of GIS. It also reflected the fact that GIS was not just a mapping system. It had become a "window" into any and all databases that contain a geographic component, such as a parcel number, street address, or street intersection. The department's name was recently changed again to the Geographic Information Systems Services Group and made a part of the Information Resources Management Division. McNoldy was promoted to director of that division.

The second key recommendation of the LIS Committee was to establish a comprehensive LIS that would serve the needs of all departments and agencies. The LIS was to be a central, comprehensive system for all land data within the county.

Conceptual LIS Design

Consultants specializing in systems planning and analysis were hired to develop the conceptual design for the LIS. They recommended that it consist of two primary components. The first component would be an administrative information system (AIS). This set of computer programs would support those county agencies that heavily relied on alphanumeric data and had little need for graphic data.

These included the finance, public works, health, and police departments. The AIS would consist of four computer programs. The first was a code enforcement module to support plan review, permit processing, and building inspections. The second module would support parcel taxation. The third would support infrastructure management, and the fourth would support a variety of analysis applications.

The second major component of the LIS would be a GIS. This system would support those departments that have a significant need for graphic data. These included the departments of planning, fire and rescue, health, and public works. The GIS would consist of a database that would integrate graphic information and nongraphic records. The GIS would support the input, management, analysis, and presentation of both graphic and attribute data.

The Office of Mapping was given the responsibility for establishing the digital base map and tax parcel overlay for the GIS. Other map themes such as the comprehensive plan, utility data, and environmental data were to be created, updated, and maintained by the user department responsible for that type of information.

Ground Control

Immediately after assuming his duties as the director of mapping, McNoldy began the process of creating a digital base map that would support all other map themes in the GIS. The first step was to establish a geodetic control network. The county had 17 existing monuments established by the National Geodetic Survey (NGS). A professional surveyor was placed under contract to expand this control network to 100 monuments spaced at approximately two-mile intervals throughout the county.

The 83 new monuments were set in the ground during the summer of 1986. The surveyor then brought in another survey firm that specialized in global positioning system (GPS) satellite surveys. This subcontractor established coordinate values for the 83 new control monuments during the winter of 1986, using eight of the existing NGS monuments as reference points. All surveying was done in accordance with the standards of the Federal Geodetic Control Committee. All computations were submitted to NGS for review and comment.

Aerial Photography and Base Mapping

An aerial photography and photogrammetric mapping firm was then selected to create the digital land base. Color aerial photographs at a scale of 1:14,400 were taken in March of 1987. Topographic mapping at a scale of 1 inch = 200 feet, with a 5-foot contour interval, was then compiled from the photography. The mapping was delivered as both traditional hard-copy maps and as digital files.

Hardware and Software

While the photogrammetric base mapping was still underway, the county chose ArcInfo from Environmental Systems Research Institute (ESRI) of Redlands, California, for its GIS software. The first GIS computer was a PRIME 2755 minicomputer located in the Office of Mapping, as it was then known. When the limits of this computer's capacity were approached, it was replaced in 1990 by a PRIME 6450. This second computer was located at the county's nearby data processing center and linked by microwave to the mapping office. The data processing center is operational around the clock and has greater security and fire protection. Moreover, the center's staff is dedicated to computer system maintenance and data security.

In 1997 the PRIME server was replaced by a Hewlett-Packard HP9000 server with 1 Gb of RAM and 100 Gb of hard disk storage, running the HP-UX operating system. GIS users have access to the system on a variety of HP X-terminals, PCs with WRQ's Reflections-X software, and Windows NT personal workstations.

GIS Division Operations

Prince William County's GIS program has evolved over the years from a mapping tool to an enterprise database that is integrated with the tax assessment and other county databases. It is used in the daily decision-making process for county business. The GIS Division in the Office of Information Technology supports user agencies throughout the county, including Fire and Rescue, Police, Public Safety Communications, Real Estate Assessments, Public Works, Economic Development, Human Services, Voter Registration, Service Authority, Park Authority, and Schools.

In its role of support to all user agencies, the GIS Division maintains the enterprise GIS database and provides technical expertise to user agencies. Monthly user group meetings provide an arena for informing county GIS users about upcoming technology upgrades and also allow discussion of data and support needs. The GIS Division has 18 staff members maintaining the core databases and supporting other user agencies.

Constructing the GIS Database

The GIS hardware and software were installed in1987, and the digital base map was delivered by the contractor in 1988. It was then possible to add the tax map overlay. This work was done by in-house personnel in order to maintain the greatest possible control of the production process. The Virginia State Department of Taxation had originally created tax parcel maps for Prince William County (at a scale of 1inch = 400 feet) in 1965.

Many discrepancies between the county's 1 inch = 200 feet topographic maps and the tax maps had been discovered. For instance, parcel map right-of-way lines sometimes fell within paved roads shown on the topographic maps, or lot lines ran through buildings. Therefore, the county decided to compile a completely new set of tax maps from recorded deed descriptions and plats.

During this parcel recompilation phase, the county found that the tax parcel maps, because they had been prepared for tax assessment purposes, showed some parcels that were never actually created by deed. Leased properties had been set up as parcels for the convenience of the tax assessment system.

On the other hand, some separate parcels never actually consolidated by deed were consolidated and shown as one parcel for tax purposes. Non-taxable parcels, such as State Corporation Commission properties, were also not shown. The county compiled its new parcel maps with strict regard to recorded legal interest. A separate layer in the GIS was established to take care of special assessment conditions.

A staff of four people began this process in 1985, and three years were required to complete it. The tax parcel overlay for the GIS was then manually digitized from this new tax map series. The digitizing was accomplished by a team of four people in two years, beginning in 1988.

The Office of Mapping and Information Resources next began automating virtually all mapping operations. Table 27-1, which follows, lists the content of the county's GIS database, including both completed topics and topics currently being added.

Table 27-1: Prince William County Digital Geographic Database

File	*Content*
Boundaries	Census areas Chesapeake Bay intensely developed areas Council of Governments areas County parks Inspection districts Political (precinct, town, city, county, state, federal)
Buildings	Building footprint Public facilities
Cadastral	Assessment splits and consolidations Comprehensive plan for future land use Easements Parcels Right-of-way Subdivision boundaries Tax levy areas Zoning

File	Content
Cultural	Cultural sites Historic districts Land use Landmarks Miscellaneous structures Recreational
Geodetic	Control monuments Virginia State plane grid (1927) Virginia State plane grid (1983)
Natural	Chesapeake Bay resources FEMA flood hazard areas Hydrography Topography Vegetation Watersheds Wetlands
Networks	Streams Street centerline
Transportation	Air traffic corridors Airfields Bridges Parking lots Railroads Roads (edge of pavement) Trails and bikeways Traffic analysis zones
Utilities	Sanitary sewer Storm water management inventory Water supply Water supply service levels

The county also developed a street centerline network file created by digitizing new centerlines from 1 inch = 200 feet county topographic maps and adding left and right address ranges for each street segment using the county's own address records. U.S. Census Bureau demographic data were attached to these centerline segments. Road centerlines are important because they serve as the boundaries for several mapping themes, including zoning, planned land use, school districts, and political districts.

The county currently uses a grid system to assign addresses to new buildings. This assignment is made by visual inspection and interpolation from an address grid map. This street address grid network has been added to the GIS.

Applications

All of the map themes that have been digitized can now be maintained, analyzed, and plotted using the GIS. Specialized thematic mapping is also possible, such as the mapping of capital improvement projects and annexation boundary changes. In addition, the GIS Services Group has been able to provide new services previously unavailable.

Using wetlands information in addition to parcel boundaries, proposed donations of land to the county have been analyzed to determine the percentage of donated land that lies within a wetland. Mailing lists for sending Gypsy Moth spraying notification letters have been prepared by digitizing and buffering the spray blocks and overlaying them with the parcels. The parcel identifier is then matched with information about the addresses of the resident and owner. The GIS parcel overlay has been analyzed to report the acreages of tax parcels and verify them against the county's tax rolls.

The Code Enforcement Module of the LIS was implemented in 1992. The GIS Services Group supplies and maintains 66 data elements on the LIS through a link to the GIS.

Today, the GIS is highly visible throughout the county, both within county offices and in private sector companies, such as engineering, surveying, and planning firms; realtors; title attorneys; and developers. Although the GIS Services Group maintains the core data layers, other departments that use the GIS are learning to maintain their own data layers. These users include the county's Fire, Police, Land Planning, Real Estate Assessment, Transportation, Watershed Management, and Social Services departments, as well as the Registrar's Office and the County Sanitary Sewer Authority. The county's most frequently requested GIS applications and products now include:

- Base Maps
  - Parcel maps
  - Planimetric maps
  - Topographic maps
  - Zoning maps
  - Addresses
  - Assessments-purpose parcels
  - Easements
  - Cemeteries

  - Cultural resources
  - Fire levy districts
- Thematic Maps
  - Police drug-free zones
  - Gypsy moth spray blocks
  - Existing and future land use
  - Magisterial district maps
  - Zoning
  - Comprehensive land use
  - County land
  - Storm water management systems
  - Economic development opportunities
- Natural Resources Maps
  - Wetlands
  - Forest types
  - Chesapeake Bay resource preservation areas
  - Water bodies
- Map Overlay and Analysis
  - Comprehensive plan land-use acreage
  - Watershed area analysis
  - Impermeable-land calculations
  - Redistricting
  - Demographic analysis
  - Facility siting
  - Fire and rescue response time analysis
  - Public safety incident and location mapping (burglaries, accidents, and so on)
  - Police street map books
  - Computer-aided dispatch streets
  - School assignment file update

- Social services client analysis
- Community services analysis
- Pre-incident planning
- Transportation planning
- School redistricting
- School bus routing

GIS Web Application

In support of its 1998 Information Technology Strategic Plan, the county looked at services that might be provided to customers using the World Wide Web. Because most county business relates to geography in some manner, and because an examination of how customers were accessing geographic information indicated that improvements were possible, it seemed appropriate to develop a web application that would provide geographic information to customers at their convenience.

The county found that even with the GIS in place, citizens were still required to purchase hard-copy maps in order to access the information. They were often sent between the Mapping Office and other offices, and the maps they purchased were quickly outdated and obsolete. Moreover, access to maps and geographic information was only available during county business hours, and customers were required to adjust their schedules to fit that time period.

Even within the county, access to the county's GIS proved formidable for many county agencies. Due to lack of funding for GIS hardware and software purchases, and the limitations of the county's network, these agencies were forced to rely on hard-copy maps that quickly became outdated. Moreover, the agencies that were able to fund access to the GIS found that a dedicated staff person was required to learn the software and to operate it. Some agencies purchased less sophisticated desktop GIS software, which gave them the information they required, but retrieval of GIS data over the network was slow, and the software was still complicated to use.

Therefore, the GIS Division developed the "County Mapper" web application to provide citizens and county staff with intranet access to the GIS data and associated information. The County Mapper is based on ESRI's Map Objects IMS. The information on County Mapper is refreshed weekly via an automated process. Because the County Mapper is maintained on a separate map server, the process reads graphic data from the county's GIS, converts it to the appropriate format, and merges it with data tables stored in the GIS and additional tables stored in the Real

Estate Assessments database. The new files are sent to the intranet map server, where they replace the existing data.

To ensure the County Mapper's usefulness, a short training session is provided to county staff. Outside customers receive instructions and support either by coming to the mapping office or by calling for help. Detailed user help notes are also available on the web pages to help users understand the application. A mapping tutorial was developed to teach school-age children about maps and about GIS. The tutorial contains four simple exercises that walk through the steps for accessing data using the County Mapper. Although the tutorial was initially created for children, it remains on the web site for anyone to access.

Prince William County's County Mapper can be viewed at *www.pwcgov.org/oit/gis/ims.htm.* The GIS Division maintains a web site at *www.pwcgov.org/oit/gis* that provides further information on its mission, activities, organization, services, and products.

Eight Benefits of Prince William County's GIS System

We have looked at the history and applications of the Prince William County GIS. But what are the benefits of GIS and the lessons learned regarding GIS implementation? According to McNoldy, the following are the main benefits from using GIS.

1. The GIS greatly increases the productivity of county employees who collect, manage, analyze, and distribute land data. They are able to accomplish more tasks because each task takes less time. They are able to tackle a greater range of tasks. Although the GIS may not eliminate the need for existing staff positions, it often erases the need for additional staff positions. Moreover, these are positions that frequently are not filled because of budget constraints. Using the GIS, the current staff becomes more productive.

   Furthermore, the products of the GIS have a higher quality and better content. The graphic quality of maps and reports generated by the GIS is more consistent because the automated mapping process virtually eliminates the individual styles and skill levels of drafters. Moreover, the content of reports and maps is improved. Because the work can be done faster, employees can spend more time checking, expanding, and improving it.

2. The GIS provides a central source for all land data. All county personnel know they can find whatever land data is available by contacting the GIS Services Group. Previously, this data was spread throughout all county agencies. Both the public and the county's employees spent a great deal of time calling the various county agencies simply to locate the data they needed.

   The GIS also provides a central location for organiziing, storing, and maintaining all address data for the county. This is very important, in that the majority of county offices deal with properties. It eliminates the need for each agency to have its own independent database of property addresses. Today, they use the database provided by the GIS Services Group. This also ensures consistency and improves coordination among agencies.

3. The GIS enables the county to perform analyses previously not feasible, usually because the data was not available, the computer software was not available, or the analysis required too much time. One example is the fire and rescue response time analysis.

4. In response to a disaster, the GIS can provide information on natural and emergency resources in a fraction of the time required using conventional maps.

5. It is much easier to update land data in the GIS than it was with previous mapping and filing systems. This favors keeping the data current. Fewer work hours and a smaller portion of the budget are required to maintain land data.

6. The GIS provides a powerful tool for resolving disputes over land ownership and boundary locations. Traditional maps do not provide enough detail or information to resolve these issues, but the GIS can pinpoint inconsistencies in parcel boundaries. These inconsistencies had been virtually hidden away when the State Department of Taxation first developed the county's tax maps 36 years ago. Because these maps were prepared at a scale of 1 inch = 400 feet, small gaps and overlaps in parcel boundary descriptions were not apparent. The GIS can more readily identify these inconsistencies because it is mapping with a precision of 1 foot.

7. All land data is registered to a common geographic reference system. This greatly improves the coordination and analysis of this information. In the past, almost every type of land data had a different geographic reference system. One department's files referenced street addresses, another referenced the state grid coordinate system, another referenced its land data to parcel numbers, and yet another referenced the property owner's name. The GIS uses a single-reference system (namely, the Virginia Coordinate System of 1983) for all map themes.

8. It is much easier to produce customized reports and maps tailored to the user's need. The GIS can produce maps at any scale, with any combination of map themes, and for any area of the county. Furthermore, it is no longer necessary to work back and forth between separate map sheets, in that the GIS can display two adjacent maps simultaneously.

Six Lessons Learned in the Prince William County Implementation

McNoldy offers the following advice to other municipalities considering a GIS program.

- Get local elected officials involved in the GIS planning process early.

McNoldy held a series of workshop meetings with the county Board of Supervisors at the outset of his GIS planning phase. Board members were told about the limitations and problems the county staff had in handling land data. The board was then given an executive-level introduction to GIS technology, concepts, and applications. They were next offered the opportunity to participate in the GIS planning process.

Through this series of workshops, these officials became very familiar with the needs and objectives of the GIS program. This has resulted in their continual, full support for the program.

- Keep elected officials advised.

McNoldy has continually informed the board of supervisors and the county executive of the progress of the GIS program, including both accomplishments and setbacks. He gives them periodic demonstrations of new GIS capabilities. This has made it far easier to gain their understanding of and support for budget expenditures.

- Establish the GIS program at the agency level.

The GIS Services Group does not report to any of the user agencies, such as planning, public works, fire and rescue, or tax assessment. Its authority equals that of its users. McNoldy believes that if his office were reporting to a user agency, it would be biased toward serving that user's needs and would not be able to look objectively at the overall GIS program and the needs of all users.

Moreover, because McNoldy reports directly to the county executive, he is able to present requests for personnel and equipment resources directly to the highest level of county management. He does not have to work through another layer of organization.

- Map at the level of parcel data.

It is important to create a GIS land base that includes tax parcels. Many GIS applications for natural resource management and land planning do not need parcel-level mapping. However, most local government agencies deal with individual parcels and addresses. GIS provides the greatest benefit for the most users when this level of detail is included in the database.

- Create and maintain a database of street addresses.

The street address is the most commonly used identifier for land records in the county. Therefore, including street addresses in the GIS database makes it possible for the greatest number of potential users to make use of data.

- Buy backlit digitizing tables.

It is far easier to read and interpret source materials being digitized on a backlit table. Often, more than one layer of source material must be taped to the table. This may include opaque material such as print paper. The backlit surface makes it easier to see the maps and the details being digitized. There is a big temptation to save money by buying digitizing tables that are not backlit. However, once these are purchased, they cannot be modified.

Summary

This chapter has shown how one Virginia municipality has successfully employed GIS. It has derived many benefits from using this new technology. There are pitfalls in using GIS, but the advice offered by experienced GIS users can help newcomers avoid them.

A Case Study in Maintaining GIS Data: Aberdeen Proving Ground, MD

How One Department of Defense Installation Uses Precision Land Surveying and GPS to Support a Sophisticated GIS Operation

In This Chapter. . .

The GIS operation at this military installation is one of the most advanced in the world. A fundamental factor in its success has been the creation and maintenance of a land base for the GIS that accurately reflects the horizontal and vertical locations of real-world features.

"The Disney World of Defense"

The May 1991 issue of *Baltimore* magazine called Aberdeen Proving Ground (APG) "The Disney World of Defense...the world's largest military theme park." In addition to testing much of the U.S. Army's glamorous equipment, APG has entertained such celebrities as the renowned scientist Edwin Hubble and author Tom Clancy, famous for *The Hunt for Red October* and other high-tech spy thrillers. It also boasts the world's largest tank museum, displaying more than 225 vehicles from many countries and tracing the evolution of tracked armored vehicles back to before World War I.

Located beside Chesapeake Bay, 25 miles north of the City of Baltimore in northern Maryland, APG is host to 55 tenant organizations representing 10 Army commands. Here the Army conducts field tests on everything from combat boots to rifles to tanks to ordnance. Testing of military equipment is a hazardous business, so the Army uses restricted, isolated locations such as APG for this work. Land and resource management is critical on these installations.

APG encompasses 72,000 acres of land and water, 510 miles of road, 189 miles of utility lines, and 389 miles of electrical distribution lines, some of which date to 1917. As with many Department of Defense (DoD) installations, the cantonment ("built up") area of APG is similar in size and density to a small city. Although many of the issues that face the APG Directorate of Public Works (DPW) are similar to those in a municipal public works department, they are considerably more complex because the Army is responsible for everything "inside the fence."

Thus, unlike a typical city, APG manages nearly *all* utilities, including electrical distribution, natural gas, telephone, and so forth. Of course, another difference between a military installation and an incorporated city is that there are no tax parcels; the government is the only landowner. APG's DPW is the landlord of more than 2,100 buildings.

On the other hand, the cantonment is just a small part of the base. The range and test areas are largely undeveloped. They include a variety of ecosystems, including the bay, rivers, wetlands, open fields, and forests. APG is home to thousands of animals in their natural habitat, as well as an important resting place for migratory birds. Moreover, activities on the range and vehicle test tracks are closely monitored and access is restricted. One mistake in range management can have disastrous results.

In addition to normal public works activities (such as managing and maintaining utilities, buildings, and grounds) the DPW supports the testing process itself. Before a tenant at APG can construct a test facility, it must receive approval from the DPW's Business Management division's Planning section. Planning must ensure compliance with environmental regulations, preserve endangered species on the base, and avoid damage to roads and utilities.

To evaluate the proposed facility, Planning must examine a variety of issues. Supporting data is retrieved from many sources to assess the tenant's requirements for acreage, utilities, and road access. These requirements are evaluated against safety and environmental regulations. It is especially important to avoid conflict with areas of known soil contamination or unexploded ordnance. Moreover, tenant requirements often conflict with one another.

Past Problems in DPW Data Management

In the past, it often took months to respond to a tenant's request for approval of a new facility. Each activity at APG had its own records of facility, natural resource, and environmental data. These were drawn on paper maps. Although the DPW generated much of this data, copies were distributed and stored in each department's filing cabinets, flat files, or hanging files. The duplication of data was enormous, as was the task of keeping everyone supplied with a current set of information.

Changes were often noted on paper drawings and maps, but not reported back to the DPW so that the master records could be changed and updated copies circulated. Where facility records were automated, they were often stored on isolated personal computers. Gathering the information required to evaluate a request for testing, or to review a proposed construction project, or to begin the design of that project, took a considerable effort. Today, APG is using a GIS to manage and coordinate this information, and the benefits have been tremendous.

The Advent of GIS

The APG GIS program began in 1991. The base was in the midst of an aggressive building program to support defense research and development, at the same time becoming more aware of the need to address environmental issues. There was deep concern that DPW did not have the facilities information it needed.

Joseph LaVoie, the Deputy Director of DPW at that time; Harry Greveris, the then chief of the DPW Engineering Plans and Services Division; and Greg Kuester, the DPW computer-aided drafting (CAD) manager, decided to find a way to centralize facilities information and solve this problem. After some research, they came to believe that a GIS, carefully managed and fully supported by DPW management, could provide the answer.

Kuester was sent on a "get smart quick" fact-finding mission, traveling to bases where GIS was already in use, going to conferences, and researching the technology to determine which system would be best for APG and how to proceed.

Preparing an Implementation Plan for GIS

The next step was to prepare a plan for the implementation of GIS at APG. DPW examined its needs, then consulted tenant activities to assess their needs. The exercise confirmed the inadequacy of the existing paper-based methods. DPW also worked with GIS consultants Dan Wheeler and Craig Huntley to help prepare the plan.

The APG GIS Implementation Plan was finished in October of 1992. It included a general discussion of GIS technology and examples of GIS products, provided an implementation schedule, defined hardware and software requirements, and presented a strategy for storing GIS data, as well as a "data dictionary," or definition of the GIS elements to be collected and maintained. The plan also called for the preparation of entirely new mapping for the base. "When I studied the system, I saw that we could not feed the old data into the new system," Kuester said. "I brought my concerns to management and it was decided that we would have to fly a new aerial survey if any system was going to work."

The implementation plan also presented a detailed cost/benefit analysis. John Willard and Barbara Wolfe of the APG Directorate of Resource Management assisted in a thorough analysis of the costs of doing business "the old way." Each cost center in DPW was reviewed to determine existing costs. This included the cost to repair utilities damaged during new construction projects or while performing routine maintenance or repair to other utility systems. It also included a tabulation of the manhours and costs of maintaining and managing the existing paper-based facilities information systems.

They then estimated what the costs would be using a GIS. The financial analysis indicated that the implementation cost of a GIS would be fully recovered within two years through savings realized from the use of the system. Little did they know that over the next several years APG would receive numerous requests for copies of its GIS Implementation Plan from military commands throughout the United States and Europe. *It had become a model for others to follow.* DPW management approved the Plan and decided to fund the GIS out of its procurement budget.

Implementing GIS

In 1991, APG first purchased UNIX-based Modular GIS Environment (MGE) software, along with a GIS data server and seven workstations from Intergraph Corporation of Huntsville, AL. In March of 1992, the base was flown for new aerial photography. College students were employed to mark 10,000 utility appurtenances (manholes, hydrants, valves, inlets, and so on) to be sure they were visible in the aerial photography and shown on the new maps.

The Combat Systems Test Activity Geodetic Section at APG supplied DPW with survey control for the mapping project, beginning a mutually beneficial information exchange. One hundred and twenty monuments referenced to the Universal Transverse Mercator (UTM) map projection were set to control aerotriangulation of the photography. Photo Science, Inc. (now Earth Data), a photogrammetric mapping contractor, prepared new digital base maps showing planimetry and topography for the entire installation. They were prepared at a scale of 1inch = 50 feet in the cantonment area, and at 1inch = 200 feet in the range areas. Digital mapping was delivered in September of 1992.

More than 8,000 existing utility as-built drawings were used to field verify existing utility locations. Survey crews headed up by Gene Burchette of the Aberdeen Test Center located the utilities based on the new survey control network.

APG also formed a GIS Steering Committee to provide overall guidance for the system, report to DPW management, and serve as a focal point for tenant organizations to receive information and express their needs. Other APG activities and tenants later purchased their own GIS workstations to be able to share GIS data interactively with DPW and add data needed for their own operations. "Everybody wants to participate once they see what the system can do," said Greveris. "Now they are willing to spend resources to integrate their information into the system and to train personnel. But most important, this effort has created a better working relationship between agencies and is improving information resources."

All of APG's GIS data has been stored in the format of the Tri-Service Spatial Data Standard (TSSDS) developed by the Tri-Service CADD/GIS Technology Center in Vicksburg, MS. The TSSDS includes both graphic standards (such as line styles, colors, layers, and symbology) and attribute data standards, such as sizes, dates, and condition. TSSDS holds out the promise of DoD-wide standardization of GIS data. Using TSSDS, one organization can read another's GIS data as if it were its own.

Similarly, one of the problems the military has had with base realignment and closure (BRAC) exercises is the various formats DoD installations and branches use to report on their assets to BRAC committees. The TSSDS can potentially eliminate this problem. The APG GIS includes metadata ("data about the data"), which identifies the source and accuracy of each data theme. Metadata is currently recorded in hard-copy form, but will be automated in the future.

The Directorate of Safety, Health and Environment was the first of the other activities to implement GIS and share data with DPW. The GIS they purchased was developed by Environmental Systems Research Institute (ESRI) of Redlands, CA. Likewise, the ATC chose to use AutoCAD software from Autodesk of Sausalito, CA, to store maps of the range facilities it managed.

Although the ESRI and Auotdesk systems were ideal for their respective uses, and although it was possible to exchange data between these and other military systems, this proved to be time consuming and resulted in duplication of data. Ultimately,

APG's commanding officer settled the issue when he issued an order that all activities standardize on Intergraph-supplied systems for both GIS and CAD. Says Kuester:

> *Before we standardized, we had a mish-mash of PCs and UNIX machines running different operating systems and different GIS packages. We found it very difficult to conduct joint projects. Now when I go to exchange data with other activities at Aberdeen, I know I won't have any problems. There's just never any data nor time lost to translations.*

Kuester adds: "We have everyone running the same operating system, office automation software, GIS, and CAD packages." Before standardizing on one GIS, DPW required up to two days to import environmental data. Now it has access to those files in minutes.

Today, over 200 persons at APG have access to the GIS data. They use a variety of PCs and personal workstations running under Microsoft Windows 95 and NT. Most of these use VistaMap, Intergraph's GIS data viewing product, for access to spatial and attribute data stored in the GIS. Microsoft Office is the standard for office automation tasks. MicroStation by Bentley Systems of Exton, PA, is the standard CAD package.

The workstations are networked over a 10BaseT Ethernet network that will be upgraded to a Fiber Distributed Data Interface (FDDI) network. GIS data is stored on two Intergraph file servers with over 100 Gb of hard disk storage running the Windows NT Server operating system. These servers store all raster and vector CAD and GIS data for the DPW.

Dan Wheeler, who helped write the plan for APG's GIS and played a key role in its implementation, today manages the system for Kuester. Al De Angelo of Michael Baker, Jr., Inc., APG's master planning consultants, manages the maintenance of the GIS database.

Numerous Uses for GIS

The APG Planning Division uses the GIS in many ways. The system is used to help planners find a suitable location for a new building, a problem that typically involves several site selection criteria. For example, an APG tenant needed a new administrative building. Its footprint would be 30 feet by 60 feet and its location had to be compatible with the installation's land use plan. Ideally, it was to be within 100 feet of a 6-inch water main, within 150 feet of a sanitary sewer, and within 50 feet of a 33-KVA transformer. The site could not encroach wetlands or other incompatible land uses.

A series of spatial queries using the GIS determined the areas that met each of these criteria. These "query sets" were then combined to determine those areas that met all criteria. This process quickly narrowed the focus of the search to just those areas, and it took a fraction of the time and effort that would have been required using paper maps and traditional manual processes.

To site a new vehicle test track, planners created a precise engineering model of the geometry of the track and its supporting facilities using MicroStation, then found a site for it using MGE. The model was shifted and rotated to make a tight fit between several restraints, including an existing airfield, a wetlands area, and several existing roadways. A precise siting analysis such as this would have been nearly impossible previous to the implementation of the GIS.

Planners also use the GIS to aid in analyzing cultural and historical resources at APG. DPW has researched the land records of neighboring Harford, Kent, and Baltimore counties to obtain deed records of properties acquired by the Army to create the installation. The GIS stores the property boundary along with attributes of each parcel. These include the grantee's name, deed and folio number, transfer date, and acreage. A series of aerial photographs of APG taken as far back as 1932 have also been scanned into the GIS. The original parcel boundaries can be overlaid with the historical aerial photography to compare them with old hedgerows, fences, and other physical features. This helps to locate old building foundations, utility lines, and other items of cultural, planning, or engineering significance.

In the DPW, the GIS has replaced the "stubby pencil" approach, says Kuester. As all data is held within databases, the system allows us to conduct a single query, where in the past we would have been forced to consult a myriad of data sources, both paper- and computer-based. The ability to fulfill the search requirement in a single query allows us to provide both tenants and prospective tenants with the space and permission they need in hours instead of weeks.

As discussed in material that follows, DPW provides design consultants for new facilities with current, digital as-built data from the GIS at the outset of the design process. As a direct result, the annual cost of repairs to underground utilities accidentally damaged during construction projects has been significantly reduced.

DPW uses the GIS to help monitor, manage, and reimburse maintenance contractors, including those who mow grass, remove snow, and re-line sewer pipelines. The system can plot maps that illustrate the desired sequence of operations, delineate areas of responsibility, compute areas and lineage of service, and show project status. Special event and traffic management maps are also produced with the GIS.

APG uses a standard Army computer system called Installation Facility System (IFS) to track buildings and other facility assets. DPW linked the GIS graphics data to the building attribute data in IFS. Now the GIS can display a map of the installation, and if the user identifies a building of interest, it will retrieve information about the building. Likewise, the GIS can color-code buildings on the map display according to their attributes, such as the building's condition, tenant, size, or current use.

The Range Operations and Control Division is also making use of the GIS. Range tower operators manage live fire exercises at APG and are responsible for firing range safety. Part of their job is to calculate and plot a surface danger zone (SDZ) for the particular combination of weapon, projectile, charge, aiming direction, and firing angle to be used in the exercise. This SDZ must be cleared before firing may commence (an active firing range is said to be "hot"). In the past, SDZs were plotted on a sheet of plastic overlaid on a 1:24,000 map using grease pencils, scaled rulers, protractors, and SDZ templates. The ranges were controlled using a light board. The operator would switch on one of the board's red lights to signify a hot range and notify down-range personnel by radio.

Although this system worked well for years, it had several drawbacks. The map was old and updated only by hand, which took too long to keep up with changes. Although tower operators were very careful, they could only be as accurate as their tools permitted. A grease pencil line on the map was 40 meters wide and a plotted firing angle was accurate only to a quarter of a degree. The horizontal location of an SDZ on the map was only accurate to ±30 meters. Tower operators compensated for these inaccuracies by being very conservative in granting downrange clearances.

Using digital maps from the DPW GIS, Range Control automated range towers. Now tower operators place SDZs with an accuracy of plus or minus several feet, enter firing angles to within 1/1,000 of a degree, and draw SDZ boundary lines only one foot thick (at the scale of the map display or plot). They can also zoom in to show more detail of an area. This results in more efficient use of the ranges.

SDZs displayed on the GIS monitor are turned on and off to control the ranges. SDZ templates for numerous combinations of weapons, projectiles, and charges are stored. They can be displayed at each firing position and at the proper firing angle with just a click and a few keystrokes. *Moreover, the latest map data is always available from the DPW.* This highlights one of the most significant benefits of GIS: real-time, real-world capabilities.

Local hunters are periodically given access to designated wildlife areas to maintain a stable animal population. Tower operators use the GIS to manage access to these areas in coordination with range operations. APG Range Control plans to integrate GPS with its GIS. GPS receivers will be placed on boats that patrol the bay, as well as land vehicles that operate downrange. These will be tracked in real time, their positions continuously monitored on the GIS display. An alarm will sound when one breaks into a hot SDZ.

Out of deference to the few residential neighbors living across the bay from APG, Range Control also uses GIS to help decide when conditions are right for firing exercises. Each morning a weather balloon is released to record atmospheric conditions, including wind speed, temperature, and relative humidity. A test charge is then fired with the resulting sound level recorded by a network of noise monitors situated around the firing range.

A standard noise modeling program computes the expected noise contours for that charge under those atmospheric conditions and plots them on the GIS display. The predicted noise levels at the monitoring stations are compared to those actually recorded following the detonation of the test charge. The standard noise prediction model is then adjusted accordingly, noise contours replotted, and a decision made whether to permit live fire exercises that day, depending on the predicted impact on civilian neighbors.

In the environmental department, GIS helps technicians respond to hazardous material spills. The GIS database has been copied to a laptop computer. When a hazardous material spill is reported, environmental technicians take it to the site, along with a GPS receiver. The spill outline is surveyed and plotted directly on the laptop, complete with notations. The data is copied into the primary GIS database back at the office. Environmental has also used this tandem GPS/GIS setup to verify natural resource data. Aerial photos and infrared satellite images are "ground truthed" in the field against the GIS database.

Precision Land Surveying Sustains an Accurate GIS Database

One of the keys to a successful GIS operation is to keep the database current, accurately reflecting all changes that occur in the real world. To achieve this goal, DPW instituted a basewide policy on all as-built drawings and surveying related to new construction. Under the regulations, the DPW is responsible for all GIS, survey control, and as-built survey standards. The ATC Engineering/Geodetic Section is the source, repository, and quality assurance group for all survey data. All design consultants and construction contractors working for APG are required by contract to follow the policy.

At the outset of a new design project, DPW supplies the consultant with existing GIS data for the project site. These data are to be used as applicable for the design. ATC also supplies data on existing survey control monuments, survey data collection codes, and brass disks for new control monuments. The design consultant sets and documents new control monuments as required, recovers and documents existing control monuments as required, and conducts field surveys to tie to the existing APG network. (Over time, the APG survey control network has been increased from the original 120 monuments to more than 500 first-order horizontal control points and over100 first-order vertical control points.)

The consultant may use either conventional or GPS survey control techniques. All survey data collected is submitted in digital form and certified by a registered professional land surveyor. All horizontal control surveys must meet or exceed third-order Class I standards, and all vertical control surveys must meet or exceed third-order standards.

As construction proceeds, the design consultant conducts as-built surveys. Starting with a copy of the APG GIS "seed file," he prepares digital as-built construction drawings in the format of the APG GIS and adhering to the TSSDS. This makes it easy for DPW to incorporate the as-built drawings into its GIS database. The as-built surveys must be performed using total stations with either "on-board" or external data collection capabilities, tied to the APG control network and datum, and certified by a professional surveyor. As-built data is collected for the following.

- Building footprints
- Pavement edges
- Fencing
- Utilities
- Other structures
- Prominent physical features
- Site grading

In addition to collecting precise, digital as-built surveys for new construction, DPW refines the accuracy of the GIS database whenever any excavation is done. When a digging permit is issued, a subsurface engineering contractor locates and surveys all underground utilities in the area. These are submitted to DPW in digital form, which then verifies the location of these utilities in its GIS database. Over time, 60 percent of the original utility data digitized from existing paper as-built drawings has been field verified and adjusted as needed. "I believe that it is most important for a land surveyor to be involved in building a GIS," Kuester says emphatically.

Gene Burchette of ATC produced the survey control system APG needed using existing monumentation and adding new monuments where he found holes in the existing network. This network was the foundation for the original aerial photography and planimetric mapping that went into the GIS. Now it provides the control for the as-built surveys that enable us to keep the GIS database accurate and up to date.

Summary

APG's use of GIS has become a model for other DoD installations. Delegations from numerous bases have visited for demonstrations, as well as many top officers and managers from the Pentagon and major DoD commands. In the near future, plans include taking advantage of the Internet. "We get a lot of calls about our data," Kuester says. "But there are too many security issues that have yet to be resolved. Once these issues are resolved, we really would like to make our data available on the World Wide Web."

APG plans to continue to develop the system, integrating as many users and databases as possible to provide even better facilities information services. According to the then Deputy Director of APG's DOW LaVoie, "We firmly believe that the centralization of information with GIS is the future of DPW. It greatly enhances efficient management of facilities in the face of dwindling resources."

This chapter was compiled from interviews with Greg Kuester and previously published articles in Government Computer News, Government Executive, *the U.S. Army Center for Public Works* PublicWorks Digest, *and the Aberdeen Proving Ground* APG News *and* ATC Times.

Appendix A
GIS Exercises Using Free GIS Software and Data

This appendix provides a series of "get acquainted" exercises with GIS. All you need is access to the Internet and you will be able to download the GIS software and data necessary to these exercises to your computer. You can then follow the instructions within the exercises toward learning how to perform the following basic GIS functions.

- Display a map
- Control the map display (i.e., zoom in and out, "fit" all map data, "fit" selected map data, and so on)
- Display a North arrow and map scale bar
- Retrieve and display map feature attributes
- Generate labels for map features
- Print out a hard-copy map
- Customize the map legend
- Create a map data query
- Create a thematic map

Part 1: Start-up Exercises

Exercises 1 through 4 of Part 1 of this appendix take you through the start-up processes of downloading software and data, installing the GeoMedia Viewer software, starting the software, and familiarizing yourself with the content and working parts of the software. Part 2, Further Exercises, contains exercises 5 through 12, which show you how to perform the other tasks of the previous list.

Exercise 1: Downloading GeoMedia Viewer and Sample Data Sets

Step 1: First, create a folder named *c:\warehouses*. You will eventually store the GIS data here. Next, open your Internet browser and go to *http://www.intergraph.com/geomedia/viewer/*.

Step 2: Select the Download FREE GeoMedia Viewer button, shown in the first of the following illustrations. Complete the upper part of the form. Check the box GeoMedia Viewer (22.9 MB), shown in the second of the following illustrations.

Download FREE GeoMedia Viewer button.

GeoMedia Viewer (22.9 MB) check box.

Download GeoMedia Viewer
Can be downloaded as one large file, or 6 small files.
☑ GeoMedia Viewer (22.9 MB)

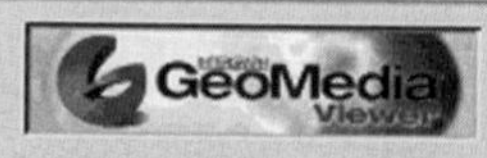

Step 3: Scroll further down the form and either download data at this time or wait until later. In either event, download the USGS GeoData data set to use with these exercises, as shown in the first of the following illustrations. You may download any of the other data sets at any time. Finally, go to the bottom of the form and click on the Go To Download button to start the process of actually copying these files to your machine, as shown in the second of the following illustrations.

USGS GeoData data set.

Choose Data Sets to Download

Please Note: *Geographical data downloaded from this site is for demonstration purposes only. Intergraph makes no claim to its accuracy, but makes every effort to ensure that the data is valid.*

☑ USGS GeoData - 1.7 MB
☐ USA -- Sample - 3.0 MB
☐ US State Metadata (hyperlinks) - 247 KB
☐ Alabama - 2.9 MB
☐ Alaska - 8.0 MB
☐ Arizona - 1.7 MB
☐ Mississippi - 2.3 MB
☐ Missouri - 2.8 MB
☐ Montana - 2.6 MB
☐ Nebraska - 1.5 MB
☐ Nevada - 993 KB
☐ New Hampshire - 750 KB

Maryland - 1.6 MB
Massachusetts - 1.8 MB
Michigan - 2.7 MB
Minnesota - 2.8 MB
Washington - 2.6 MB
West Virginia - 2.8 MB
Wisconsin - 2.4 MB
Wyoming - 1.2 MB

Go To Download

Download started with the Go To Download button.

Step 4: You will next see the Download GeoMedia Viewer and Sample Data Sets form, shown in the following illustration. Simply click on a file to download it to your system. You will use the GeoMedia Viewer file, *gmview20.exe*, to install GeoMedia Viewer later.

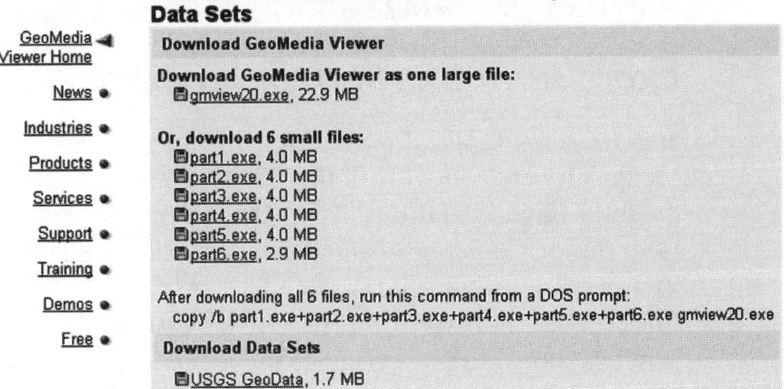

Download GeoMedia Viewer and Sample Data Sets

GeoMedia Viewer Home
News
Industries
Products
Services
Support
Training
Demos
Free

Download GeoMedia Viewer

Download GeoMedia Viewer as one large file:
gmview20.exe, 22.9 MB

Or, download 6 small files:
part1.exe, 4.0 MB
part2.exe, 4.0 MB
part3.exe, 4.0 MB
part4.exe, 4.0 MB
part5.exe, 4.0 MB
part6.exe, 2.9 MB

After downloading all 6 files, run this command from a DOS prompt:
copy /b part1.exe+part2.exe+part3.exe+part4.exe+part5.exe+part6.exe gmview20.exe

Download Data Sets

USGS GeoData, 1.7 MB

Download GeoMedia Viewer and Sample Data Sets form.

Step 5: The USGS GeoData you download will be delivered in a "zipped" file format. The actual file downloaded will be named *cape_gir.zip*. Save this file in a temporary folder, as you can delete it later. Next, extract the content of *cape_gir.zip* using your zip file utility (e.g., PKUNZIP or WinZIP) to the folder *c:\warehouses*. The GIS data will then be in the file *CAPE GIRADEAU.mdb*, a Microsoft ACCESS database file.

Step 6: The text file *Readme* contains background information and descriptions of all files. It will be necessary to locate the **.mdb* files after you have started the GeoMedia Viewer. Finally, check each *.mdb* file and remove or turn off the "Read-only" attribute (right click on the file in Windows Explorer, click on Properties, and uncheck the attribute "Read-only"). This is very important.

Exercise 2: Installing GeoMedia Viewer

Step 1: If you downloaded the all-inclusive file *gmview20.exe*, simply double click on it to start the installation process. If you downloaded the six individual files, you will have to use them to create the *gmview20.exe* file. To do this, run the following command from a DOS command window.

```
copy /b part1.exe+part2.exe+part3.exe+part4.exe+part5.
exe+part6. exe gmview20.exe
```

This will create the *gmview20.exe* file.

Step 2: Once you have the *gmview20.exe* file (it should be 23,496 Kb in size when viewed in Windows Explorer), click on it to start the setup process. Installation is easy. Take all defaults suggested by the setup program and you will be ready to start using GeoMedia Viewer.

Exercise 3: Starting GeoMedia Viewer

Step 1: Go to the Windows START button and select Programs, and then select GeoMedia Viewer. You will see the GeoMedia menu selection list, shown in the first of the following illustrations. All of these items are discussed in the GeoMedia Viewer Help topics. Click on this menu item to start the Help function. To start GeoMedia Viewer, click on the blue GeoMedia symbol or on "GeoMedia Viewer" in the menu. Click anywhere (except on the Learn More button) on the first form that appears, which is shown in the second of the following illustrations. This will start GeoMedia Viewer.

GeoMedia menu selection list.

First form presented by GeoMedia Viewer.

When GeoMedia Viewer starts, it will display the GeoMedia Map window and a legend, shown in the following illustration. The legend is empty because you have not chosen any features to display and have not performed any "queries," which also display features. The Map window displays the graphic characteristics of the GIS features. The menu bar and toolbar that access GeoMedia commands are at the top of the Map Window. In the lower left corner you see that the current map scale is 1:3,779,527. The active coordinate units being displayed are on the lower right corner of the Map window.

Map window and legend.

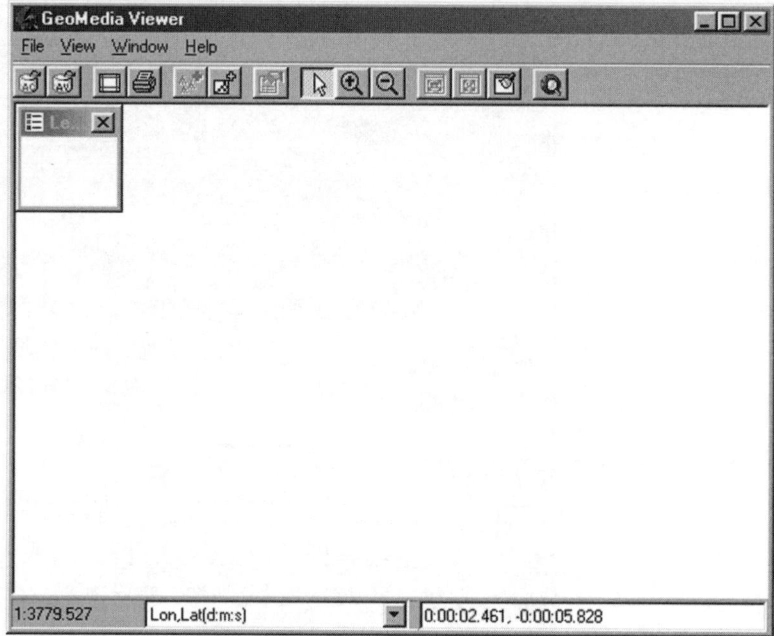

Step 2: Once GeoMedia Viewer is started, you need to select a GIS data set. To do this, click on the File menu, shown in the following illustration, and select either a Microsoft ACCESS Warehouse or an ArcView data set. For this exercise, use the Open Access Warehouse form, browse to the folder *c:\warehouses* (where you stored the *CAPE GIRARDEAU.mdb* file), and select it. Although nothing seems to happen, the Cape Girardeau database is now the "active" warehouse and you can display or query its features.

File menu.

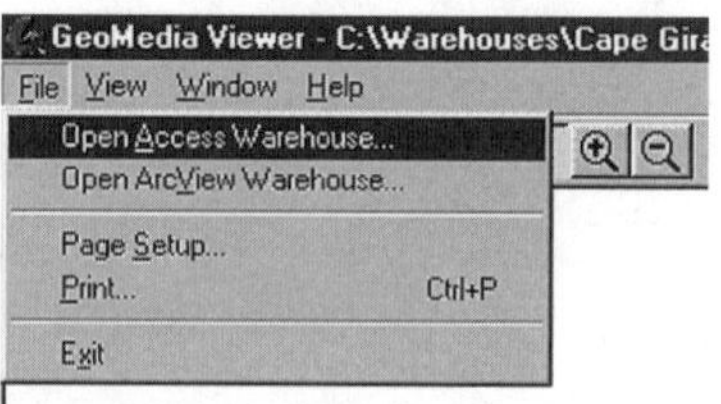

Step 3: To view these features, click on the View menu and select Map Features, shown in the first of the following illustrations. This will display the View Map Features form (see the second of the following illustrations), which is a listing of the geographic features available in the data set. The CAPE GIRARDEAU data set contains a number of point, line, and area features. Table A-1, which follows, describes these features.

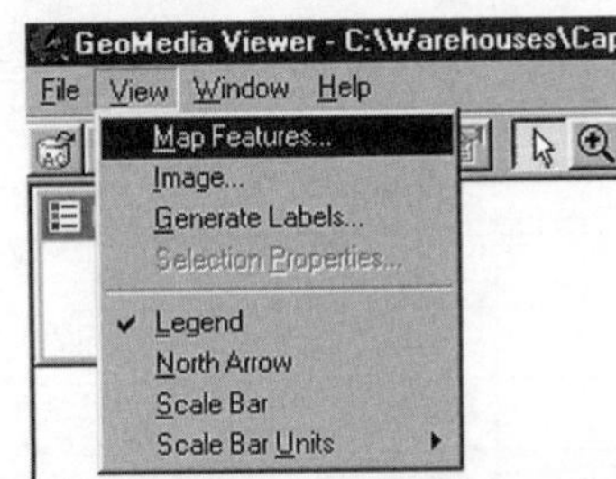

View menu showing Map Features option.

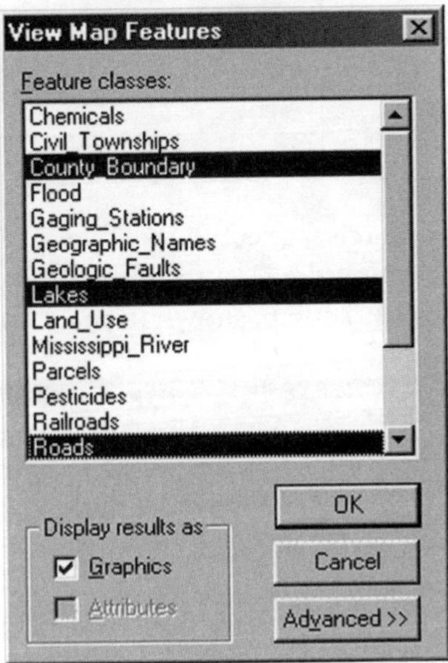

View Map Features form.

Table A-1: View Map Features

Feature Name	**Feature Type**
Chemicals	Point
Civil_Townships	Area
County_Boundary	Area
Flood	Area
Gaging_Stations	Point
Geographic_Names	Point
Geologic_Faults	Line
Lakes	Area
Land_Use	Area

Feature Name	Feature Type
Mississippi_River	Area
Parcels	Area
Pesticides	Area
Railroads	Line
Roads	Line
School_Districts	Area
Section_Lines	Line
Streams	Line
Voting_Districts	Area
Zip_Codes	Area

Step 4: For this exercise, select the following features: County Boundary, Lakes, and Roads. After selecting the features to be displayed, click on the OK button and you have created your first map, shown in the following illustration.

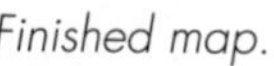

Finished map.

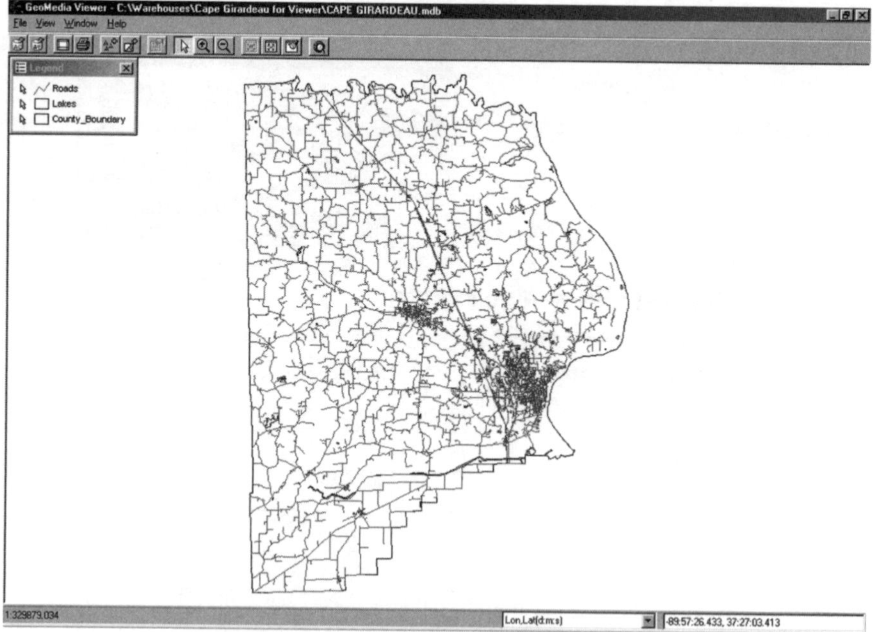

Exercise 4: Exploring User Controls in GeoMedia Viewer

GeoMedia Viewer provides pull-down menus and toolbar buttons for controlling the mapping environment. Some features are only available from the menu, such as controlling display of the North arrow or scale bar. Other controls available from the toolbar are shown in the following illustration.

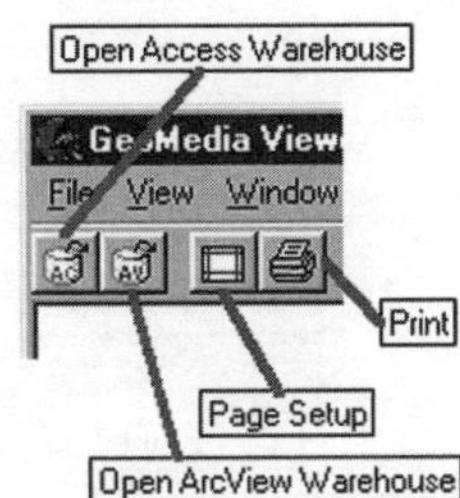

Toolbar control options.

The first two buttons are used to connect to an ACCESS data warehouse or an ArcView data warehouse, respectively. Because ArcView data does not provide any method of identifying their coordinate system, it is necessary to create a *.csf* (Coordinate System File) file using the tools provided with GeoMedia Viewer before it can be displayed.

Page Setup allows the user to set print and paper size, margins, print scale, and other characteristics required for printing a copy of a map. Print sends the current map to the printer.

The next set of buttons controls the display of features, raster images, and attributes of features currently selected in the Map window, shown in the following illustration. You have already seen how View Map Features works for vector data. In addition, GeoMedia Viewer can display GeoTIFF-format raster images properly aligned to the coordinate system of the currently active warehouse.

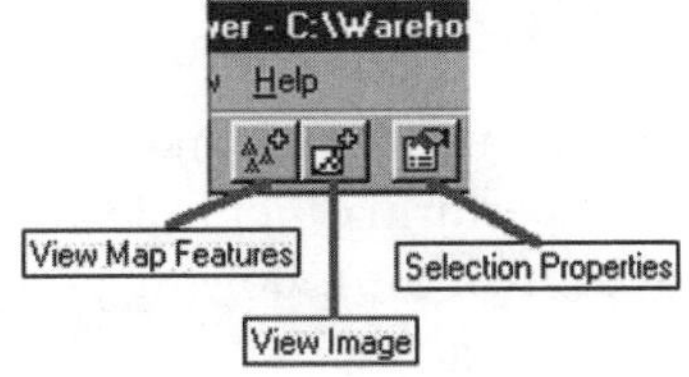

Map window features.

The View Image button will start the View GeoTIFF Image form for selecting the image. Last, Selection Properties will display a list of the attributes associated with the currently selected feature. The same action can also be accomplished by double clicking on any feature displayed in the map view.

The last set of toolbar buttons on the right controls the display of the Map window, as shown in the following illustration. After clicking on the Selection Properties button, the user can select any feature displayed that has the LOCATABLE characteristic set (discussed later in this appendix). After a feature has been selected, you can display its attribute values (NAME, ADDRESS, and so on) by clicking on the View Map Features button.

Toolbar options for controlling the Map window.

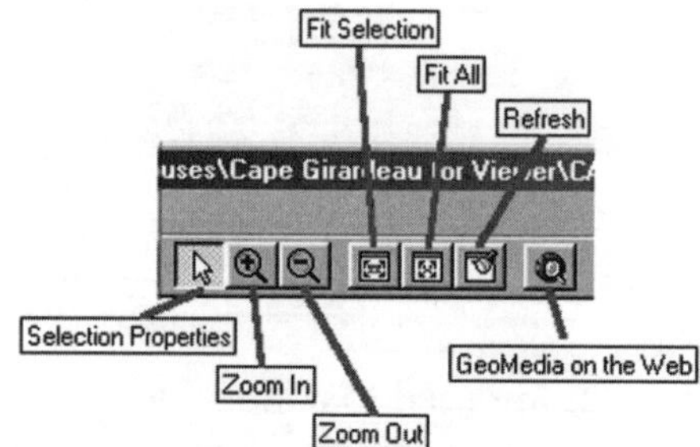

If a feature is currently selected, clicking on the Fit Selection button will change the Map window so that the selected feature fills the window. Zoom In and Zoom Out work just like a camera lens to reduce or enlarge the area of features displayed. The Fit All button will fit all features currently displayed in the legend within the window. Use it if you ever loose sight of some map features. Refresh will redisplay the features at the current window settings.

Although not a display control, the GeoMedia on the Web button will access the Internet, if it is available, and point your browser to a locally installed web page (i.e., on your computer). This web page contains links to the GeoMedia product center at Intergraph.

Part 2: Further Exercises

Now that you have created a map and are familiar with the user controls of GeoMedia Viewer, you can perform other tasks with the software. The eight exercises in this section take you through the following processes.

- Adding map objects
- Displaying feature attributes
- Displaying attributes in table format

- Labeling features
- Printing a map
- Querying a map
- Modifying style definitions
- Creating a multi-colored thematic map

Exercise 5: Adding Map Objects

Step 1: Using the View menu, you can add a North arrow or a scale bar to your map. In the following illustration, only the Legend option has been chosen. Clicking on the North Arrow and Scale Bar options will display those Map Objects. These are called "Map Objects" because they are more than graphic symbols. The North arrow "knows" where north is, and therefore always points to true north, regardless of the map projection.

Adding a North arrow and scale bar.

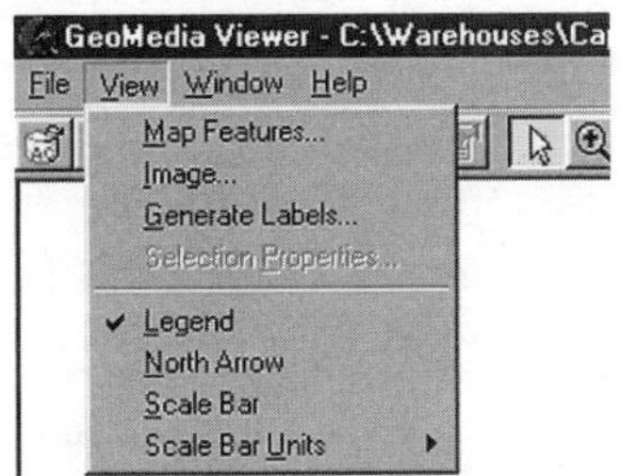

Likewise, the scale bar "knows" what the current map display scale is, and therefore will automatically adjust to display reasonable values. Scale bar units can be set to kilometers, meters, miles, or feet, depending on your preference, as shown in the following illustration.

Setting scale bar units.

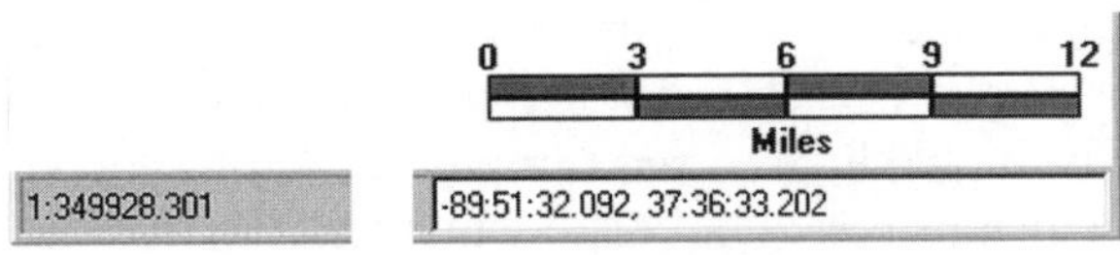

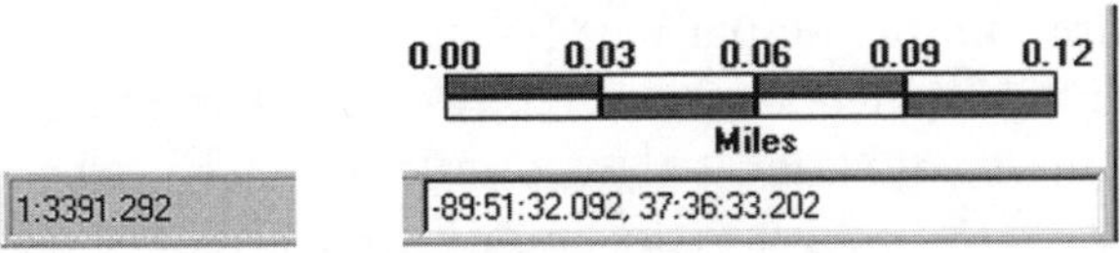

Exercise 6: Displaying Feature Attributes

As previously mentioned, GeoMedia Viewer provides a number of tools for determining the non-geographic attributes (e.g., NAME, ADDRESS, and INCOME) of features. Double clicking on one feature will produce a pop-up display of the attributes attached to that feature, an example of which is shown in the following illustration.

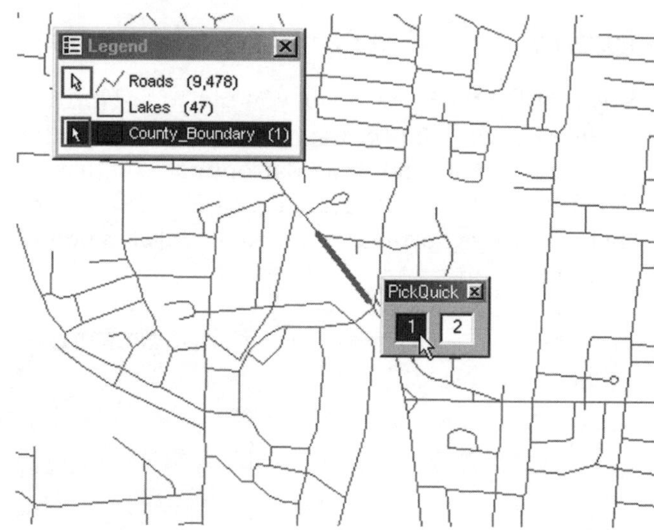

Display of attached attributes.

Step 1: In this exercise, you have two features on the left that are "locatable" and that are overlaid: Roads and County_Boundary. Note the image of the selection arrow beside these features in the legend surrounded by the red boxes. This symbol in the legend indicates that these features are both locatable and can be selected.

Step 2: To select one of these features, hold the selection arrow over the desired feature (choose the road segment in red). A PickQuick box appears, with a separate cell for every locatable feature. Move the selector arrow from cell to cell in the PickQuick box. As each cell is activated, the corresponding feature in the Map window will be highlighted in red (emphasized in the previous illustration) so that you can see which feature is linked to which cell.

Step 3: When the desired feature is highlighted, click on the corresponding cell in the PickQuick box and that feature will be "selected." Click on the Selection Properties button on the toolbar (see the following illustration), and the attributes associated with the selected feature will be displayed. In this case, this segment is a part of Kingshighway, and is located in county 31 (CNTY_NO). The county on the left is county 31 (COUNTYL), as is the county on the right (COUNTYR).

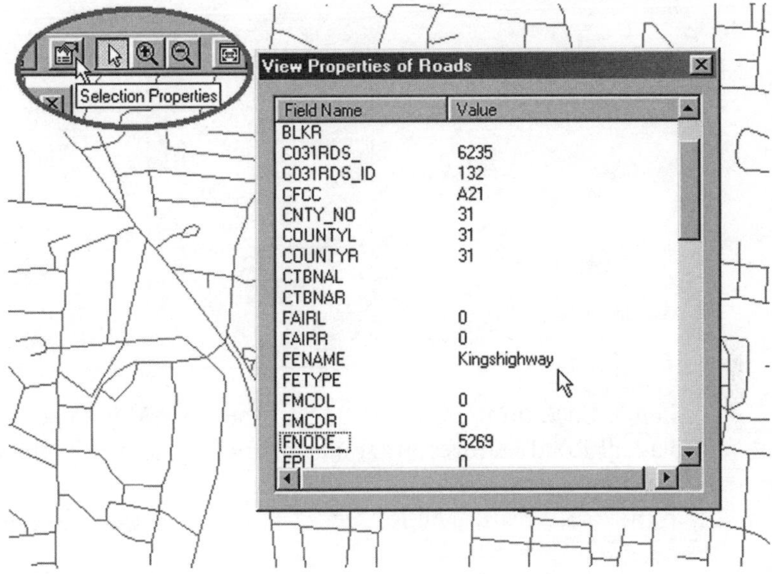

Selection Properties button.

Step 4: Normally, you would turn off the LOCATABLE property for features that are not the subject of a current search (discussed further later in this appendix). For this exercise, simply double click on the feature to display the view properties of <Feature Name>.

Exercise 7: Displaying Attributes in Table Format

Step 1: All map feature attributes may be displayed in a table format for review and comparison. To do this, begin the same way you did for viewing the graphic representation of features: click on the View Map Features button, shown in the following illustration.

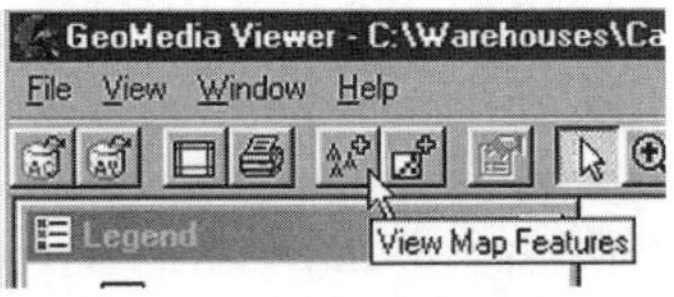

View Map Features button.

Step 2: Here you can display the attributes in the View Map Features form, shown in the following illustration. Select the desired feature. Select Roads. Next, uncheck the "(Display results as) Graphics" box, which is the default, and check the "(Display results as) Attributes" box. Do this because the feature Roads is already shown as a graphic. Note that both boxes can be checked at the same time to display both the graphics and the tabular format of the data.

View Map Features form.

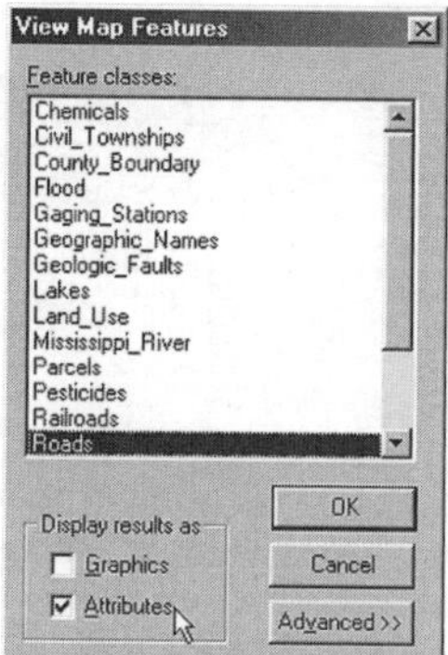

Step 3: Click on the OK button and the View Map Feature Attributes form will display all Road features in tabular format, as shown in the following illustration. By scrolling down or across, you will see all rows and columns, respectively, contained in the Roads feature table.

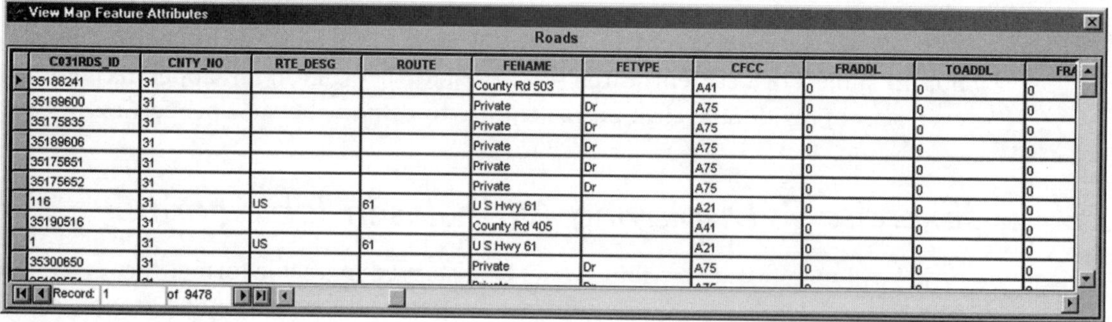

View Map Feature Attributes

Roads

C031RDS_ID	CNTY_NO	RTE_DESG	ROUTE	FENAME	FETYPE	CFCC	FRADDL	TOADDL	FRA
35188241	31			County Rd 503		A41	0	0	0
35189600	31			Private	Dr	A75	0	0	0
35175835	31			Private	Dr	A75	0	0	0
35189606	31			Private	Dr	A75	0	0	0
35175651	31			Private	Dr	A75	0	0	0
35175652	31			Private	Dr	A75	0	0	0
116	31	US	61	U S Hwy 61		A21	0	0	0
35190516	31			County Rd 405		A41	0	0	0
1	31	US	61	U S Hwy 61		A21	0	0	0
35300650	31			Private	Dr	A75	0	0	0

Record: 1 of 9478

Road features displayed in tabular format.

Exercise 8: Labeling Features

Step 1: Labeling features, even large numbers of features, is easy. From the View pull-down menu (there is no corresponding button), select Generate Labels. This will display the Generate Labels form, shown in the following illustration.

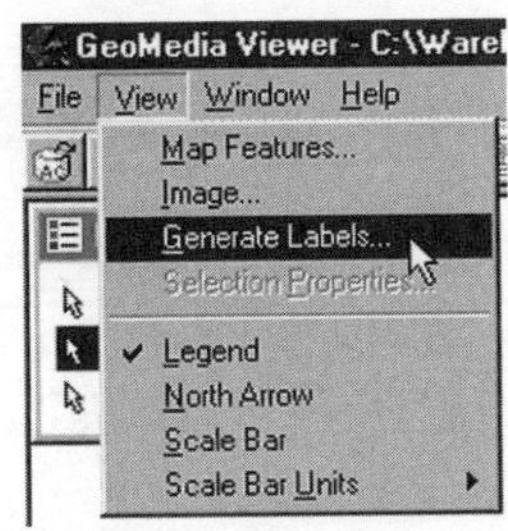

Generate Labels form.

Step 2: Using this form, select the Feature Class to be labeled, and then select the Attribute to be used for a label. Here, you are going to create a label with the name of the county, as shown in the following illustration. Although this data set contains just one county, this same process will create a label for each county contained in the feature table. It is just as easy to create labels for all 3,000-plus counties in the nation as it is for one.

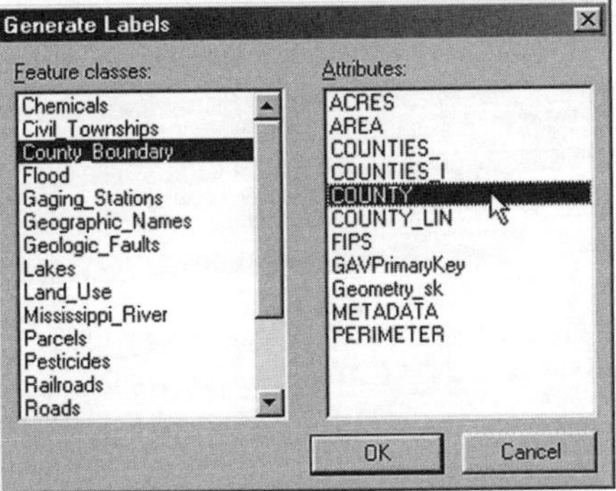

Label showing county name.

Step 3: After selecting the County_Boundary feature class and the COUNTY attribute, which contains the name of the county, click on the OK button to display the label. However, it will appear at first that no label was created. This is because the default label font size and view scale are such that the label will not display until you zoom in several times. This is corrected by adjusting the style of the feature display graphics.

The map legend now shows a new entry, County_Boundary Labels (1) (the "1" indicates that only one label was placed in the map view), and a capital A inside a small box, as shown in the following illustration.

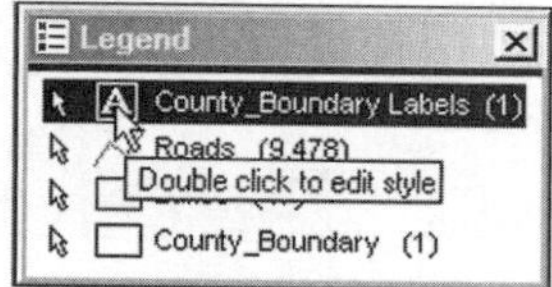

Map legend showing the letter A within a box.

Step 4: Double click on the "A" in the box. This will bring up the Style Definition form, shown in the following illustration, which is used to modify the graphic characteristics (i.e., color, weight, area fill, and so on) of the features displayed. Different characteristics are presented for different types of geometry. In the case of labels, the geometry is "letters and numbers," and therefore the form presents a font and its associated font characteristics (i.e., color, size, and so on).

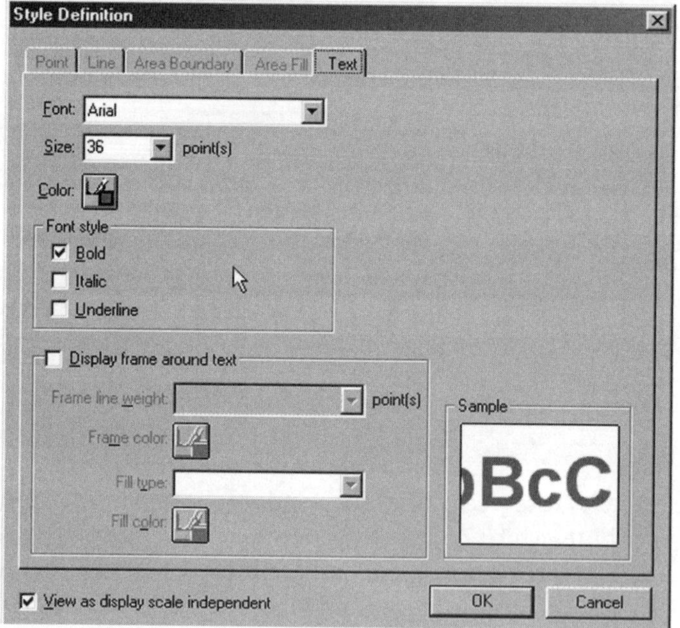

Style Definition form.

Step 5: Of particular importance is the check box on the bottom left side of the Style Definition form, "View as display scale independent." Select this option. The font characteristics will display as if the map view were not reduced from a scale of 1 to 1. Set the font size to 36 to achieve a legible display of the name of the county (do not forget to check the "View as display scale independent" box). The following illustration shows the map view after applying this style definition to the county name label.

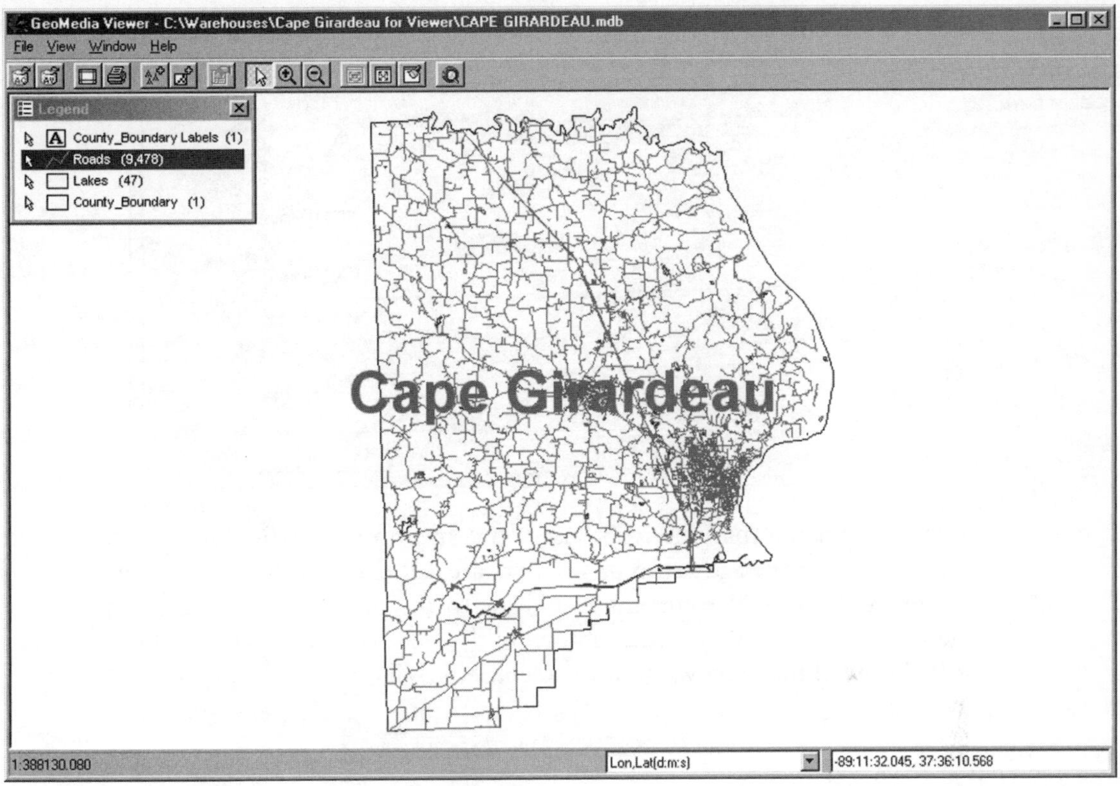

Map view after scale type and font size specified.

Exercise 9: Printing a Map

Step 1: Before printing a map, you first need to set the output scale and orientation. From the File pull-down menu, shown in the first of the following illustrations, select Page Setup to display the Map Window Page Setup form, shown in the second of the following illustrations. To create a map of "Cape Girardeau County Roads and Lakes," select the Landscape orientation, the standard Letter sized paper, and "Fit to 1 page(s) wide by 1 tall" in the Print Scale box on the form.

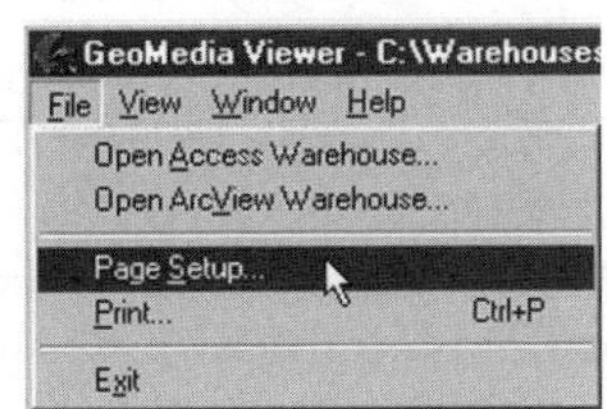

File pull-down menu Page Setup selection.

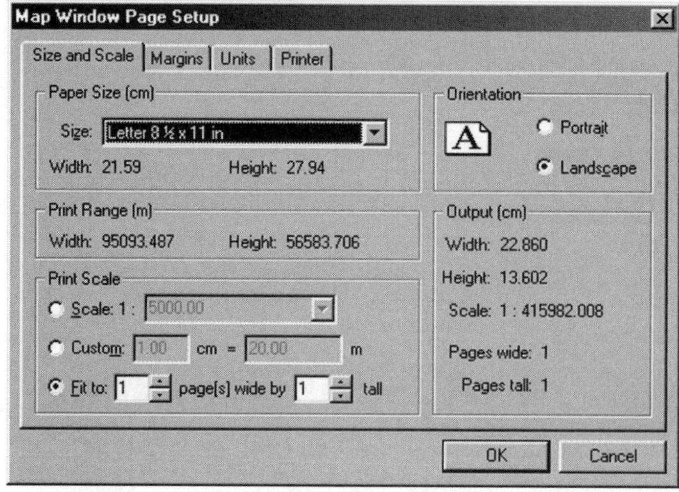

Map Window Page Setup form.

Step 2: Select the margins to be used, the size units (English or Metric), and the printer. Once the Page Setup parameters are set, click on the OK button to dismiss the form. To send the current map view to the printer, select the File pull-down menu, and then select Print, as shown in the following illustration. This will print a scaled view of the map, without the GeoMedia Viewer menu and toolbar buttons.

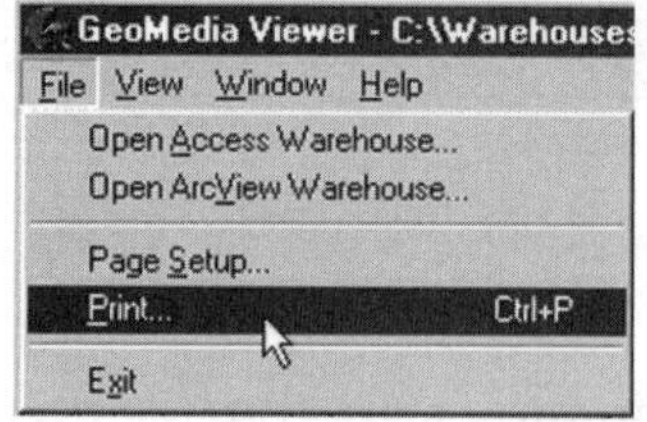

Selecting Print in the File menu.

Now you can move on to some more sophisticated exercises in GIS; namely, creating a map "query" (i.e., asking questions about the map data), and creating a thematic map (i.e., a map symbolized according to a data theme).

Exercise 10: Querying a Map

Step 1: Start by displaying some map features that will provide a reference for your work area. From the menu bar, select the File menu item, and then select Open Access Warehouse. Using the Open Access Warehouse form, browse to the folder where you stored the *CAPE GIRARDEAU.mdb* file and select it. Although nothing seems to happen, the Cape Girardeau data set is now the active warehouse, and you can display or query its features.

Step 2: To view some map reference features, click on the View menu item and select Map Features. This will display the View Map Features form listing the geographic features available in the data set. Select County_Boundary, Civil_Townships, and Parcels (see the first of the following illustrations). Selecting these features with the Graphics box checked will produce the map shown in the second of the following illustrations.

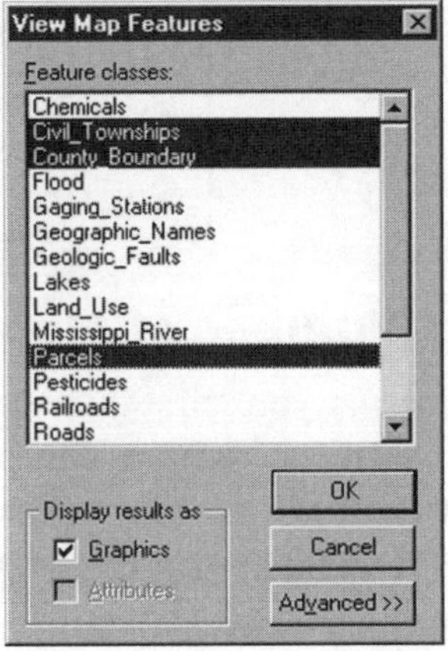

Selecting data set geographic features.

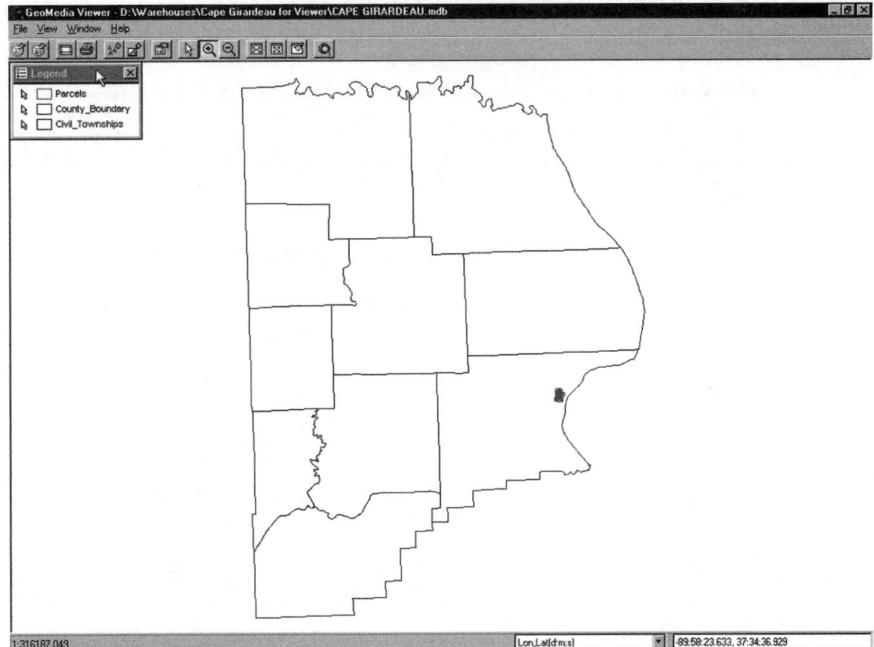

Resultant map.

The Cape Girardeau data set contains a number of point, line, and area features. You can display these features using "default" colors and line styles chosen by GeoMedia Viewer. However, you can adjust these colors and line styles, as well as their order of display in the legend (see the following illustration). The adjustment of map feature symbology is discussed in material that follows. The steps of this exercise continue in the following section.

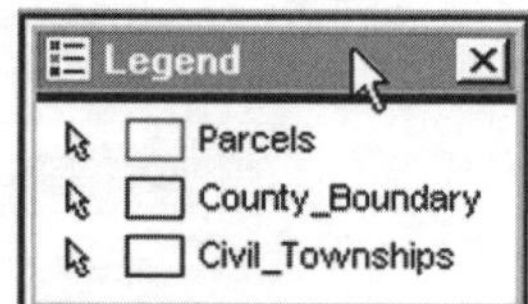

Adjustment of styles and display order in the legend.

Determining Number of Map Features Displayed

Before you try creating your first query, you need to understand the importance of being aware of how many items, or "instances," of a particular map feature are displayed in the Map window. A query will always create a legend entry. However, you may be confused if there are no matches to the query and, therefore, no map features are displayed. Selecting the Show Statistics option of the legend will indicate the number of instances displayed for each type of map feature. If the query produces no instances of a map feature, the legend will display the number zero (0).

Step 3: The Show Statistics option is inactive by default. To select this option, first place the cursor inside the legend area, but not on any of the map features. Right click your mouse in this area to display the Legend menu. Click on Properties at the bottom of the list to open the Legend Properties form, shown in the following illustration.

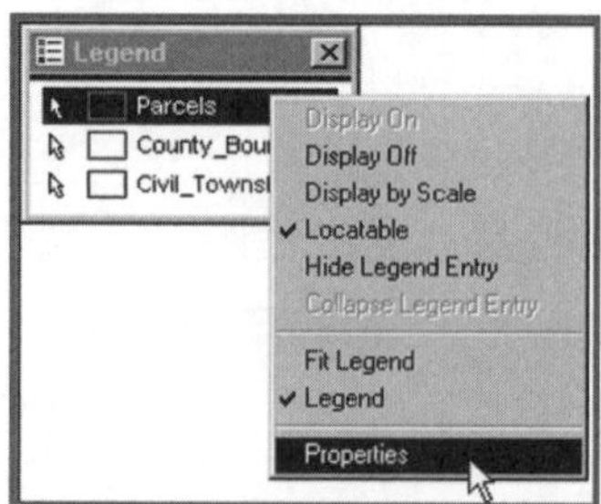

Legend Properties form.

Step 4: Click on the General tab in the upper left corner of the form (see the first of the following illustrations). Now check the Show Statistics box. Finally, click on OK to dismiss this form. You will now see the number of features displayed in the map: 1 County_Boundary, 10 Civil_Townships, and 584 Parcels (see the second of the following illustrations). The steps of this exercise continue in the following section.

General tab of the Legend Properties form.

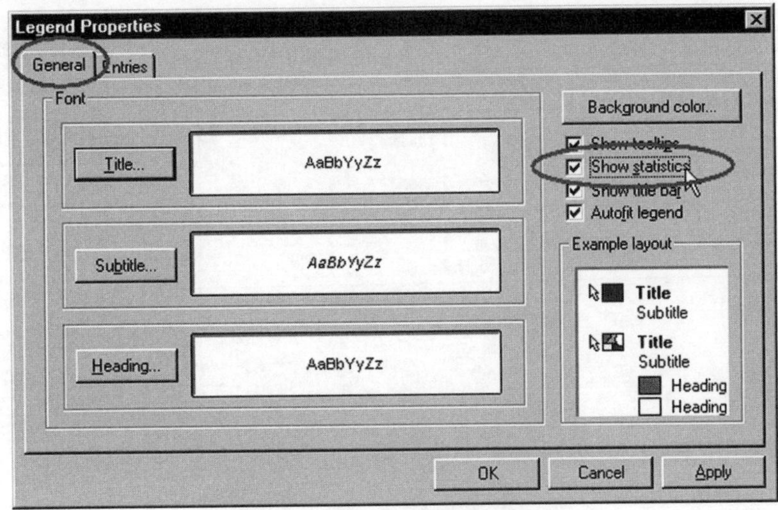

Display of number of each specified feature.

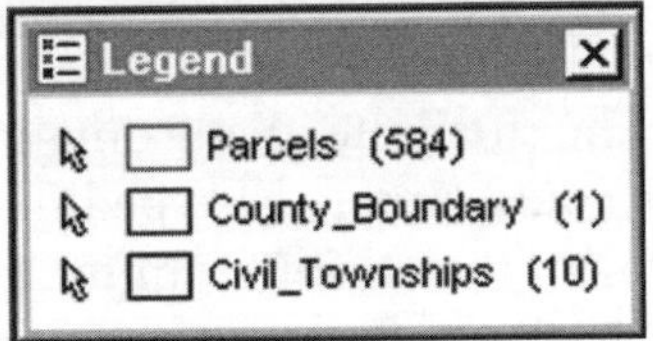

A Simple Query Regarding Parcels

You are now ready to construct the query. Although it is easy to show all 584 parcels on the map, you might want to display only those parcels, for example, that satisfy a certain condition, or conditions. For example, say you want to locate only those parcels owned by Bruce Wayne. In database parlance, this query would be phrased as "Show me all parcels where OWNER = 'Bruce Wayne.'"

When you selected a map feature earlier, you were, in effect, executing the first part of this query, "Show me all parcels." As you saw, the result was all 584 parcels (instances) in the database being displayed. The following is how you execute the second part of the statement, the "where" clause.

Step 5: First, use the Zoom In tool to fill the Map window with the parcels. This will make it easier to see the results of the query. Next, select the type of map feature to be displayed in the Map window (click on the View menu and select Map Features). The View Map Features form appears. Click on the Parcels option, and then click on the Advanced button. This will expand the form to display the Attribute Filter portion of the form, shown in the following illustration.

Attribute Filter field of the View Map Features form.

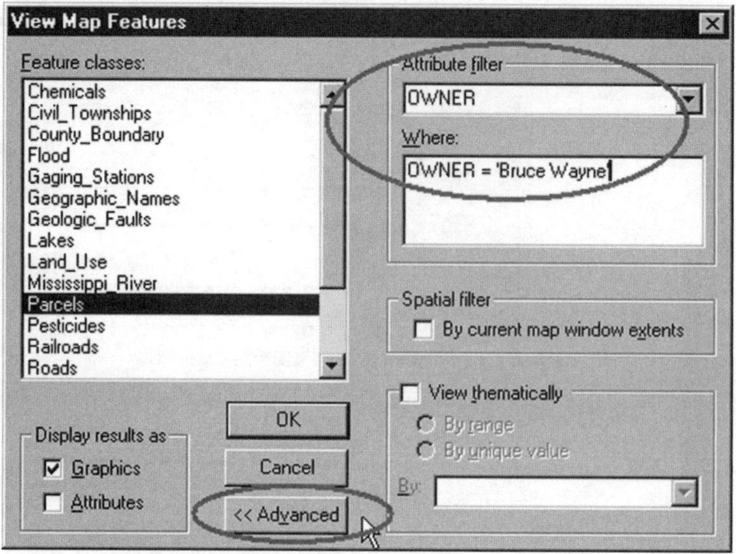

Step 6: Select the attribute that contains the names of the parcel owners. This attribute name could simply be typed into the Where input field. However, if you are unfamiliar with the attributes of the parcel features, you can select the attribute by clicking on the down-facing triangle on the right of the Attribute Filter field. This will display a "pick-list" of the attributes associated with the map feature Parcels. Click on the attribute Owner, which will then be copied into the Where field.

Step 7: To complete the query, click inside the Where field just to the right of Owner, and then type in = *'Bruce Wayne'*, as shown in the following illustration. When finished, click on OK to dismiss this form and display the parcels owned by Bruce Wayne, a total of six (see the following illustration).

Typing in the "where" clause of the query.

At first it may not be clear which six parcels belong to Bruce Wayne. This is because the default color for the six parcels owned by Bruce Wayne is similar to the default color used to display the 584 parcels. Changing the display characteristics, or style definitions, for the individual legend entries will correct this problem.

Exercise 11: Modifying Style Definitions

There are two methods of accessing the Style Definition forms. The first method is to place your cursor inside the Legend window and right click the mouse, as you did previously. This will produce the Legend menu. Here, you select Properties, select the feature to be modified, and then click on Style and modify the style definition.

This method permits access to other legend properties, such as Scale (features will only display when the scale of the Map window is within a certain range) and Display Priority (determines which features display on top of other features, and whether a feature is displayed on the map or displayed in the legend). These are all useful options when creating a map that conveys a particular theme.

Step 1: There is also a shortcut that allows you to quickly change a map feature's style. First, double click on the style symbol in the legend. For this exercise, this is the box in the legend, as indicated in the following illustration.

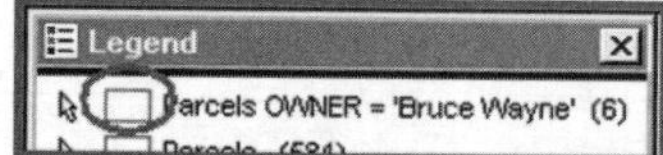

Style symbol to be selected.

Each type of map feature's geometry has a different style symbol. Parcels and County_Boundary are area features and are represented by a small box in the legend. The box in the legend displays the same style characteristics as the features displayed in the Map window. The following illustration shows several examples of styles for map features.

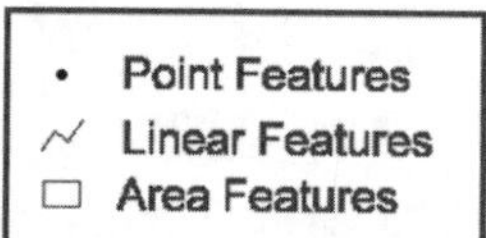

Map feature styles.

Step 2: To change the symbology of Bruce Wayne's parcels, double click on these parcels' respective style symbol (i.e., the green box in the legend). This will activate the Style Definition form. Because this is an area map feature, you can change the style of the Area Boundary, Area Fill, or both. In this case, color-fill Bruce Wayne's parcels to distinguish them from the other 584 parcels. To do this, click on the Area Fill tab at the top of the form.

Step 3: Click on the Type field in the Primary Fill box and select Solid. Finally, click on the square button next to the Color label. This will display the Color Picker form, shown in the following illustration.

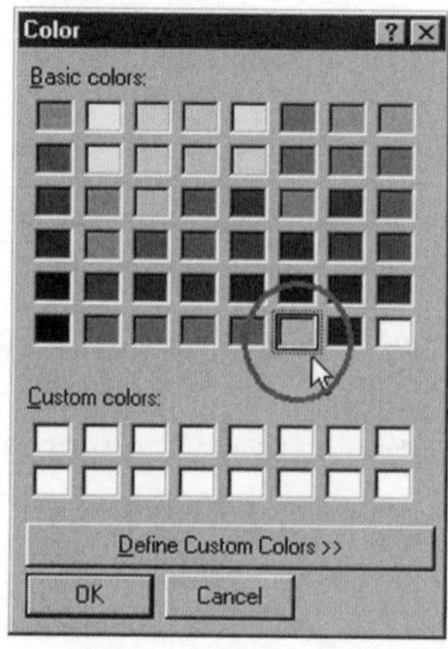

Color Picker form.

Step 4: Select light gray (6th row, 6th column), and then click on OK to dismiss the form. Note that the button next to the Color label has now changed to match the Area Fill color you selected. Your Style Definition form should now look like that shown in the following illustration.

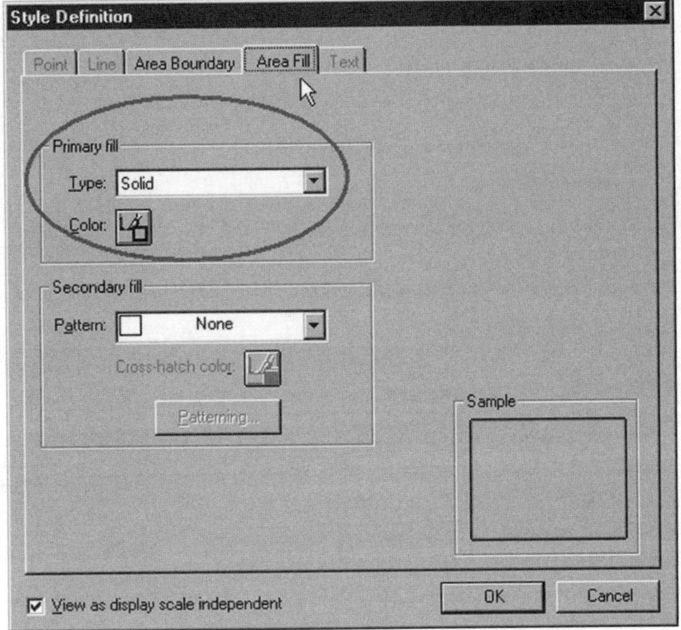

Specifications in the Style Definition form.

The Area Boundary line style and the Area Fill style you have chosen will be displayed in the Sample box in the Style Definition form. The last step is to dismiss the Style Definition form by clicking on the OK button. GeoMedia changes the symbology of the parcels found in the OWNER = 'Bruce Wayne' query to the style you have just selected. Note that the Style Symbol in the legend also changes to match that of the map, as shown in the following illustration.

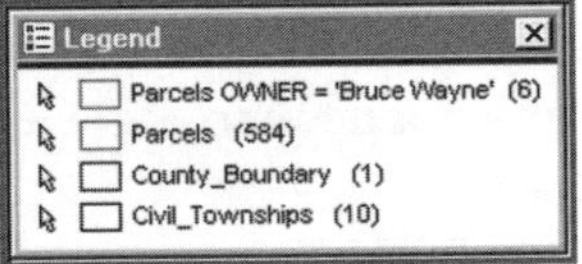

Revised legend.

Exercise 12: Creating a Multi-colored Thematic Map

To this point, you have used only one type of symbology to display all instances of a certain type of map feature. (In the exercises, an instance of the map feature Parcels is an individual map parcel.) Another method of representing map features is through use of the Thematic Map display, which can distinguish the instances of a map feature according to their attribute data. This can be done using colors and other types of symbology.

In this exercise you will create a map showing all parcels having an area between 500 and 1,000 units, with changes in symbology incremented by 100 units. To do this, you will again use the Advanced button on the View Map Features form; in this case to combine an Attribute Filter with a Thematic Map display.

Step 1: Note in the following illustration that a query has been specified for Parcels where "AREA>=500 and Area<1,000" (i.e., AREA is both greater than or equal to 500 and less than 1,000). Note also that the "View thematically" box has been checked in the View Map Features form. Access the View Map Features form, establish this query, and make sure the "View thematically" box is checked.

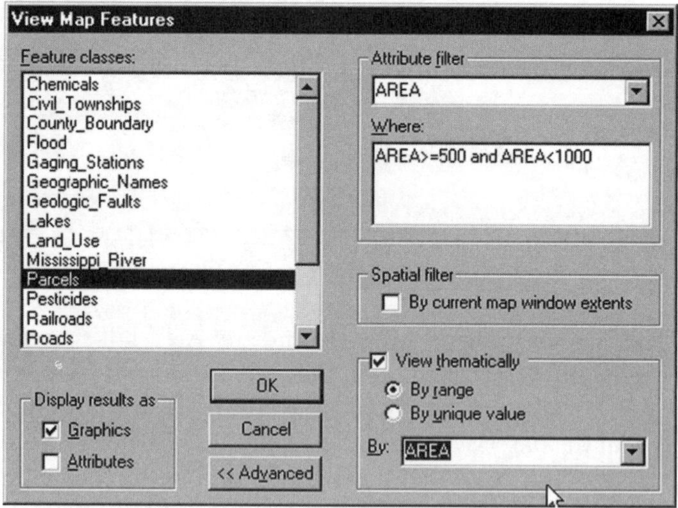

Query and "View thematically" specified in View Map Features form.

Thematic map displays can be viewed "By range" or "By unique value." "By range" is the better choice for working with the Parcel features. "By unique value" might be a better choice for a thematic map of the Civil_Townships because there are only 10 instances of that feature.

Step 2: Select "By Range," and then select AREA in the By selection box. Note that the attribute used for the "where" clause of the query must also be used in the By selection box. Click on OK. The result should look like the map display shown in the following illustration.

Resultant map display.

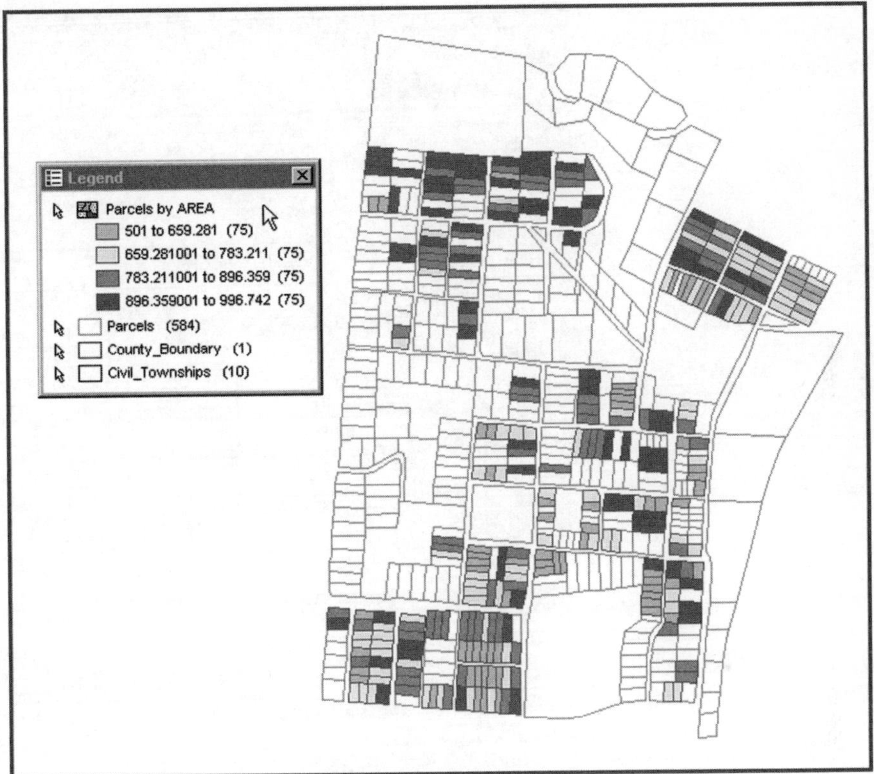

Note that the number of changes in symbology and ranges are "default" values at this point, and therefore the map symbology is not exactly what you want. Modify the Legend entry to produce the correct symbology.

Step 3: First, place your cursor on the map symbology in the legend and right click the mouse. This will bring up the Map By Ranges form, shown in the following illustration, which is used to modify the criteria for map display.

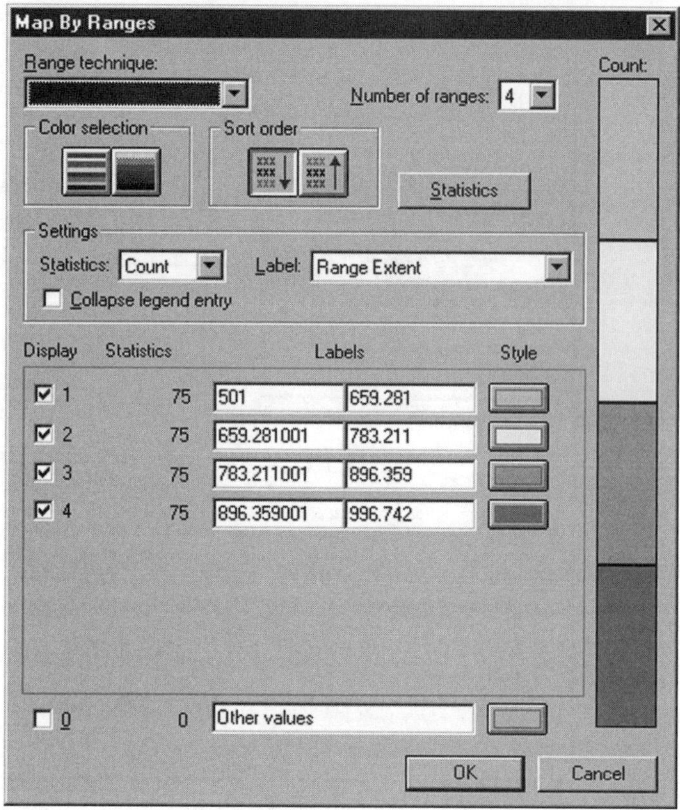

Map By Ranges form.

Step 4: You want five breaks in map symbology, not four. Change this by modifying the "Number of ranges" box in the upper right side of the Map By Ranges form to 5. This will also generate a new Color Table for ranges. Note, however, that the range values are still not what you want, because the default values are taken from actual values in the attribute data table. Adjust them by editing the values in the Labels boxes in the center of the form. The following illustration shows the correct range values for this exercise.

Desired range values.

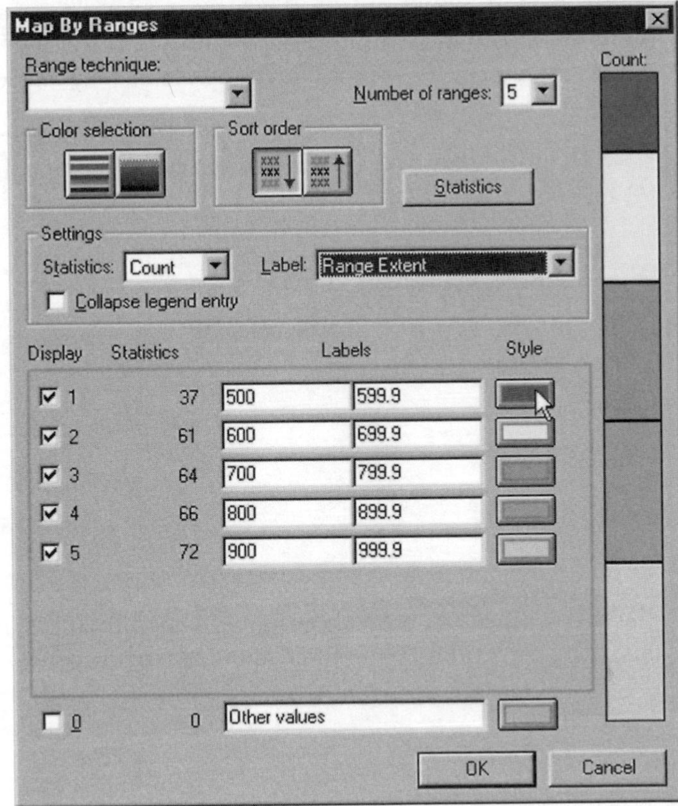

Step 5: Count is the default setting for the Statistics option in the Map by Ranges form. This option divides the ranges so that an equal number of attribute records fall in each range. Note that as you edit the range values the number of feature instances in each range changes. For example, there are 37 parcels with areas between 500 and 599.9 units. Keep in mind that the range values must not overlap. For instance, avoid making the first range "500 to 599.9" when the second range starts at 587. Establish the range values.

Step 6: You may also want to modify a range's color display using the Map by Ranges form. For instance, you could emphasize the "500 to 599.9" range on the map and in the legend. Click on the Color Button under the Style option to display the Color Picker form. In the same way you used the Color Picker form previously, change this range to bright red. Likewise, modify Ranges 4 and 5 to softer colors.

Step 7: Make one last change to the legend, this time without changing the map display. Hide all other entries except Parcels by AREA. To hide the Parcels legend entry without removing that feature from the map display, click on the Parcels entry with the right-hand mouse button. This will display a drop-down list of Legend Display Switches, shown in the following illustration. Select the Hide Legend Entry option to hide the legend entry without affecting the display of those features in the map.

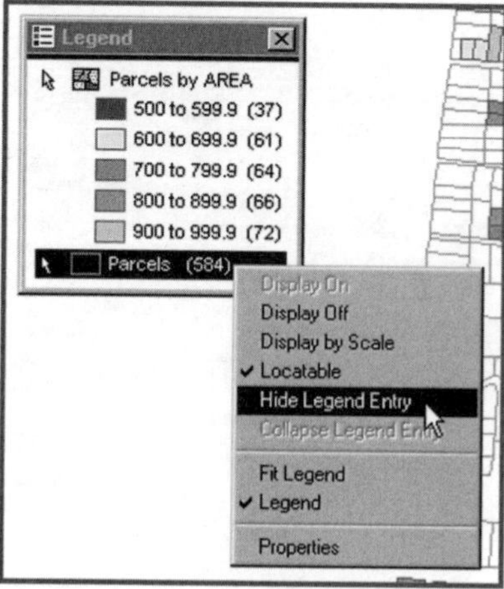

Legend Display Switches drop-down list.

The following illustration shows the finished thematic map. If your map looks like this, congratulations. You receive an A for the course!

Finished thematic map.

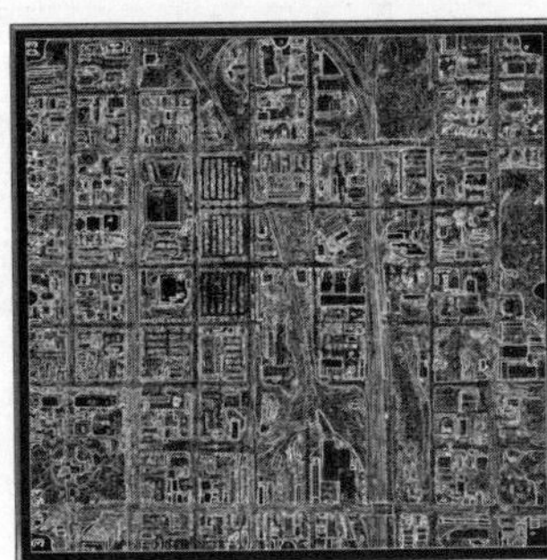

Appendix B
Resources for GIS Information and Training

This appendix contains lists of suggested resources for learning more about, and staying abreast of, developments in GIS. These resources comprise six categories:

- Associations
- Events
- Books
- Periodicals
- Educational institutions

These lists are preceded by information on education and training in GIS. Obviously there are a great many resources available for learning about GIS. Books and schools can be extremely useful for both introductory and advanced training in GIS. The author also recommends joining an association that focuses on GIS, subscribing to a GIS periodical, and attending annual GIS conferences to stay abreast of developments in this rapidly changing field. One should also contact software developers, resellers, and systems integrators for information on their training programs for specific GIS products. This information is usually posted on their web sites.

Education and Training in GIS

There are a variety of ways of learning about GIS. These include degree and certificate programs that focus on GIS as a general discipline, principally offered by colleges and universities. More specific GIS training may also be obtained from GIS software developers and resellers, technical institutes, systems integrators, consultants, and GIS conferences. "Self-paced" GIS training materials are also available.

Finally, many organizations that use GIS often establish their own "in-house" training programs. This section summarizes each of these options. The section ends with a brief discussion of how land surveyors can take advantage of these sources.

Colleges and Universities

Although many of their GIS programs may include training in specific aspects of GIS applications, colleges and universities are concentrating on educating students in GIS concepts and theory. Instead, training in particular GIS software is being done by an increasing number of community colleges. Many of these local schools have already developed courses for training in specific word processing, spreadsheet, and computer-aided drafting (CAD) programs, and are adding classes for specific GIS packages. Roughly 1,000 schools teach GIS courses around the world.

Over a hundred colleges and universities in the United States offer GIS studies, including schools in almost all fifty states. Most undergraduate degrees in geography now include a heavy emphasis on GIS. GIS can be a minor degree option for undergraduate geography majors, or a graduate school major. Comprehensive GIS study programs include classes in mapping and geodesy, spatial data modeling, software programming and customization, attribute data management, data input and editing, data quality, data analysis, analytical modeling, and data output, as well as organizational and management issues.

These classes usually include the use of one or more GIS software and database programs, as well as hands-on use of computers and peripheral devices, such as digitizing tables, scanners, and plotters. Moreover, college GIS studies are often offered as part of the program for a related discipline, such as agriculture, civil engineering, computer sciences, earth sciences, environmental and natural resources, forestry, geography, land planning, land surveying, oceanography, and many others.

Software Developers and Resellers

All of the leading GIS software developers offer comprehensive training programs for their products. Most of their distributors and product resellers also offer training classes. These training classes not only cover the software products themselves, but related topics. These include techniques in data input and output, software programming and customization, database management systems (DBMS), data networks, system management and administration, and GIS management issues. These classes are usually offered at the vendor's headquarters and major branch offices, but on-site training at a customer's office is usually also available.

Classroom instruction at the vendor's location is the most common method chosen by customers because of its low cost, but individual student training, usually done "on the job," can be arranged as well. Although most classes follow a predetermined curriculum, customized training is also usually available at a somewhat greater cost. "Webcasts" by software developers are a more recent development. Customers visit a site on the World Wide Web at a prearranged time to view a short class or special demonstration using their web browser. This is an example of "distance learning" that is now becoming, and will undoubtedly continue to become, increasingly popular and prevalent in the next few years.

Technical Institutes

Many private technical training institutes also offer training in GIS. They are usually different from software resellers in that they do not offer the products for sale. Many of these have developed from technical schools that initially offered training in technical areas such as drafting, building trades, and CAD. The training is usually focused on specific GIS software and applications, and typically results in a certificate of completion.

Systems Integrators

A systems integrator is a company that offers general consulting services in computer systems, data networks, and/or telecommunications. This general systems capability usually distinguishes it from a software reseller, which is typically focused on a particular computer application, such as CAD or GIS. However, the systems integrator often resells computer software and hardware as well. Therefore, it may offer many of the same types of GIS training classes described under the previous section, "Software Developers and Resellers." However, a systems integrator will often emphasize customized and specialized training programs for either classroom or individual instruction.

Consultants

There are many companies and individuals that offer general, "independent" GIS consulting services; that is, without bias toward one GIS software product or another. This typically means the consultant does not resell GIS software, in order to remain "neutral," when making recommendations regarding product selection and use. Although the consultant may offer GIS training, it is usually limited to general GIS concepts, work processing, and management issues.

Self-paced Training

There are some books on the market that provide not only an overview of specific GIS products but an in-depth review of product capabilities, as well as "tips and tricks" for users. These books include training exercises the reader can follow at his own pace. For example, see the OnWord Press line of GIS books. In addition, most GIS software products are delivered on a CD or downloaded over the Internet, and self-paced user instruction is usually included. Self-paced training is obviously the least expensive in terms of cash outlays, but the results are less certain, and the process can take much longer than a comparable formal training program.

Conferences

There are a number of major conferences that cover topics in GIS and closely related fields. These conferences typically offer a variety of opportunities to learn about GIS. Most of these opportunities come in the form of conference sessions. Conference sessions cover a wide range of topics, including technical, management, and productivity issues; lessons learned; case studies; and standards developments.

Conferences often include "hands-on" sessions in which attendees can work with the software in a classroom setting. In addition, these conferences usually include exhibits by GIS vendors. Vendor exhibits typically include product demonstrations, usually highlighting their latest features. Finally, conferences often include "user group" sessions, in which persons with an interest in one particular GIS software product or application area can meet and discuss topics of common interest.

In-house Training Programs

For many employees, GIS introduces a very different way of doing work. Typically, they have little exposure to computer systems in general, and none to GIS in particular. Most employees sincerely want to do well in their work. However, to do well in their jobs, they need to feel confident about what they are doing, and they need to have all of the knowledge and skills necessary to do it well. Trying to work with a new technology such as GIS can destroy employee confidence and morale.

This can only be overcome through a comprehensive training program. The software vendor, a system integrator, or a local training center often provides the initial GIS user training. Subsequent training can also be purchased, but many organizations set up in-house training programs, run by employees with previous experience. Moreover, an ongoing training program is needed for new GIS users, and all users will need to learn to operate any new software and equipment.

Different levels of training are usually required, depending on the type of GIS user the employee happens to be. Most organizations have a large number of "casual" GIS users that need only to learn the functions for basic GIS data viewing. The in-house GIS support staff can usually train them in one relatively short session. Most organizations also have a few GIS "analysts," who need to use a more sophisticated GIS product and functions. Once again, the GIS support staff is usually capable of providing this training. A GIS analyst's training could require as much as 40 hours of the employee's time, which may be spread over a number of sessions.

However, the GIS support staff itself will most often take advantage of vendor training classes because they need to master all of the GIS products in use, many of which are very complex. In-house training programs rarely offer comparable instruction. System managers may require one to two weeks of additional training to learn system management, network management, user account management, and more advanced GIS and database functions the average user will not need.

GIS Training for Land Surveyors

You might work for organizations that create and maintain GIS databases of their assets. This includes military bases, utilities, municipalities, state transportation agencies, and so forth. Other readers are with private companies that could, or already do, offer GIS services. I believe the following comments pertain to both.

To get started in GIS, many organizations designate a GIS "champion" to provide overall coordination and direction for the GIS program. If this person is an existing employee, all of the options previously discussed may be used to provide him with training in GIS, perhaps with the exception of a college degree program that takes several years to complete.

A "champion" with prior GIS experience hired from the outside may only need some additional training to tune his skills to the needs of the organization's GIS. After the GIS is in place, these organizations can still take advantage of all of the options for training previously discussed, including the hiring of recent graduates from GIS study programs.

Associations

The following associations are either devoted principally to GIS users and GIS issues, or are organizations devoted to a related field that deals extensively with the use of GIS. These groups play a valuable role in the promotion and use of GIS. They deal with important legal and policy issues, as well as offer valuable educational programs for their members. The list includes the name, address, and telephone and fax numbers of each association. In some cases an Internet address is also included.

Association of American Geographers (AAG)

1710 Sixteenth St. N.W.
Washington D.C. 20009-3198
202/234-1450; fax: 202/234-2744
www.aag.org

American Congress on Surveying and Mapping (ACSM)

5410 Grosvenor Lane, Suite 100
Bethesda, MD 20814-2144
301/493-0200; fax: 301/493-8245
www.survmap.org

American Public Works Association (APWA)

2345 Grand Boulevard, Suite 500
Kansas City, MO 64108-2641
816/472-6100; fax: 816/472-1610
www.pubworks.org

American Society for Photogrammetry and Remote Sensing (ASPRS)

5410 Grosvenor Lane, Suite 210
Bethesda, MD 20814-2160
301/493-0290; fax: 301/493-0208
www.asprs.org

Geographic Information and Technology Association (GITA)

(Formerly AM/FM International)
14456 E. Evans Ave.
Aurora, CO 80014
303/337-0513; fax: 303/337-1001
www.gita.org

Management Association for Private Photogrammetric Surveyors

1760 Reston Parkway, Suite 515
Reston, VA 20190
703/787-6996; fax 703/787-7550
www.mapps.org

National Association of Counties (NACo)

440 First Street, N.W., Suite 800
Washington, D.C. 20001
202/393-6226; fax: 202/393-2630
www.naco.org

National States Geographic Information Council (NSGIC)

1225 New York Avenue N.W., Suite 350
Washington D.C. 20005-6156
202/628-9724; fax: 202/628-9744
www.nsgic.org

Urban and Regional Information Systems Association (URISA)

1460 Renaissance Dr., #305
Park Ridge, IL 60068
847/824-6300; fax: 847/824-6363
www.urisa.org

GIS Events

The events listed are a sampling of annual North American events and conferences either devoted to GIS or that deal extensively with GIS issues. These events typically include meetings and presentations on topics of interest to GIS users, as well as exhibits by GIS software, hardware, and service vendors.

Note that the month and place for the event may vary from year to year. You will need to verify the exact date and location of the upcoming conference with its sponsor. See the previous list of Associations for contact information when not provided below.

In addition to these association events, GIS software vendors sponsor both local and national user groups and conferences. Local meetings typically occur several times each year, whereas national conferences are usually held annually. These conferences are usually beneficial and highly recommended.

A/E/C Systems
Contact: Penton Media
Exton, Pennsylvania
www.aecsystems.com

- American Association of Geographers (AAG) Annual Meeting
- American Congress on Surveying and Mapping (ACSM) Annual Convention and Exposition
- American Public Works Association (APWA) International Public Works Congress & Exposition
- American Society for Photogrammetry and Remote Sensing (ASPRS) Annual Conference
- Geographic Information and Technology Association (GITA) Annual Conference XXIII

Books

The books listed cover a wide variety of topics related to GIS use, and include several introductory volumes. The list includes the title, publisher, year of publication, ISBN number, and suggested retail price of the book. The list is alphabetized by author.

Geographic Information Systems: A Guide to the Technology
J. Antenucci, K. Brown, P. Croswell, M. Kevany, with H. Archer
Van Nostrand Reinhold
1991 0-442-00756-6 $59.95

Practical Handbook of Digital Mapping: Terms and Concepts
Sandra L. Arlinghaus, Editor
CRC Press
1994 0-8493-0131-9 $61.95

Geographic Information Systems: A Management Perspective
Stan Aronoff
WDL Publications
1989 0-921804-00-8 $57.00

Marketing Government Geographic Information: Issues and Guidelines
William Bamberger and Nora Sherwood Bryan, Editors
Urban and Regional Information Systems Association
1993 0-916848-00-0 $35.00

Introduction to Environmental Remote Sensing
Eric C. Barrett and Leonard F. Curtis
Chapman & Hall
1992 (Third Edition) 0-412-37170-7 $46.50

Integrating GIS & CAMA, Volumes 1 and 2
Bruce A. Belon
Urban and Regional Information Systems Association (URISA)
$55.00 each

Geographic Information Systems
Tor Bernhardsen
VIAK IT
1992 82-991928-3-8 $42.50

Beyong Mapping: Concepts, Algorithms, and Issues in GIS
Joseph K. Berry
GIS World
1993 0-9625063-6-2 $29.95

Spatial Reasoning for Effective GIS
Joseph K. Berry
GIS World
1995 1-882610-14-8 $29.95

Intelligent GIS : Location Decisions and Strategic Planning
Mark Birkin, Graham Clarke, Martin Clarke, and Alan Wilson
John Wiley & Sons
1996 0470236140 $64.95

Geographic Information Systems for Geoscientists: Modelling with GIS
Graeme F. Bonham-Carter
Pergamon/Elsevier Science Publications
1994 0-08-042420-1 $43.00

Introduction to Remote Sensing
James B. Campbell
The Guilford Press
1987 0-89862-776-1 $49.95

GIS and Organizations
Heather Campbell and Ian Masser
Taylor & Francis
1995 $36

Geographical Information Systems in Assessing Natural Hazards
Alberto Carrarra and Fausto Guzzetti, Editors
Kluwer Academic Publishers
1995 0-7923-3502-3 $178.00

Introduction to Integrated Geo-Information Management
Seppe Cassettari
Chapman & Hall
1993 0-412-48900-7 $39.00

Profiting from a Geographic Information System
Gilbert H. Castle
GIS World
1993 0-9625063-7-0 $49.95

Exploring Spatial Analysis in GIS
Yue-Hong Chou
OnWord Press
1997 1-56690-119-7 $52.90

Exploring Geographic Information Systems
Nicholas Chrisman
John Wiley & Sons
1997 0471108421 $61.95

Getting Started With Geographic Information Systems
Keith C. Clarke
Prentice-Hall
1998 0139238891 $69.33

International Symposium on the Spatial Accuracy of Natural Resource Data Bases
Russell G. Congalton, Editor
American Society for Photogrammetry and Remote Sensing
1994 1-57083-008-8 $65.00

Introduction to Remote Sensing
A.P. Cracknell and L.W.B. Hayes
Taylor & Francis
1991 0-85066-335-0 $36.00

Statistics for Spatial Data
Noel A.C. Cressie
John Wiley & Sons
1993 (Revised Edition)
0-471-00255-0 $99.95

Annotated Bibliography on GIS-Related Standards
Peter Croswell, Editor
Urban and Regional Information Systems Association (URISA)

Spatial Information Technology Standards and System Integration
Peter Croswell
Urban and Regional Information Systems Association (URISA)
$55.00

Geographic Information Systems: A Visual Approach
Bruce Davis
OnWord Press
1-56690-098-0 $42.95

The Added Value of Geographical Information Systems in Public and Environmental Health
Marion de Lepper, Henk Scholten, and Richard Stern, Editors
Kluwer Academic Publishers
1995 0-7923-1887-0 $139.50

Fundamentals of Geographic Information Systems
Michael N. Demers
John Wiley & Sons
1996 0471142840 $84.95

Environmental GIS Applications to Industrial Facilities
William J. Douglas
Lewis Publishers
1995 0-87371-991-3 $59.95

Innovations in GIS 2
Peter Fisher, Editor
Taylor & Francis
1995 0-7484-0269-1 $45.00

Environmental Remote Sensing from Regional to Global Scales
Giles Foody and Paul Curran, Editors
John Wiley & Sons
1994 0-471-94434-3 $89.95

Spatial Analysis and GIS
Stewart Fotheringham and Peter Rogerson, Editors
Taylor & Francis
1994 0-7484-0104-0 $39.50

Accuracy of Spatial Databases
Michael Goodchild and Sucharita Gopal
Taylor & Francis
1989 0-85066-847-6 $85.00

Environmental Modeling with GIS
Micael F. Goodchild, Bradley O. Parks, and Louis T. Steyaert, Editors
Oxford University Press
1993 0-19-508007-6 $65.00

Bringing Geographical Information Systems Into Business
David J. Grimshaw
Longman Scientific & Technical
1994 0-470-23426-1 $34.95

Landscape Ecology and GIS
Roy Haines-Young, David R. Green, and Stephen H. Cousins, Editors
Taylor & Francis
1994 0-7484-0252-7 $37.50

GIS and Site Design
Karen C. Hanna and R. Brian Culpepper
John Wiley & Sons
1998 0471163872 $54.95

Visualization in Geographical Information Systems
Hillary M. Hearnshaw and David J. Unwin, Editors
John Wiley & Sons
1994 0-471-94435-1 $105.00

GIS Data Conversion: Strategies, Techniques, and Management
Pat Hohl, Editor
OnWord Press
1998 1-56690-175-8 $49.90

An Introduction to Urban Geographic Systems
William E. Huxhold
Oxford University Press
1991 0-19-506535-2 $34.00

Managing Geographic Information Projects
William E. Huxhold and Allan G. Levinsohn
Oxford University Press
1995 0-19-507869-1 $45.00

The Local Government Guide to GIS: Planning and Implementation
International City/County Management Association
International City/County
1991 0-87326-919-5 $32.00

Geographic Information Systems & Mapping: Practices and Standards
A. I. Johnson, C. B. Petterson, and J. L. Fulton
American Society for Testing & Materials
1992 0-8031-1471-0 $75.00

The Global Positioning System and GIS: An Introduction
Michael Kennedy
Ann Arbor
1996 $60

An Intensive Comparison of Triangulated Irregular Networks (TINs) and Digital Elevation Models (DEMs)
Mark P. Kumler
University of Toronto Press, Inc.
1995 $20.00

Time in Geographic Information Systems
Gail Langran
Taylor & Francis
1992 0-7484-0003-6 $39.95

Land Registration and Cadastral Systems
Gerhard Larsson
John Wiley & Sons
1991 0-470-21798-7 $43.95

Fundamentals of Spatial Information Systems
Robert Laurini and Derek Thompson
Academic Press Limited
1992 0-12-438380-7 $49.95

Remote Sensing and Image Interpretation
Thomas M. Lillesand and Ralph W. Kiefer
John Wiley & Sons
1994 (Third Edition) 0-471-57783-9 $78.95

Spatial Analysis : Modelling in a GIS Environment
Paul Longley (Editor), Michael Batty (Editor), and Mike Batty (Contributor)
John Wiley & Sons
1997 0470236159 $90.00

Some Truth with Maps: A Primer on Symbolization & Design
Alan M. MacEachren
Association of American Geographers
1994 $15.00

Visualization in Modern Cartography
Alan M. MacEachren and D. R. Fraser Taylor, Editors
Pergamon
1994 0-08-042415-5 $43.00

Geographic Information Systems: Principles and Applications
David J. Maguire, Michael Goodchild, and David W. Rhind, Editors
John Wiley & Sons
1991 0-470-21789-8 $295.00

Geographic Information Systems and Their Socioeconomic Applications
David Martin
Routledge
1991 0-415-05697-7 $19.95

Handling Geographic Information: Methodology and Potential Applications
Ian Masser and Michael Blakemore, Editors
John Wiley & Sons
1991 0-470-21792-8 $99.00

Diffusion and Use of Geographic Information Technologies
Ian Masser and Harland J. Onsrud, Editors
Kluwer Academic Publishers
1993 0-7923-2190-1 $136.00

Generalization in Digital Cartography
Robert B. McMaster and K. Stuart Shea
Association of American Geographers
1992 8-89291-209-X $15.00

Human Factors in Geographical Information Systems
David Medyckyj-Scott and Hilary M. Hearnshaw, Editors
Belhaven Press
1993 1-85293-262-7 $79.95

Drawing the Line: Tales of Maps and Cartocontroversy
Mark Monmonier
Henry Holt and Company, Inc.
1995 0-8050-2581-2 $27.50

How to Lie with Maps
Mark Monmonier
The University of Chicago Press
1991 0-226-53414-4 $14.95

GIS Data Conversion Handbook
Glenn E. Montgomery and Harold C. Schuch
GIS World
1993 0-9625063-4-6 $44.95

Raster Imagery in Geographic Information Systems
Stan Morain and Shirley López Baros, Editors
OnWord Press
1996 1-56690-097-2 $59.90

Directory of Colleges and Universities Offering Geographic Information Systems Courses
John M. Morgan III and Genevieve R. Bennett
American Congress on Surveying and Mapping
1990 $25.00

GIS and Generalization: Methodology and Practice
Jean-Claude Muller, Jean-Philipe Lagrange, and Robert Weibel, Editors
Taylor & Francis
1995 0-7484-0319-1 $45.00

Managing Geographic Information Systems
Nancy J. Obermeyer and Jeffrey K. Pinto
The Guilford Press
1994 0-89862-005-8 $31.50

Geographic and Land Information Systems for Practicing Surveyors
Harlan J. Onsrud and David W. Cook, Editors
American Congress on Surveying and Mapping
1990 $45

Digital Parcel Mapping
Karen Parrish
Urban and Regional Information Systems Association (URISA)
$55

Simple Computer Imaging and Mapping
Micha Pazner, Nancy Thies, and Roberto Chavez
ThinkSpace, Inc.
1994 1-896671-00-4 $39.95

GIS Technologies for the Transportation Industry
Hilary Perkins, Editor
Urban and Regional Information Systems Association (URISA)
$69.00

Terrain Modeling in Surveying and Civil Engineering
G. Petrie and T. J. M. Kennie, Editors
McGraw-Hill, Inc.
1991 0-07-049683-8 $72.00

Introductory Readings in Geographical Informations Systems
Donna J. Peuquet and Duane F. Marble, Editors
Taylor & Francis
1990 0-85066-857-3 $39.50

Ground Truth: The Social Implications of GIS
John Pickles, Editor
The Guilford Press
1995 $19

GIS Online: Information Retrieval, Mapping, and the Internet
Brandon Plewe
OnWord Press
1997 1-56690-137-5 $46.95

Mountain Environments and Geographic Information Systems
Martin F. Price and D. Ian Heywood, Editors
Taylor & Francis
1994 0-7484-0088-5 $99.00

Three-Dimensional Applications in Geographic Information Systems
Jonathan Raper, Editor
Taylor & Francis
1989 0-85066-776-3 $75.00

Applications of Spatial Data Structures
Hanan Samet
Addison-Wesley
1990 0-201-50300-x $45.25

The Design and Analysis of Spatial Data Structures
Hanan Samet
Addison-Wesley
1990 0-201-50255-0 $43.25

Development Management Systems
Dennis A. Sandquist, AICP
Urban and Regional Information Systems Association (URISA)
1998 $19.00

Geographical Information Systems for Urban and Regional Planning
Henk J. Scholten and John C. H. Stillwell, Editors
Kluwer Academic Publishers
1990 0-7923-0793-3 $130.50

Geographic Information Systems: An Introduction
Jeffrey Star and John Estes
Prentice-Hall
1990 0-13-351123-5 $55.00

Geographic Information Systems: The Microcomputer and Modern Cartography
D. R. Fraser Taylor, Editor
Pergamon Press
1991 0-08-040277-1 $47.00

Geographic Information Systems and Cartographic Modeling
C. Dana Tomlin
Prentice-Hall
1990 0-13-350927-3 $68.00

Three-Dimensional Modeling with Geoscientific Information Systems
A. Keith Turner, Editor
Kluwer Academic Publishers
1992 0-9625063-7-0 $167.50

GIS Database Concepts
Urban and Regional Information Systems Association (URISA)
1998 $19.00

GIS Guidelines for Assessors
Urban and Regional Information Systems Association (URISA)
1999 $55.00

Marketing Government Geographic Information: Issues & Guidelines
Urban and Regional Information Systems Association (URISA)
$45.00

Implementation of Land Information Systems in Local Government
Stephen J. Ventura
Wisconsin State Cartographer's Office
1991 $15

Enterprise GIS
Nancy R. von Meyer and R. Scott Oppman
Urban and Regional Information Systems Association (URISA)
$55.00

Data Modeling
Michael D. Walls
Urban and Regional Information Systems Association (URISA)
1998 $19.00

GIS in Cities & Counties: A Nationwide Assessment
Lisa Warnecke, Jeff Beattie, Cheryl Kollin, Winifred Lyday, and Steve French
Urban and Regional Information Systems Association (URISA)
$55.00

Contouring: A Guide to the Analysis and Display of Spatial Data
David F. Watson
Pergamon Press
1992 0-08-040286-0 $120.00

Advances in GIS Research
T. C. Waugh and R. G. Healey, Editors
Taylor & Francis
1994 0-7484-0315-9 $125.00

The Power of Maps
Denis Wood
The Guilford Press
1992 0-89862-493-2 $17.95

Innovations in GIS 1
Michael F. Worboys, Editor
Taylor & Francis
1994 0-7484-141-5 $44.00

Spatial Analysis and Spatial Policy Using Geographic Information Systems
Lew Worrall, Editor
Bellhaven Press
1991 1-85293-141-8 $83.95

Periodicals

The periodicals listed are either principally devoted to GIS issues or cover a related field and treat GIS issues extensively. The list of periodicals includes the title of the publication, and the publisher's name, address, and telephone and fax numbers. The list is alphabetized by publication title.

ACSM Bulletin

Cartography and Geographic Information Science
Surveying and Land Information Systems
American Congress on Surveying and Mapping (ACSM)
5410 Grosvenor Lane, Bethesda, MD 20814
301/493-0200; fax: 301/493-8245

http://www.acsm.net/publist.html

American Demographics

American Demographics, Inc.
P.O. Box 68, Ithaca, NY 14851
1-800-828-1133; fax: 607/273-3196

http://www.demographics.com/

AM/FM/GIS Networks

Energy IT Magazine
Geospatial Information & Technology Association (GITA)

(Formerly AM/FM International)
14456 E. Evans Avenue, Aurora, CO 80014
303/337-0513; fax: 303/337-1001

http://www.gita.org

Business Geographics

GeoTec Media
2101 S. Arlington Heights Road Suite 150
Arlington Heights, IL 60005
847/427-9512; fax: 847/427-2079

http://www.gisworld.com/bg/2000/0500/

Design & Drafting News

American Design Drafting Association
P.O. Box 11937
Columbia, SC 29211
803/771-0008; fax: 803/771-4272

http://www.adda.org/

Earth Observation Magazine

EOM, Inc.
4901 E. Dry Creek Road
Suite 170
Littleton, CO 80122
303/713-9500; fax: 303/713-9944

http://www.eomonline.com/

FGDC Newsletter

Federal Geographic Data Committee
590 National Center, USGS
Reston, VA 20192
703/648-5740

http://fgdc.er.usgs.gov/publications/publications.html

Geo Information Systems

Advanstar Communications, Inc.
859 Willamette St.
Eugene, OR 97401
503/344-3514

http://www.advanstar.com/index_allpubs.html

GEOWorld

GeoTec Media
2101 S. Arlington Heights Road, Suite 150
Arlington Heights, IL 60005
847/427-9512; fax 847/427-2079

http://www.gisworld.com/gw/

GPS World

Advanstar Communications, Inc.
859 Willamette St.
Eugene, OR 97401
503/344-3514

http://www.advanstar.com/index_allpubs.html

ICMA Newsletter

International City Management Association (ICMA)
777 North Capitol Street, NE
Suite 500
Washington, DC 20002
202/289-4262

http://icma.org/

International Journal of Geographic Information Systems

Taylor & Francis, Ltd.
325 Chestnut Street
8th Floor
Philadelphia, PA 19106
215/625-8900; fax: 215/625-2940

http://www.journals.tandf.co.uk/

PE & RS (Photogrammetric Engineering & Remote Sensing)

American Society for Photogrammetry and Remote Sensing
5410 Grosvenor Lane #210
Bethesda, MD 20814-2160
301/493-0290; fax: 301/493-0208

http://www.asprs.org/publications.html

P.O.B.

Business News Publishing
755 West Big Beaver
Suite 1000
Troy, MI 48084
248/362-3700; fax: 248/362-0317

http://www.pobonline.com/

Professional Surveyor

Professional Surveyors Publishing Company, Inc.
1713-J Rosemont Avenue
Frederick, MD 21702-4170
301/682-6101; fax: 301/682-6105

http://www.profsurv.com/

URISA Journal

Urban & Regional Information Systems Association (URISA)
1460 Renaissance Drive
#305 Park Ridge, IL 60068
847/824-6300; fax: 847/824-6363

http://www.urisa.org/journal.htm

URISA NEWS

Urban & Regional Information Systems Association (URISA)
1460 Renaissance Drive
#305Park Ridge, IL 60068
847/824-6300; fax: 847/824-6363

http://www.urisa.org/NEWS/Newset.htm

Educational Institutions

The following schools offer training or study in GIS. The list includes the school's name, address, and telephone and fax numbers, as well as the department principally responsible for GIS instruction. It includes schools in North America and is arranged alphabetically by state.

You should note that many community colleges also offer GIS training. Moreover, GIS software vendors offer training in the use of their particular GIS products, as do many private companies. These companies include privately owned GIS and CADD training centers, computer system integrators, value-added resellers (VARs), software developers, and data conversion service bureaus that use or resell GIS software.

Although many of their GIS programs may include training in specific aspects of GIS applications, colleges and universities are concentrating on educating students in GIS concepts and theory. Moreover, training in particular GIS software is being done by an increasing number of community colleges.

Alabama

Auburn University
Geography
2190 Haley Center
Auburn, AL 36849-5224
205/844-3420; fax: 205/844-2378

Jacksonville State University
Geography/Anthropology
237 Martin Hall
Jacksonville, AL 36265
205/782-5232; fax: 205/782-5228

Samford University
History/Political, Science/Geography
800 Lakeshore Drive
Birmingham, AL 35229
205/870-2109; fax: 205/870-2384

University of North Alabama
Geographic Research Center
Geography
Box 5064
Florence, AL 35632-0001
205/760-4640; fax: 205/760-4329

Arkansas

University of Arkansas
Geography
108 Ozark Hall
Fayetteville, AR 72701
501/575-6159; fax: 501/575-3846

University of Central Arkansas
Geography
201 Donaghey
Conway, AR 72035
501/450-3164; fax: 501/450-5185

California

California State University, Fullerton
P.O. Box 34080
Geography
Fullerton, CA 92634
714/773-3161; fax: 714/773-2209

California State University, Northridge
Geography
18111 Nordhoff Street
Northridge, CA 91330-8249
818/885-3532; fax: 818/885-2723

California State University, Sacramento
Geography
6000 J. Street
Sacramento, CA 95819-6003
916/278-6109; fax: 916/278-5787

Humboldt State University
Natural Resources Planning and
Interpretation Department
1 Harpst Street
Arcata, CA 95521-8299
707/826-3438; fax: 707/826-4145

San Diego State University
Geography
San Diego, CA 92182
619/594-5466; fax: 619/594-4938

San Francisco State University
Geography
1600 Holloway Avenue
San Francisco, CA 94044
415/338-2983; fax: 415/338-1980

San Jose State University
Geography and Environmental Studies
One Washington Square, WSQ 118
San Jose, CA 95192-0116
408/924-5475; fax: 408/924-5477

University of California, Los Angeles
Geography, Bunche Hall
405 Hilgard Avenue
Los Angeles, CA 90024-1524
310/825-3525 or 825-1071; fax: 310/206-5976

University of California, Riverside
Earth Sciences
Riverside, CA 92521
714/787-3434; fax: 714/787-4324

University of California, Santa Barbara
Geography
Santa Barbara, CA 93106-4060
805/893-8224; fax: 805/893-8617

Colorado

Colorado State University
Forest Sciences
Colorado State University
Fort Collins, CO 80523
303/491-6911; fax: 303/491-6754

University Corporation for Atmospheric Research (UCAR)
University NAVstar Consortium (UNAV-CO)
P.O. Box 3000
Boulder, CO 80307-3000
303/497-8020; fax: 303/497-8028

University of Colorado, Colorado Springs
Geography and Environmental Studies
1420 Austin Bluffs Parkway
Colorado Springs, CO 80933-7150
719/593-3166; fax: 719/593-3362

University of Southern Colorado
Civil Engineering Technology
2200 Bonforte Boulevard
Pueblo, CO 81001
719/549-2683; fax: 719/549-2519

Connecticut

University of Connecticut
Natural Resources Management/
Engineering U-87
Room 308
1376 Storrs Road
Storrs, CT 06269-4087
203/486-2840; fax: 203/486-5408

University of New Haven
Biology and Environmental Sciences
300 Orange Avenue
West Haven, CT 06516
203/932-7108; fax: 203/932-2036

Florida

Florida Atlantic University
Geography
P.O. Box 3091
Boca Raton, FL 33431-0991
407/367-3250; fax: 407/367-2744

University of Florida
Urban & Regional Planning
Room 431, Arch. Building
College of Architecture
Gainesville, FL 32611
904/392-0997; fax: 904/392-7266

University of Florida
Geography
3141 Turlington Hall
P.O. Box 117315
Gainesville, FL 32611-7315
904/392-0494; fax: 904/392-3584

University of South Florida
Geography
SOC 107
Tampa, FL 33620-8100
813/974-2386; fax: 813/974-2668

Georgia

Georgia State University
Geography
University Plaza
Atlanta, GA 30303-3083
404/874-9642

University of Georgia
Institute of Government
201 N. Milledge Avenue
Athens, GA 30602
404/542-2736; fax: 404/542-9301

Hawaii

University of Hawaii
Geography
2424 Maile Way
Honolulu, HI 96822
808/956-8465; fax: 808/956-3528

Idaho

University of Idaho
Department of Geography
University of Idaho
Moscow, ID 83843
208/885-6216; fax: 208/885-5724

Illinois

Augustana College
Geography
New Science Building
639 38th Street
Rock Island, IL 61201
309/794-7325; fax: 309/794-7422

Northern Illinois University
Geography
Davis Hall 118
DeKalb, IL 60115-2854
815/753-6827; fax: 815/753-6872

Northwestern University
Civil Engineering
Evanston, IL 60208
708/491-4338; fax: 708/491-4011

Western Illinois University
Geography Department
Macomb, IL 61455
309/298-1764; fax: 309/298-2400

Indiana

Ball State University
Geography
Cooper Science Building
Muncie, IN 47306
317/285-1776; fax: 317/285-2351

Purdue University
Agricultural Engineering
1146 AGEN
West Lafayette, IN 47907-1146
317/494-1198; fax: 317/496-1115

Purdue University
Civil Engineering
1284 Civil Engineering Building
West Lafayette, IN 47907-1146
317/494-2157; fax: 317/496-0395

Iowa

University of Iowa
Geography
316 Jessup Hall
Iowa City, IA 52242-1316
319/335-0153; fax: 319/335-2725

Kansas

Kansas State University
Geography
Dickens Hall
Manhattan, KS 66506-0801
913/532-6727; fax: 913/532-7310

University of Kansas
Geography
213 Lindley Hall
Lawrence, KS 66045-2121
913/864-5143; fax: 913/864-5276

Kentucky

Eastern Kentucky University
Geography and Planning
Roark 201
Richmond, KY 40475-3129
606/622-1418; fax: 606/622-1020

Murray State University
Mid-America Remote Sensing Center
Lowry Center
Murray, KY 42071
502/762-2148; fax: 502/762-4417

Louisiana

University of Southwestern Louisiana
Center for Analysis of Spatial & Temporal Systems (CASTS)
P.O. Box 43730
Lafayette, LA 70504-3730
318/231-5813; fax: 318/231-6688

Maryland

Towson State University
Geography and Environmental Planning
Linthicum Hall, Room 30
Baltimore, MD 21204-7097
410/830-2964; fax: 410/830-3888

University of Maryland at College Park
Geography
Lefrak Hall, University of Maryland
College Park, MD 20742
301/405-4050; fax: 301/314-9299

University of Maryland
Baltimore County Campus
Geography
5401 Wilkins Avenue
211 Social Sciences
Baltimore, MD 21228
410/455-3149; fax: 410/455-1056

Michigan

Central Michigan University
Geography
296A Dow Science Building
Mt. Pleasant, MI 48859
517/774-3323; fax: 517/774-3537

Grand Valley State University
Biology and Natural Resources Management
226 Mackinac
Allendale, MI 49401
616/895-2470; fax: 616/895-3506

Michigan State University
Geography
315 Natural Science Building
E. Lansing, MI 48824
517/355-4651; fax: 517/336-1076

Michigan Technological University
School of Technology-Surveying
1400 Townsend Drive
Houghton, MI 49931-1295
906/487-2259; fax: 906/487-2583

The University of Michigan-Flint
Resource Science
516 Murchie Science Building
Flint, MI 48502-2186
313/762-3355; fax: 313/762-3687

Minnesota

Saint Mary's College of Minnesota
Biology
700 Terrace Heights
Winona, MN 55987
507/457-1544; fax: 507/457-1633

St. Cloud State University
Geography
720 S. 4th Avenue, SH 317
St. Cloud, MN 56301-4498
612/255-2170; fax: 612/654-5198

University of Minnesota
Natural Resources Institute
5013 Miller Trunk Highway
Duluth, MN 55811
218/720-4269; fax: 218/720-4219

University of Minnesota
Forest Resources
1530 N. Cleveland Avenue
St. Paul, MN 55108
612/624-9271; fax: 612/625-5212

University of Minnesota
Geography
414 Social Science Tower
Minneapolis, MN 55455
612/625-6080; fax: 612/624-1044

University of St. Thomas
Geography
LOR 306, 2115 Summit Avenue
St. Paul, MN 55105-1096
612/962-5560; fax: 612/962-6410

Missouri

Northwest Missouri State University
Computer Science and Information Systems
113 Garrett-Strong Hall
Maryville, MO 64468
816/562-1600; fax: 816/562-1900

University of Missouri-Columbia
Geography
Room 3, Stewart Hall
Columbia, MO 65211
314/882-8370; fax: 314/884-4239

Montana

Montana State University
Earth Sciences
200 Traphagen
Bozeman, MT 59717-0348
406/994-3331; fax: 406/994-6923

Nebraska

University of Nebraska at Omaha
Geography
60th & Dodge
Omaha, NE 68182
402/554-4805; fax: 402/554-3518

University of Nebraska-Lincoln
Center for Advanced Land Management
Information Technologies
113 Nebraska Hall
Lincoln, NE 68588-0517
402/472-7531; fax: 402/472-2410

Nevada

University of Nevada-Las Vegas
Civil and Environmental Engineering
P.O. Box 454015
Las Vegas, NV 89154-4015
702/895-3701; fax: 702/895-3936

New Hampshire

Dartmouth College
Geography
6017 Fairchild Hall
Hanover, NH 03755
603/646-1309

New Jersey

Glassboro State College
Geography & Anthropology
Robinson Building
201 Mullica Hill Road
Glassboro, NJ 08028
609/863-7311; fax: 609/863-6165

New Jersey Institute of Technology
Civil Engineering
323 King Boulevard
Newark, NJ 07102
201/596-5808; fax: 201/242-1823

Rutgers University
Cook College Remote Sensing Center
Natural Resources
College Farm Road
New Brunswick, NJ 08903
908/932-9631; fax: 908/932-8644

New Mexico

New Mexico State University
Geography
P.O. Box 30001-Department 3MAP
Las Cruces, NM 88003-0001
505/646-3509; fax: 505/646-7430

University of New Mexico
Geography/Technology Application Center
Albuquerque, NM 87131-6031
505/77-3622; fax: 505/277-3614

New York

Cornell University
Natural Resources
Fernow Hall
Ithaca, NY 14853
607/255-9423

Pace University
Biological Sciences
861 Bedford Road
Pleasantville, NY 10570
914/773-3563; fax: 914/773-3541

Syracuse University
343 H. B. Crouse Hall
Syracuse, NY 13244-1160
315/443-2605; fax: 315/443-4227

University at Albany
State University of New York
Information Science PhD Program
Draper 113, University at Albany
135 Western Avenue
Albany, NY 12222
518/442-3306; fax: 518/442-5232

North Carolina

Organization for Tropical Studies
410 Swift Avenue
P.O. Box 90630
Durham, NC 27708
919/684-5774; fax: 919/684-5661

University of North Carolina at Chapel Hill
City & Regional Planning
CB #3140 New East Building
Chapel Hill, NC 27599
919/962-3983; fax: 919/962-5206

University of North Carolina-Charlotte
Geography and Earth Sciences
Charlotte, NC 28223
704/547-4247; fax: 704/547-3182

Western Carolina University
Natural Resources Management Program
218-B Stillwell
Cullowhee, NC 28723
704/227-7367; fax: 704/227-7647

Ohio

Bowling Green State University
Geography
305 Hanna Hall
Bowling Green, OH 43403
419/372-2925; fax: 419/372-8600

Bowling Green State University
Geology
Overman Hall
Bowling Green, OH 43403
419/372-2490; fax: 419/372-7205

Kent State University
Geography
413 McGilvrey Hall, KSU Campus
P.O. Box 5190
Kent, OH 44242-0001
216/672-2045; fax: 216/672-4304

Ohio State University
Geodetic Science and Surveying
1958 Neil Avenue
Columbus, OH 43210
614/292-6753; fax: 614/292-2957

Ohio State University
School of Natural Resources
210 Kottman Hall, 2021 Coffey Road
Columbus, OH 43210-1085
614/292-2265; fax: 614/292-7432

Ohio State University
Geography
1036 Derby Hall
190 N. Oval Mall
Columbus, OH 43210
614/292-2514; fax: 614/292-6213

Ohio University
Geography
122 Clippinger Labs
Athens, OH 45701-2979
614/593-1140; fax: 614/593-1139

University of Akron
Geography and Planning
Carroll Hall 306
Akron, OH 44325-5005
216/972-7620; fax: 216/972-6080

University of Cincinnati
Geography
714 Swift Hall, ML 0131
Cincinnati, OH 45221-0131
513/556-3421; fax: 513/556-3370

Wittenberg University
Geography
Springfield, OH 45501
513/327-7515; fax: 513/327-6340

Oklahoma

Northeastern State University
Geography & Sociology
Seminary Hall
Tahlequah, OK 74464
918/456-5511, ext. 3525; fax: 918/458-2193

Oklahoma State University
Geography
GEOG Building 308
Stillwater, OK 74078
405/744-9173; fax: 405/744-5620

University of Oklahoma
Geography
Sarkeys Energy Center, Room 684
100 E. Boyd Street
Norman, OK 73019
405/325-5325; fax: 405/325-3148

University of Oklahoma
College of Architecture
Regional and City Planning
and Landscape Architecture
Gould Hall, Room 162
Norman, OK 73019-0385
405/325-3871; fax: 405/325-7558

University of Oklahoma
School of Civil Engineering
and Environmental Science
202 W. Boyd Street, Room 334
Norman, OK 73019
405/325-5911; fax: 405/325-4217

Oregon

Oregon Institute of Technology
Surveying
3201 Campus Drive
Klamath Falls, OR 97601-8801
503/885-1511; fax: 503/885-1777

Oregon State University
Geosciences
Wilkinson Hall 104
Corvallis, OR 97331-5506
503/737-1201; fax: 503/737-1200

Pennsylvania

Indiana University of Pennsylvania
Geography & Regional Planning
IUP, 2 Leonard Hall
Indiana, PA 15705
412/357-2250; fax: 412/357-6213

Kutztown University
Geography
Grim Science Building
Kutztown, PA 19530
215/683-4364; fax: 215/683-1352

Pennsylvania State University
Geography
302 Walker Building
University Park, PA 16802
814/865-3433; fax: 814/863-7943

Shippensburg University
Geography-Earth Science
1871 Old Main Drive
Shippensburg, PA 17257-2299
717/532-1399; fax: 717/532-1273

Slippery Rock University
Geography & Environmental Studies
Slippery Rock, PA 16057
412/738-2048; fax: 412/738-2188

Temple University
Office of Undergraduate Admissions
Broad Street and Montgomery Avenue
Philadelphia, PA 19122
215/204-7200; fax: 215/204-5694

University of Pennsylvania
Landscape Architecture and Regional Planning
Graduate School of Fine Arts, Meyerson Hall
Philadelphia, PA 19104
215/898-6591; fax: 215/898-9215

Rhode Island

University of Rhode Island
Natural Resources Science
210B Woodward Hall
Kingston, RI 02881
401/792-2495; fax: 401/792-4561

South Carolina

University of South Carolina
Geography
Columbia, SC 29208
803/777-8976; fax: 803/777-4972

South Dakota

South Dakota State University
Geography
Scobey Hall, Box 504
Brookings, SD 57007-0648
605/689-4511; fax: 605/688-6386

Tennessee

Memphis State University
Geography
Johnson Hall, Room 107
Memphis, TN 38152
901/678-2386; fax: 901/678-3299

Middle Tennessee State University
Geography and Geology
P.O. Box 9
Murfreesboro, TN 37132
615/898-2726; fax: 615/898-5538

University of Tennessee at Chattanooga
Sociology, Anthropology, and Geography
615 McCallie Avenue
Chattanooga, TN 37403
615/755-4435; fax: 615/755-4279

Texas

Texas A&M University
Forest Science
Remote Sensing/GIS Laboratory
Texas A&M University
College Station, TX 77845-2135
409/845-5069; fax: 409/845-6049

Utah

University of Utah Geography
OSH 270
Salt Lake City, UT 84112
801/581-8218; fax: 801/581-6957

University of Utah
Geography, DIGIT Lab
270 Orson Spencer Hall
Salt Lake City, UT 84112
801/581-3612 or 581-3613; fax: 801/581-6957

Utah State University
Department of Geography & Earth Resources
College of Natural Resources
Logan, UT 84322-5240
801/750-1292; fax: 801/750-4048

Utah State University
Agriculture and Irrigation Engineering
Room EC216
Logan, UT 84322-4105
801/750-2785; fax: 801/750-1248

Virginia

George Mason University
Spatial Decision Support Systems Laboratory
Graduate House
Fairfax, VA 22030
703/993-3351; fax: 703/993-2284

George Mason University
Geography and Earth Systems Science
4400 University Drive
Room 2067, King Hall
Fairfax, VA 22030
703/993-1210; fax: 703/993-1359

Virginia Commonwealth University
Urban Studies and Planning
812 W. Franklin Street
Richmond, VA 23284-2008
804/367-1134; fax: 804/367-6681

Virginia Tech
Geography
301 Patton Hall
Blacksburg, VA 24061-0115
703/231-6886; fax: 703/231-6367

Washington

Central Washington University
GIS Lab
Geography and Land Studies
Ellensburg, WA 98926-7500
509/963-1447; fax: 509/963-1047

Wisconsin

University of Wisconsin-Madison
Environmental Remote Sensing Center
1225 W. Dayton Street, 12th Floor
Madison, WI 53706
608/262-1585; fax: 608/262-5964

University of Wisconsin-Madison
Geography
550 N. Park Street
384 Science Hall
Madison, WI 53706-1491
608/262-4846; fax: 608/265-3991

University of Wisconsin-Madison
Landscape Architecture
LICGF (Land Info. & Computer Graphics Facility)
25 Agriculture Hall
Madison, WI 53706
608/263-5534; fax: 608/262-2500

University of Wisconsin-Milwaukee
Urban Planning
2131 E. Hartford
Milwaukee, WI 53211
414/229-4014; fax: 414/229-6976

University of Wisconsin-Milwaukee
Center for Continuing Engineering Education
GIS Technology and Education Center
929 N. Sixth Street
Milwaukee, WI 53203
414/227-3115; fax: 414/227-3119

University of Wisconsin-Milwaukee
Geography
P.O. Box 413
Milwaukee, WI 53201
414/229-4866; fax: 414/229-3981

University of Wisconsin-Stevens Point
Geography/Geology
Stevens Point, WI 54481
715/346-2629; fax: 715/346-3624

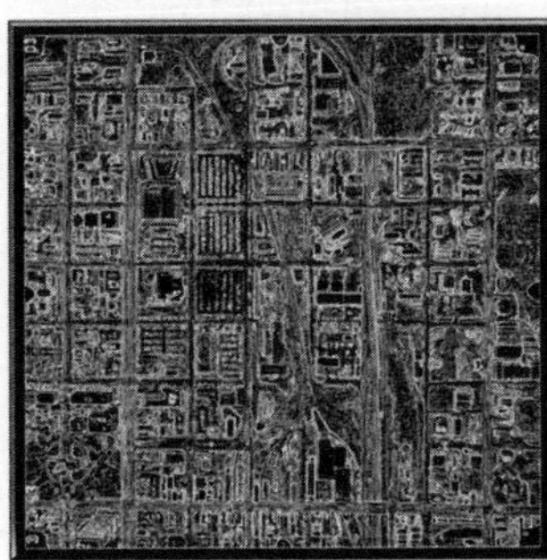

Appendix C
GIS Data Sources

Roughly two-thirds of the total cost of implementing a GIS involves building its database. This cost is often the biggest obstacle to justifying a GIS program. Therefore, GIS users are constantly searching for ways to build the GIS database at a lower cost. One way is to buy GIS data that someone else has collected. Existing digital map data can usually be purchased for much less than the cost of creating it. This is especially true if the seller has many buyers. The seller can recover the initial investment over a larger number of sales; therefore, the seller can charge a lower price on each purchase.

A typical county or large city may have to spend hundreds of thousands of dollars to obtain new digital topographic mapping for a GIS base map. On the other hand, it may be able to purchase data containing the same data themes (e.g., roads, drainage, topography, buildings, and vegetation) and covering the same area for a few hundred dollars. "What's the catch?" Well, it's a big one. The horizontal and vertical accuracy of the less expensive existing data is likely to be much lower than that of the new topographic mapping. It is also likely to provide far fewer details on the map features. (See Chapter 21 for a more detailed discussion of map accuracy.)

However, many GIS programs are beginning with existing digital map data because of its lower cost. The plan may be to supplement and improve this data over time with more accurate and more detailed data, so in the long run the total cost may be greater than simply buying the more accurate data to begin with. Nonetheless, this incremental approach to GIS database construction is gaining in popularity.

Moreover, this approach does not just lower the initial GIS investment threshold. It also helps the GIS program produce visible results more quickly. Building the GIS database "from scratch" can take a very long time. Therefore, when the GIS database is built with existing digital map data, users can get on to initial applications sooner. Showing some benefits in short order is a good way to secure management support and additional funding for the GIS program.

This appendix presents a table of federal sources of existing GIS data. The list is taken from the Manual of Federal Geographic Data Products published by the Federal Geographic Data Committee (FGDC). Paper and microfiche copies of the manual are available for sale from the National Technical Information Service, 5285 Port Royal Road, Springfield, VA 22161, telephone 703-605-6000. To see the latest FGDC information on federal GIS data sources, it would be best visit their web site at *http://www.fgdc.gov/data/data.html.*

The FGDC web site links to another excellent web site for public GIS data sources maintained by the Center for Advanced Spatial Technology (CAST) of the University of Arkansas. It is located at *http://www.cast.uark.edu/local/hun*t.

State and local governments, as well as local and regional utilities, also collect and sell GIS data. They too usually charge only a nominal fee. Of course, the data will be limited to the geographic area for which the data supplier is responsible. The format of the data will depend on the GIS the agency or utility has chosen, and this may not be compatible with your GIS. But they should be able to provide the data in a popular data transfer format, such as the USGS DLG or Autodesk DXF formats. In the case of state agencies and utilities, the available data themes are usually only those related to their "business."

Local governments that have implemented GIS typically offer data related to land value, use, and ownership, as well as public works, topography, flood plains, and so forth. Many local governments are also responsible for utility services, including the water, storm, and sanitary systems. It is far beyond the scope of this book to attempt to list all of these local sources, but you should only need to make a few phone calls to a state agency, a utility, or local government in order to determine what GIS data is available and how to obtain it.

GIS data is also available from commercial suppliers. These data sets were often initially obtained from federal government agencies, then repackaged and/or enhanced by the vendor. The web site *http://www.geoplace.com* is one of several that list such companies. Select the "GeoDirectory," then select "Data Sources," and finally click on "Find Company" to see a list of commercial GIS data suppliers.

	Atmospheric				Boundaries						Socioecon.			
GIS DATA SUPPLIERS	Climate	Radiation	Temperature	Weather	Administrative	Census Geography	County	International	Local Government	State	Demographic	Economic	Mortality	Natality
FEDERAL														
Department of Agriculture														
Agricultural Stabilization & Conservation Service 801/524-5013 FAX: 801/524-5244														
Forest Service 202/205-1760					|		|							
Soil Conservation Service 817/334-5559 FAX: 817/334-5469														
Department of Commerce														
Bureau of the Census 301/763-4100 FAX: 301/763-4794					|	|	|		|	|	|	|		
Bureau of Economic Analysis 202/606-9900							|			|	|	|		
National Climatic Data Center 704/271-4800			|	|										
National Env. Satellite Data & Info. Service 301/763-8400 FAX: 301/763-8443	|	|	|	|										
National Geodetic Survey 301/713-3242 FAX: 301/713-4172														
National Oceanographic Data Center 202/606-4549 FAX: 202/606-4586														
Department of Defense														
Defense Mapping Agency 800/826-0342								|						
Department of Health and Human Services														
Centers for Disease Control National Technical Information Service (NTIS) 703/487-4763							|			|	|		|	|
Department of the Interior														
Bureau of Land Management 303/236-6376 FAX: 303/236-6564														
Bureau of Mines 202/501-9597														
Bureau of Reclamation 303/236-6741														
Minerals Management Service 303/236-5825					|			|		|				

GIS DATA SUPPLIERS	Atmospheric				Boundaries						Socioecon.			
	Climate	Radiation	Temperature	Weather	Administrative	Census Geography	County	International	Local Government	State	Demographic	Economic	Mortality	Natality
National Park Service (NPS) 303/969-2590					✓									
U.S. Fish and Wildlife Service 703/491-6255														
United States Geological Survey (USGS) 703/648-5920 FAX: 703/648-5548	✓	✓	✓	✓	✓	✓	✓	✓	✓	✓	✓	✓	✓	
Department of Transportation														
Bureau of Transportation Statistics 202/554-3564														
Federal Emergency Management Agency														
Federal Map Distribution Center 800/358-9616														
National Aeronautics & Space Administration														
National Space Science Data Center 301/286-6695 FAX: 301/286-1771														
Tennessee Valley Authority (TVA)														
TVA Map Sales 615/751-6277					✓		✓		✓	✓	✓			
COMMERCIAL														
American Digital Cartography 414/733-6678 FAX: 414/734-3375					✓	✓	✓	✓	✓		✓			
Earth Satellite Corporation 301/231-0660 FAX: 301/231-5020	✓	✓	✓	✓										
Environmental Research Institute of Michigan 313/994-1200 FAX: 313/665-6559														
EOSAT 301/552-0500 FAX: 301/552-3762		✓	✓											
ETAK, Inc. 800/765-0555 FAX: 415/328-3148					✓	✓	✓	✓	✓	✓				
Geographic Data Technology, Inc. 800/331-7881 FAX: 603/643-6808						✓				✓	✓			
SPOT Image Corporation 703/715-3100 FAX: 703/648-1813														
GeoSystems 717/293-7500 FAX: 717/293-7577														
MapInfo Corporation 800/327-8627 FAX: 518/285-6070					✓	✓	✓	✓			✓	✓		

GIS DATA SUPPLIERS	Geodetic			Geophys.			Hydrologic									
	Global Positioning System	Horizontal Control	Vertical Control	Gravity	Magnetics	Seismic	Coastal	Flood	Floodplain	Hydrography	Hydrologic Units	Ice	Snow	Water	Waterways	Wetlands
FEDERAL																
Department of Agriculture																
Agricultural Stabilization & Conservation Service 801/524-5013 FAX: 801/524-5244																
Forest Service 202/205-1760										✓						
Soil Conservation Service 817/334-5559 FAX: 817/334-5469										✓	✓			✓		
Department of Commerce																
Bureau of the Census 301/763-4100 FAX: 301/763-4794							✓			✓				✓	✓	
Bureau of Economic Analysis 202/606-9900																
National Climatic Data Center 704/271-4800																
National Env. Satellite Data & Info. Service 301/763-8400 FAX: 301/763-8443				✓	✓	✓	✓			✓		✓				
National Geodetic Survey 301/713-3242 FAX: 301/713-4172	✓	✓	✓													
National Oceanographic Data Center 202/606-4549 FAX: 202/606-4586							✓			✓	✓					
Department of Defense																
Defense Mapping Agency 800/826-0342							✓								✓	
Department of Health and Human Services																
Centers for Disease Control National Technical Information Service (NTIS) 703/487-4763																
Department of the Interior																
Bureau of Land Management 303/236-6376 FAX: 303/236-6564																✓
Bureau of Mines 202/501-9597																
Bureau of Reclamation 303/236-6741															✓	
Minerals Management Service 303/236-5825							✓							✓		
National Park Service (NPS) 303/969-2590										✓						

GIS DATA SUPPLIERS	Geodetic			Geophys.			Hydrologic									
	Global Positioning System	Horizontal Control	Vertical Control	Gravity	Magnetics	Seismic	Coastal	Flood	Floodplain	Hydrography	Hydrologic Units	Ice	Snow	Water	Waterways	Wetlands
U.S. Fish and Wildlife Service 703/491-6255																│
United States Geological Survey (USGS) 703/648-5920 FAX: 703/648-5548		│	│	│	│	│	│	│	│	│	│	│	│	│	│	│
Department of Transportation																
Bureau of Transportation Statistics 202/554-3564																
Federal Emergency Management Agency																
Federal Map Distribution Center 800/358-9616								│	│							
National Aeronautics & Space Administration																
National Space Science Data Center 301/286-6695 FAX: 301/286-1771													│			│
Tennessee Valley Authority (TVA)																
TVA Map Sales 615/751-6277		│	│								│			│	│	
COMMERCIAL																
American Digital Cartography 414/733-6678 FAX: 414/734-3375		│	│							│				│	│	│
Earth Satellite Corporation 301/231-0660 FAX: 301/231-5020						│	│	│	│	│			│	│		│
Environmental Research Institute of Michigan 313/994-1200 FAX: 313/665-6559																
EOSAT 301/552-0500 FAX: 301/552-3762							│	│	│	│		│	│	│	│	│
ETAK, Inc. 800/765-0555 FAX: 415/328-3148																
Geographic Data Technology, Inc. 800/331-7881 FAX: 603/643-6808															│	
SPOT Image Corporation 703/715-3100 FAX: 703/648-1813		│	│				│	│	│	│	│	│	│	│	│	│
GeoSystems 717/293-7500 FAX: 717/293-7577																
MapInfo Corporation 800/327-8627 FAX: 518/285-6070																

GIS DATA SUPPLIERS	Land Ownership							Rem.Sens.			
	Cadastral	County	Federal	Private	State	Subsurface	Public Land Survey System	Aerial Photography	Orthophoto Quads	Radar	Satellite
FEDERAL											
Department of Agriculture											
Agricultural Stabilization & Conservation Service 801/524-5013 FAX: 801/524-5244								✓			
Forest Service 202/205-1760			✓								
Soil Conservation Service 817/334-5559 FAX: 817/334-5469											
Department of Commerce											
Bureau of the Census 301/763-4100 FAX: 301/763-4794											
Bureau of Economic Analysis 202/606-9900											
National Climatic Data Center 704/271-4800										✓	✓
National Env. Satellite Data & Info. Service 301/763-8400 FAX: 301/763-8443											✓
National Geodetic Survey 301/713-3242 FAX: 301/713-4172											
National Oceanographic Data Center 202/606-4549 FAX: 202/606-4586								✓			✓
Department of Defense											
Defense Mapping Agency 800/826-0342											
Department of Health and Human Services											
Centers for Disease Control National Technical Information Service (NTIS) 703/487-4763											
Department of the Interior											
Bureau of Land Management 303/236-6376 FAX: 303/236-6564	✓		✓	✓	✓	✓	✓	✓			
Bureau of Mines 202/501-9597											
Bureau of Reclamation 303/236-6741											
Minerals Management Service 303/236-5825	✓		✓	✓	✓						

	Land Ownership							Rem.Sens.			
GIS DATA SUPPLIERS	Cadastral	County	Federal	Private	State	Subsurface	Public Land Survey System	Aerial Photography	Orthophoto Quads	Radar	Satellite
National Park Service (NPS) 303/969-2590											
U.S. Fish and Wildlife Service 703/491-6255											
United States Geological Survey (USGS) 703/648-5920 FAX: 703/648-5548		|	|	|	|	|	|	|	|	|	|
Department of Transportation											
Bureau of Transportation Statistics 202/554-3564											
Federal Emergency Management Agency											
Federal Map Distribution Center 800/358-9616											
National Aeronautics & Space Administration											
National Space Science Data Center 301/286-6695 FAX: 301/286-1771										|	|
Tennessee Valley Authority (TVA)											
TVA Map Sales 615/751-6277	|		|				|	|	|		
COMMERCIAL											
American Digital Cartography 414/733-6678 FAX: 414/734-3375							|				|
Earth Satellite Corporation 301/231-0660 FAX: 301/231-5020			|					|	|	|	|
Environmental Research Institute of Michigan 313/994-1200 FAX: 313/665-6559								|	|	|	|
EOSAT 301/552-0500 FAX: 301/552-3762										|	|
ETAK, Inc. 800/765-0555 FAX: 415/328-3148											
Geographic Data Technology, Inc. 800/331-7881 FAX: 603/643-6808		|			|						
SPOT Image Corporation 703/715-3100 FAX: 703/648-1813											|
GeoSystems 717/293-7500 FAX: 717/293-7577											
MapInfo Corporation 800/327-8627 FAX: 518/285-6070											

	Subsurface					Surface Features									
GIS DATA SUPPLIERS	Bathymetry	Geology	Marine	Minerals	Soils	Airports	Land Cover	Land Use	Pipelines	Ports	Railroads	Roads	Structures	Transmission lines	Vegetation
FEDERAL															
Department of Agriculture															
Agricultural Stabilization & Conservation Service 801/524-5013 FAX: 801/524-5244															
Forest Service 202/205-1760												✓			
Soil Conservation Service 817/334-5559 FAX: 817/334-5469					✓		✓	✓							
Department of Commerce															
Bureau of the Census 301/763-4100 FAX: 301/763-4794						✓			✓		✓	✓			
Bureau of Economic Analysis 202/606-9900															
National Climatic Data Center 704/271-4800															
National Env. Satellite Data & Info. Service 301/763-8400 FAX: 301/763-8443	✓	✓	✓				✓	✓							✓
National Geodetic Survey 301/713-3242 FAX: 301/713-4172															
National Oceanographic Data Center 202/606-4549 FAX: 202/606-4586	✓	✓						✓							
Department of Defense															
Defense Mapping Agency 800/826-0342						✓	✓			✓	✓	✓		✓	
Department of Health and Human Services															
Centers for Disease Control National Technical Information Service (NTIS) 703/487-4763															
Department of the Interior															
Bureau of Land Management 303/236-6376 FAX: 303/236-6564															
Bureau of Mines 202/501-9597				✓											
Bureau of Reclamation 303/236-6741								✓							
Minerals Management Service 303/236-5825				✓											

GIS DATA SUPPLIERS	Subsurface					Surface Features									
	Bathymetry	Geology	Marine	Minerals	Soils	Airports	Land Cover	Land Use	Pipelines	Ports	Railroads	Roads	Structures	Transmission lines	Vegetation
National Park Service (NPS) 303/969-2590												✓	✓		
U.S. Fish and Wildlife Service 703/491-6255							✓								
United States Geological Survey (USGS) 703/648-5920 FAX: 703/648-5548	✓	✓	✓	✓	✓	✓	✓	✓	✓	✓	✓	✓	✓	✓	✓
Department of Transportation															
Bureau of Transportation Statistics 202/554-3564						✓					✓	✓			
Federal Emergency Management Agency															
Federal Map Distribution Center 800/358-9616															
National Aeronautics & Space Administration															
National Space Science Data Center 301/286-6695 FAX: 301/286-1771					✓			✓							✓
Tennessee Valley Authority (TVA)															
TVA Map Sales 615/751-6277	✓	✓		✓	✓	✓	✓	✓	✓	✓	✓	✓		✓	✓
COMMERCIAL															
American Digital Cartography 414/733-6678 FAX: 414/734-3375						✓	✓	✓	✓	✓	✓	✓		✓	
Earth Satellite Corporation 301/231-0660 FAX: 301/231-5020	✓	✓	✓	✓	✓		✓	✓	✓			✓		✓	✓
Environmental Research Institute of Michigan 313/994-1200 FAX: 313/665-6559	✓						✓	✓							
EOSAT 301/552-0500 FAX: 301/552-3762	✓	✓	✓	✓	✓	✓	✓	✓	✓	✓	✓	✓		✓	✓
ETAK, Inc. 800/765-0555 FAX: 415/328-3148						✓					✓	✓			
Geographic Data Technology, Inc. 800/331-7881 FAX: 603/643-6808						✓					✓	✓		✓	
SPOT Image Corporation 703/715-3100 FAX: 703/648-1813	✓	✓	✓	✓	✓	✓	✓	✓	✓	✓	✓	✓	✓	✓	✓
GeoSystems 717/293-7500 FAX: 717/293-7577															
MapInfo Corporation 800/327-8627 FAX: 518/285-6070						✓					✓	✓			

GIS DATA SUPPLIERS	Topo		Other						
	Contours	Elevations	Ecological	Environmental	Geographic Names	Navigation	Recreation	Travel	ZIP Code Boundaries
FEDERAL									
Department of Agriculture									
Agricultural Stabilization & Conservation Service 801/524-5013 FAX: 801/524-5244									
Forest Service 202/205-1760		✓			✓		✓	✓	
Soil Conservation Service 817/334-5559 FAX: 817/334-5469									
Department of Commerce									
Bureau of the Census 301/763-4100 FAX: 301/763-4794					✓				✓
Bureau of Economic Analysis 202/606-9900									
National Climatic Data Center 704/271-4800				✓					
National Env. Satellite Data & Info. Service 301/763-8400 FAX: 301/763-8443	✓	✓	✓	✓					
National Geodetic Survey 301/713-3242 FAX: 301/713-4172									
National Oceanographic Data Center 202/606-4549 FAX: 202/606-4586	✓	✓		✓		✓			
Department of Defense									
Defense Mapping Agency 800/826-0342	✓	✓			✓	✓			
Department of Health and Human Services									
Centers for Disease Control National Technical Information Service (NTIS) 703/487-4763									
Department of the Interior									
Bureau of Land Management 303/236-6376 FAX: 303/236-6564			✓				✓	✓	
Bureau of Mines 202/501-9597									
Bureau of Reclamation 303/236-6741							✓		
Minerals Management Service 303/236-5825									

	Topo		Other						
GIS DATA SUPPLIERS	Contours	Elevations	Ecological	Environmental	Geographic Names	Navigation	Recreation	Travel	ZIP Code Boundaries
National Park Service (NPS) 303/969-2590							I		
U.S. Fish and Wildlife Service 703/491-6255									
United States Geological Survey (USGS) 703/648-5920 FAX: 703/648-5548	I	I	I		I	I	I		
Department of Transportation									
Bureau of Transportation Statistics 202/554-3564									
Federal Emergency Management Agency									
Federal Map Distribution Center 800/358-9616									
National Aeronautics & Space Administration									
National Space Science Data Center 301/286-6695 FAX: 301/286-1771									
Tennessee Valley Authority (TVA)									
TVA Map Sales 615/751-6277	I	I				I	I		
COMMERCIAL									
American Digital Cartography 414/733-6678 FAX: 414/734-3375	I	I			I				
Earth Satellite Corporation 301/231-0660 FAX: 301/231-5020	I	I	I	I		I			
Environmental Research Institute of Michigan 313/994-1200 FAX: 313/665-6559									
EOSAT 301/552-0500 FAX: 301/552-3762	I	I	I	I		I			
ETAK, Inc. 800/765-0555 FAX: 415/328-3148						I	I	I	
Geographic Data Technology, Inc. 800/331-7881 FAX: 603/643-6808									I
SPOT Image Corporation 703/715-3100 FAX: 703/648-1813	I	I	I	I		I	I		
GeoSystems 717/293-7500 FAX: 717/293-7577									
MapInfo Corporation 800/327-8627 FAX: 518/285-6070									

Glossary of Commonly Used GIS Terms

Accuracy— The degree of correctness of a measurement, or degree of conformity with a standard. Accuracy relates to the quality of a result, and is distinguished from precision, which relates to the quality of the operation by which the result is obtained.

Accuracy standards— Specific standards to which a finished product must adhere.

Aggregation— The process of combining smaller areas, and the data they contain, into larger areas by dissolving common boundaries.

Alphanumeric string— Continuous segment of information consisting of both numbers and letters. May include symbols such as punctuation marks and mathematical symbols.

AM/FM— Automated mapping and facility management.

Annotation— Text, numbers, or symbols added to a map to enhance its information content.

Arc data— Location data representing linear features or the borders of polygon features.

Area— A 2D defined space expressed as a spatial measurement. A polygon on the earth as projected onto a horizontal plane is an example of an area.

Aspatial query— A question asked of a GIS database on the basis of non-spatial attributes.

Aspect— Horizontal direction toward which a slope faces. Commonly expressed as the direction clockwise from north.

Attribute— Descriptive characteristic of a feature. An attribute value is a measurement assigned to an attribute for a feature instance.

Attribute tagging— The process of assigning an attribute to a particular feature.

Automatic clipping/joining— System capability for copying small portions of a database for movement and placement elsewhere in the database without operator intervention.

Automatic polygon centroid calculation— System capability for determining the center of a polygon area without operator intervention.

Automatic snapping— System capability for completing a line segment whose end falls within a predefined proximity to an intersection or node, without operator intervention.

Azimuth— Horizontal direction of a line measured clockwise from a reference plane, usually the meridian.

Base data— Basic level of map data on which other information is placed for purposes of comparison or geographical correlation. Is often in the form of a base map.

Base line— A starting point to which future changes will be compared.

Base map— A map portraying background reference information onto which other information is placed. Base maps usually show the location and extent of natural earth surface features and permanent manmade objects.

Bearing— Horizontal angle at a given point measured clockwise from a specific reference datum to a second point.

Benchmark— A series of tests for ensuring that hardware and/or software meets user needs.

Boolean analysis— Strategy for searching and retrieving information based on the use of the logical operators AND, OR, and NOT to represent symbolic relationships.

Boolean operator— An operator based on Boolean algebra.

Browsing— System capability for finding an undefined feature or set of features in a database.

Buffer— A polygon enclosing an area within a specified distance from a point, line, or other polygon.

Cartesian coordinates— A system that locates a point by referring to its distance from axes measured at right angles, usually represented as a grid on a map.

Cartography— The profession of drawing maps; the study of maps.

Centroid— A point within a polygon whose coordinates are the averages of the corresponding coordinates for all points included in the given area.

Clustering operations— Processes that agglomerate (cluster) individual items or features into groups.

Compression— A series of techniques that reduces space, bandwidth, cost, transmission, generating time, and storage of data. These techniques are designed for the elimination of repetition, for the removal of irrelevant data, and for the use of special coding techniques, such as run-length encoding.

Computer-aided cartography— Software that assists an operator in standard mapping operations.

Computer-aided design and drafting (CADD)— Software that assists an operator in engineering and architectural design and drafting operations.

Computer-aided drafting (CAD)— Software that assists an operator in drafting operations.

Connectivity analysis— Analytical technique for determining whether or not points (nodes) or lines are connected to each other.

Contiguity analysis— Analytical technique for determining whether or not areas (polygons) are situated next to each other. Sometimes referred to as adjacency analysis.

Contour— An imaginary line on the ground, all points of which are at the same elevation above or below a specified datum surface, usually mean sea level.

Control point— Any station in a horizontal or vertical control network identified in a data set or photograph and used for correlating the data shown in that data set or photograph.

Coordinate pair— Set of Cartesian coordinates describing the 2D location of a point, line, or area (polygon) feature in relation to the common coordinate system of the database.

Coordinate system— A particular type of reference frame or system, such as plane rectangular coordinates or spherical coordinates, that uses linear or angular quantities to designate the position of points within that particular reference frame or system.

Corner joins— The location at which three or more contiguous map sheets come together.

CPU time— Actual computational time necessary to process a set of instructions in the arithmetic and logic units of a computer.

Dangling arc (or line)— An arc or line feature connected to other lines at only one end.

Data— Observations made of the real world, and collected as facts or evidence, which may be processed to give them meaning and to convey information.

Database— A collection of information related by a common fact or purpose.

Database creation— The process of bringing data into the electronic environment of a database for later use.

Database development— The process of determining what elements will be included in a database and their internal relationships.

Database management system (DBMS)— Software designed to access and structure a database.

Data accuracy— The extent to which an estimated data value approaches its true value.

Data bias— The systematic variation of data from reality.

Data capture— Series of operations required to encode data in a computer-readable digital form (digitizing).

Data category (layer)— Data having similar characteristics and contained in the same data set (for example, roads and rivers). Usually information contained within a data category is related and is designed to be used with other categories.

Data conversion— The process of converting data from one format to another, whether from paper format to digital format or from digital format to digital format.

Data dictionary— Repository of information about the definition, structure, and use of data. It does not contain the actual data.

Data editing— The process of correcting errors in GIS data.

Data encoding— To apply a code, frequently consisting of binary numbers, to represent individual or groups of data.

Data error— The difference between the real world and its representation in GIS data.

Data input— The process of converting data into a format that can be used by a GIS.

Data manipulation— The performance of those data-processing chores common to most users, such as sorting, input/output operations, and report generation.

Data precision— The recorded level of detail of data.

Data quality— The degree of excellence exhibited by data in relation to its portrayal of actual phenomena.

Data reduction— The process of transforming masses of raw data into useful, ordered, or simplified intelligence.

Data set— Collection of similar and related data records recorded for use by a computer.

Data structure— The organization of data, particularly the reference linkages among data elements.

Data topology— The order or relationship of specific items of data to other items of data.

Decompression— The process by which compressed data are expanded to their former file size.

Differential GPS— The process of using two GPS receivers to obtain highly accurate and precise position fixes (see Global Positioning System).

Digital data— Of, or relating to, data in the form of digits; data displayed, recorded, or stored in binary notation.

Digital elevation model (DEM)— A file with terrain elevations recorded at the intersections of a fine grid and organized by quadrangle to be the digital equivalent of the elevation data on a topographic base map.

Digital orthophotography— See orthophotography.

Digital terrain analysis— The process of analyzing a digital terrain model to determine volumes, cross-sectional areas, ground profiles, lines of sight, and the like.

Digital terrain model (DTM)— A land surface represented in digital form by an elevation grid or lists of 3D coordinates.

Digitization— The process of converting an analog image or map into a digital format usable by a computer.

Digitizing table— A flat surface (table) with an embedded electronic grid and equipped with a cursor. Used to trace (digitize) hard-copy maps and drawings into a GIS or CADD database.

Disaggregation— The reverse of "aggregation." (See Aggregation.)

Dissolve— The process of removing shared common attributes by eliminating the shared boundaries when merging two or more polygons.

Distributed database— Database with unique components in geographically dispersed locations linked through a telecommunications network.

DXF— Drawing eXchange format (developed by Autodesk, Inc.) to facilitate the conversion of digital data from one format to another.

Edge matching— The comparison and graphic adjustment of features to obtain agreement along the edges of adjoining map sheets.

Editing— Inserting, deleting, and changing attribute and geometric elements to correct and/or update a model or database.

Electronic Distance Measuring (EDM)— A theodolite combined with an optical rangefinder for accurate distance measurements. (See also Theodolite.)

Electrostatic plotter— An output device that generates hard copies of computer graphics files using electrically generated static charges to plot a raster image.

Elements— Entities of graphic data that represent map features.

Entities— Items about which information is stored. Items may be tangible or intangible, and are further defined by attributes.

Export— The process of transferring data or software from one system to another.

Feature— A set of data with common attributes and relationships. The concept of feature encompasses both entity and object.

Feature attribute— Also called a feature object or feature code. An element used to represent the non-positional aspects of an entity.

Feature merge— The process of combining area features using selected rules.

Filtering— The selective process of removing certain spatial data points to enhance features in an image.

Format— Predetermined arrangement of characters, fields, lines, punctuation, page numbers, and the like.

Format conversion— Converting data in one format into a format usable by another system.

Generalization— Reduction in detail in a map or data model to selectively remove information in order to simplify a pattern without distorting overall content.

Geographic information system (GIS)— System of computer hardware, software, and procedures designed to support the capture, management, manipulation, analysis, modeling, and display of spatially referenced data for solving complex planning and management problems.

Global Positioning System (GPS)— A system of orbiting satellites used for navigating and capable of providing highly accurate geographic coordinates to portable devices on the ground.

Graphic entities— Entities graphically portrayed as geometric shapes or symbols on a source document.

Graphical user interface— A computer interface that employs intuitive symbols and prompts (such as menus, icons, and pick lists) to represent computer commands. Distinguished from older "command line" computer interfaces that required the entry of text-based commands.

Grid— A network of uniformly spaced horizontal and vertical lines that encloses an area (a cell), with an associated value assigned.

Grid cell— An element of a raster data structure.

Grid cell analysis— The process of analyzing grid cell data.

Hard copy— Printed paper or film copy of machine output in a visually readable form such as printed reports, listings, graphs, drawings, maps, or summaries.

Horizontal control— Network of stations of known geographic or grid positions referred to a common horizontal datum, which controls the horizontal positions of mapped features with respect to parallels and meridians (or northing and easting grid lines) on a map.

Hypertext Markup Language (HTML)— Standardized codes, or "tags," used to define the structure of information in web pages on the World Wide Web (WWW). HTML defines several aspects of a web page, including heading levels, bold, italics, images, paragraph breaks, and hypertext links to other resources.

Hypertext Transfer Protocol (HTTP)— The protocol that governs the transfer of Hypertext Markup Language (HTML) documents between two or more computers, usually over the Internet or an intranet.

IDENTITY overlay— Polygon-on-polygon overlay corresponding to the Boolean OR and AND overlays. The resulting map will contain all polygons from the first map layer, and all polygons from the second map layer that fall entirely within those of the first layer.

Image processing— The various operations that can be applied to image format data. These include, but are not limited to, image compression, image restoration, image enhancement, image rectification, preprocessing, quantization, spatial filtering, and other image pattern recognition techniques.

Import— The process of bringing data or software from another system into a system.

Interactive system— A system allowing two-way electronic communication between user and computer.

Internet— A global network of computers using the same communications protocol to exchange data, originally designed by the U.S. Department of Defense to provide secure communications for military and intelligence purposes.

INTERSECT overlay— Polygon-on-polygon overlay corresponding to the Boolean AND overlay. The resulting map will contain only those polygons that cover areas common to both sets of input polygons.

Intersection— The coexistence of endpoints at a specific geographic location; the set of all objects common to two or more intersecting sets.

Islands— Polygons completely enclosed within another polygon.

Isoline— A line joining points of equal value.

Join— Area at which two or more adjacent maps or images are brought together to form a continuous model.

Labeling— The process of assigning attributes to polygons.

Layers— Various "overlays" of data, each of which normally deals with one thematic topic. These overlays are registered to each other by the common coordinate system of the database.

Least-squares adjustment— Method of adjusting observations in which the sum of the squares of all deviations or residuals derived in fitting the observations to a mathematical model is minimized. Such an adjustment is based on the assumption that blunders and systematic errors have been removed from the data, and that only random errors remain.

Line— A one-dimensional object having a length and direction, and connecting at least two points. Examples are roads, railroads, telecommunication lines, and streams.

Line-in-polygon overlay— The process of overlaying a line map over a polygon map to determine which lines cross which polygons.

Lineage— Information about a data source (particularly the original scale) and its accuracy.

Line string— A line with more than two points.

Local area network (LAN)— A data communications system linking computers and devices, usually within the same building or campus.

Management information system— System of computer hardware, software, and procedures designed to support the capture, management, manipulation, analysis, and display of financial, personnel, logistical, operational, and other management data.

Manual digitizing— The process of digitizing maps and aerial photographs using a digitizing table.

Map overlay— The process of combining the data from two or more maps to generate a new data set or map.

Map projection— Systematic drawing of lines of a plane surface to represent the parallels of latitude and the meridians of longitude of the earth.

Menu tablet— Small, movable, flat surface (tablet) with a command menu, an embedded electronic grid, and cursor. Used to input command instructions by pointing and clicking with the cursor.

Merge— To take two or more maps or data sets and combine them into a single, coherent map or database without redundant information.

Multimedia— Digital media presented in multiple forms, including text, pictures, sound, and video.

Network— (1) A collection of lines in a GIS or AM/FM system used to represent interconnected lines in the real world, such as streets or utilities. (2) A data communications system connecting computers and devices to facilitate the transfer of data.

Network analysis— Analytical techniques concerned with the relationships between locations on a network, such as the calculation of optimal routes through road networks, capacities of network systems, or best location for facilities along networks.

Node— Zero-dimensional topological entity representing the start or endpoint of an edge, the position of a point feature, or both.

Object-oriented GIS— An approach to organizing spatial data as discrete objects.

Off-line processing— Transmission of information between a computer and a peripheral unit before or after, but not during, processing.

On-line processing— Transmission of information between a computer and a terminal or display device while processing is occurring.

Operand— A query set operated on.

Operator— Term in a query string that specifies an operation on the resultant content of an operand.

Orthographic— The representation of related views of an object as if they were all geometrically projected upon a plane with a point of projection at infinity.

Orthophotography— A rectified photographic image that has no relief displacement or radial distortion inherent in aerial photos. Digital orthophotography refers to aerial photography that has been digitized.

Overlay— Data layer, usually dealing with only one aspect of related information, which is used to supplement a database. Overlays are registered to the base data by a common coordinate system.

Overshoot— A topological error in which a polygon is unclosed and two of its bounding arcs intersect.

Photogrammetrically digitized— Digitized from aerial photographs and geodetic control data by means of photogrammetric instruments, providing 3D coordinates.

Photogrammetry— The art and science of obtaining reliable measurements through the use of photographs.

Pixel— Short for "picture element." The smallest discrete element that makes up an image.

Planimetric base mapping— Map prepared from aerial photographs by photogrammetric methods, as a guide or base for contouring.

Planimetric data— Spatial data that do not take topographic relief information into account for establishing position.

Plotter— An output device that generates hard copies of computer graphics files.

Point— A level of spatial definition referring to an object that has no dimension. Map examples include wells, weather stations, and navigational lights.

Point-in-polygon overlay— The process of laying a point map over a polygon map to determine which points fall within which polygons.

Polygon dissolve— The process of merging adjacent area polygons having the same attributes, combining them into a single, larger area.

Polygon eliminate— The process of merging very small area polygons with larger, adjacent area polygons.

Polygon net— A collection of polygons having the same attribute.

Polygon-on-polygon overlay— The process of overlaying two polygon maps to determine how the two sets of polygons overlap. There are three main types of polygon overlay: UNION, INTERSECT, and IDENTITY.

Polygonization— Process of connecting linear feature information into polygons.

Positional accuracy— Term used in evaluating the overall reliability of the positions of cartographic features relative to their true position, or to an established standard.

Projection change— Procedure for transferring features from one map projection surface to the corresponding position on another map projection surface by graphical or analytical methods.

Proximity analysis— Analytical technique used to determine the spatial relationship between a selected point and its neighbors.

Query— A set of conditions or questions that form the basis for the retrieval of information from a database.

Query set— The result of a question or query posed to a topological file. The results from such a query can be used to generate reports, graphic displays, and new map files. Query sets are groups of elements and/or features in a topological file that meet a given criterion.

Query statement— Queries posed to a topological file, expressing spatial and/or attribute criteria.

Query string— See Query statement.

Raster data model— A system of rectangular cells in which individual cells are used to create images of point, line, and area map features.

Rasterization— The process of converting vector map data to raster data format.

Reclassification— Procedure for changing the classification of existing data.

Rectification— The process of projecting a tilted or oblique image onto a reference plane.

Relational database— A computer database employing an ordered set of values or records (known as tuples) grouped into 2D tables (called relations).

Relations— Two-dimensional tables of data in a relational database.

Remote sensing— The science of observation without touching. Often used to refer to Earth observation from satellites using electromagnetic sensors.

Resolution— Measure of the ability of a display system to distinguish detail under specific conditions. The measure of this ability is normally expressed in lines per millimeter, meters per pixel, dots per inch, and the like.

Route tracing— Application of network analysis for tracing the route of flows through a network from origin to destination.

Rubber sheeting— Topological process of stretching or shrinking a subarea or portion of a map or image to fit in registration with selected control points.

Satellite image— A digital image of the earth taken using electromagnetic sensors on a satellite.

Scale— Ratio between the distance on a map, chart, or photograph and the corresponding distance on the surface of the earth.

Scanning— The process of using an electronic input device to convert analog information such as maps, photographs, or overlays into a digital format usable by a computer.

Scrubbing— The process of preparing data for input to a GIS, intended to eliminate errors and questionable data in advance.

Sequential— Refers to data files in a serial order; that is, one file after another.

Set query— The comparison or combination of query sets in a logical way.

Slivers— Polygons formed when two adjacent polygons do not abut along a single common line and leave a small space between.

Slope— Rate of rise or fall of a quantity against horizontal distance expressed as a ratio, decimal, fraction, percentage, or the tangent of the angle of inclination. Also called gradient.

Smoothing— The selective process of removing certain spatial data points to enhance the appearance of features in an image.

Source material— Data of any type required for the production of mapping, charting, and geodesy products. Source material includes, but is not limited to, ground-control aerial and terrestrial photographs, sketches, maps, and charts; topographic, hydrographic, hypsographic, magnetic, geodetic, oceanographic, and meteorological information; intelligence documents; and written reports pertaining to natural and manmade features of the area to be mapped or charted.

Spatial analysis— Analytical techniques associated with the study of the location of geographical entities together with their spatial dimensions. Also referred to as quantitative analysis.

Spatial data— Data pertaining to the location of geographical entities together with their spatial dimensions. Spatial data are classified as point, line, area, or surface.

Spatial operator— Term in a query string that specifies location constraints in an operation.

Spatial query— A query that includes criteria for which selected features must meet location conditions.

State plane grid— A Cartesian coordinate mapping system covering a state in the United States.

Stereo model— A mathematical model that relates two stereophotographs to a photographed object or area.

Stereo pair— Two photographs having sufficient perspective overlap of detail to make possible stereoscopic examination of an object or an area common to both.

Structured Query Language (SQL)— A standard computer language developed to facilitate the query of relational databases.

Surface— A level of spatial measurement referring to a 3D defined space. Examples include contours, isolines, and bathymetry.

Tabular data— Data in a row and column format.

Temporal data— Data that can be linked to a certain time, or to a period between moments in time.

Terrain model— A surface model of terrain (see Digital terrain model and Triangulated irregular network).

Thematic maps— Maps pertaining to one particular theme or subject.

Thematic topics— Mapping categories, consisting of a single type of data, that are intended to be used with base data.

Theodolite— A survey instrument for measuring horizontal and vertical angles using a rotating and tilting telescope.

Thinning— The process whereby a linear feature is generalized through the use of a series of rules that reduces the number of data points while maintaining the basic shape of the feature.

Three-dimensional (3D) data— Volumetric data representing measurements in three dimensions and as angular or linear measures such as latitude-longitude-elevation.

Topographic analysis— Analysis of the configuration of a surface, including its relief and the position of streams, roads, cities, and the like. Usually subdivided into hypsography (the relief features), hydrography (the water and drainage features), culture (manmade features), and vegetation.

Topological— Refers to such properties of geometric figures as adjacency that are not altered by distortion as long as the surface is not torn.

Topological error checking— The process of ensuring that the logical consistency of data is intact; that is, all polygons are closed, all arcs are connected to nodes, and so on.

Topological relationships— The relationships of data elements within a database. Topologically related refers to how a change to one element affects other elements.

Topology— A branch of geometrical mathematics concerned with order, contiguity, and relative position, rather than actual linear dimensions.

Total station— A theodolite combined with a data recording device and coordinate geometry (COGO) software.

Transformation— The process of converting data from one mapping coordinate system to another.

Triangulated irregular network (TIN)— A data structure that describes a 3D surface as a series of irregularly shaped triangles. Usually used in connection with terrain modeling.

Tuples— Individual records (rows) in a relational database.

Two-dimensional (2D) data— Areal data in two dimensions, such as northing-easting or latitude-longitude.

Undershoot— A type of topological error in which an arc is unclosed.

UNION overlay— Polygon-on-polygon overlay corresponding to the Boolean OR overlay. The resulting map will contain a composite of all polygons in both input map layers.

User interface— Method and means by which a human operator communicates with database and applications modules.

Vector— Directed line segment, with magnitude commonly represented by a pair of Cartesian (x,y) coordinates for endpoints. Vector data refers to data in the form of an array with one dimension.

Vector data model— A spatial data model using 2D Cartesian (x,y) coordinates to store the shape of spatial entities.

Vectorization— The process of converting data from raster to vector format.

Viewshed— A polygon map resulting from an analysis of the locations visible from a specified point.

Web page— A single unit of information, often called a document, that is available on the World Wide Web (WWW).

Wide area network (WAN)— A data communications system linking two or more LANs (see Local area network).

Window— Rectangular frame with a specified size and location on the screen of an interactive graphics system, and within which a rectangular portion, or window, of the map is displayed.

World Wide Web (WWW)— A global network of computers and software using the Internet and the Hypertext Transfer Protocol (HTTP) standard to access and exchange digital information and multimedia.

Index